Viruses and the Nucleus

Viruses and the Nucleus

Editor

Julian A. Hiscox
School of Biochemistry and Microbiology, University of Leeds, UK

John Wiley & Sons, Ltd

Other Wiley Editorial Offices

John Wiley & Sons Inc., 111 River Street, Hoboken, NJ 07030, USA

Jossey-Bass, 989 Market Street, San Francisco, CA 94103-1741, USA

Wiley-VCH Verlag GmbH, Boschstr. 12, D-69469 Weinheim, Germany

John Wiley & Sons Australia Ltd, 33 Park Road, Milton, Queensland 4064, Australia

John Wiley & Sons (Asia) Pte Ltd, 2 Clementi Loop #02-01, Jin Xing Distripark, Singapore 129809

John Wiley & Sons Canada Ltd, 22 Worcester Road, Etobicoke, Ontario, Canada M9W 1L1

Wiley also publishes its books in a variety of electronic formats. Some content that appears in print
may not be available in electronic books.

Library of Congress Cataloging in Publication Data

Viruses and the nucleus / edited by Julian A. Hiscox.
      p.   ;   cm.
   Includes bibliographical references and index.
   ISBN-13: 978-0-470-85112-8 (alk. paper)
   ISBN-10: 0-470-85112-0 (alk. paper)
   1. Molecular virology.   I. Hiscox, Julian A.
   [DNLM: 1. Cell Nucleus—virology.   2. Cytopathogenic Effect, Viral.
   QW 160 V82137 2005]
   QP389.V57 2005
   616.9′101—dc22   2005028183

British Library Cataloguing in Publication Data

A catalogue record for this book is available from the British Library

ISBN-13 978-0-470-85112-8 (Hardback) 978-0-470-85113-5 (Paperback)
ISBN-10 470-85112-0 (Hardback) 0-470-85113-9 (Paperback)

Typeset in 10.5/12.5pt Times by Integra Software Services Pvt. Ltd, Pondicherry, India
Printed and bound in Great Britain by Antony Rowe Ltd, Chippenham, Wiltshire
This book is printed on acid-free paper responsibly manufactured from sustainable forestry in which
at least two trees are planted for each one used for paper production.

Contents

Contributors

G. Eric Blair   Molecular Cell Biology Research Group, School of Biochemistry and Microbiology, Faculty of Biological Sciences, University of Leeds, Leeds LS2 9JT, UK

Gavin Brooks   Cardiovascular Research Group, School of Animal and Microbial Sciences, University of Reading, PO Box 228, Whiteknights, Reading, Berkshire RG6 6AJ, UK

Paul Digard   Division of Virology, Department of Pathology, University of Cambridge, Tennis Court Road, Cambridge CB2 1QP, UK

John Dimmock   Department of Biological Sciences, University of Warwick, Coventry CV4 7AL, UK

Brian Dove   School of Biochemistry and Microbiology, University of Leeds, Leeds LS2 9JT, UK

Debra Elton   Division of Virology, Department of Pathology, University of Cambridge, Tennis Court Road, Cambridge CB2 1QP, UK

Warner C. Greene   Gladstone Institute of Virology and Immunology, University of California, PO Box 419100, San Francisco, CA 94141-9100, USA

Kurt E. Gustin   Department of Microbiology, Molecular Biology and Biochemistry, University of Idaho, Moscow, ID 83844, USA

Crisanto Gutierrez   Centro de Biologia Molecular 'Severo Ochoa', Consejo Superior de Investigaciones Cientificas, Universidad Autonoma de Madrid, Cantoblanco, 28049 Madrid, Spain

Jane V. Harper   Cardiovascular Research Group, School of Animal and Microbial Sciences, University of Reading, PO Box 228, Whiteknights, Reading, Berkshire RG6 6AJ, UK

Julian A. Hiscox   School of Biochemistry and Microbiology, University of Leeds, Leeds LS2 9JT, UK

Dean A. Jackson   Faculty of Life Sciences, University of Manchester, Sackville Street, PO Box 88, Manchester M60 1QD, UK

Nicola James   Molecular Cell Biology Research Group, School of Biochemistry and Microbiology, Faculty of Biological Sciences, University of Leeds, Leeds LS2 9JT, UK

Keith N. Leppard   Department of Biological Sciences, University of Warwick, Coventry CV4 7AL, UK

David A. Matthews   Division of Virology, Department of Pathology and Microbiology, School of Medical Sciences, University of Bristol, Bristol BS8 1TD, UK

Carlos de Noronha   Gladstone Institute of Virology and Immunology, University of California, San Francisco, CA 94141, USA

Peter Sarnow   Department of Microbiology and Immunology, Stanford University School of Medicine, Stanford, CA 94305, USA

1 The Nucleus – An Overview

Dean A. Jackson

Faculty of Life Sciences, University of Manchester,
Sackville Street, Manchester, UK

1.1 Introduction

Eukaryotic cells, unlike those of prokaryotic bacteria and archaea, possess a membrane-bound organelle called the nucleus. In general, nuclei are the most conspicuous organells in somatic animal cells (see Figure 1.1), occupying roughly one third of the cell volume. The major function of the nucleus is to act as a repository for the genetic information, encoded in DNA. Somatic mammalian cells generally carry two copies of the DNA genome, one inherited from the female parent and the other from the male. In humans, this information is held within 46 chromosomes, two copies each of 22 autosomes and two sex chromosomes – XY in males and XX in females. Nuclei contain the machinery needed to ensure the fidelity (repair) and duplication (replication) of the genetic code. Nuclei also house the complex machinery needed to ensure that the genetic information can be selectively decoded so that the requisite genes are expressed in the many different cell types present in multicellular eukaryotes. Activating programmes of gene expression is a highly complex process that relies on the interplay between activating transcription factors, the synthetic machinery and different chromatin states that arise through the action of a variety of chromatin remodelling systems. Some chromatin states are permissive for gene expression and allow transcription factors to access DNA so that the programme of events that lead to RNA synthesis can be initiated. Other types of chromatin have a structure that is operationally immune from the action of transcription factors. Many factors are known to influence and stabilize these chromatin states. Local and long-range features of chromosome architecture in particular play a fundamental role in defining both chromatin or gene domains and nuclear compartments where

Viruses and the Nucleus   Edited by Julian A. Hiscox
© 2006 John Wiley & Sons, Ltd

transcription is permitted or excluded. As well as establishing chromatin compartments, nuclei are also believed to support a wide variety of other compartments that probably influence the efficiency of chromatin function. Understanding the importance of these specialized nuclear regions, how they work and the purpose they serve could be of fundamental importance to understanding the critical subtleties of chromatin function.

Much of our present understanding of nuclear structure and function in eukaryotic cells has evolved as a result of experiments performed over the past few decades. Viruses, in contrast, have experienced many millions of years of co-evolution with these cells, during which time they have grown to utilize and subvert their hosts. This volume is dedicated to understanding how viruses use host cells to meet their own needs. In many cases viruses are selfish and the infected cell is destined to die. Nevertheless, it is interesting to investigate how viruses use the host machinery to their own ends, and in particular to consider what studies of viral infection are able to teach us about the fundamental principles of nuclear architecture in eukaryotic cells. This chapter sets the scene and explains what confronts a virus as it enters the eukaryotic cell nucleus.

1.2   The gene expression pathway in mammalian cells

In mammals, some 250 distinct cell lineages can be defined by the specialized roles that each performs. Different cell lineages have quite distinct properties and appearance, even though almost all of these cells contain the same genetic information. The phenotype of each of these lineages is determined by patterns of protein synthesis that in turn reflect levels of gene expression. Patterns of gene expression are determined by the dynamic interplay between chromatin, chromatin remodelling machinery, transcription factors and synthetic machinery that drives RNA synthesis. Three different RNA polymerase complexes perform RNA synthesis in the nuclei of mammalian cells. In most cells, RNA polymerase II (pol II) is the major activity, transcribing all protein-coding genes to generate patterns of gene expression that determine cell type. Synthesis is performed by a ~ 4 MD holoenzyme that contains the pol II core enzyme associated with other activities required during RNA synthesis and processing. RNA polymerase I (pol I) is dedicated to the synthesis of the repeated ribosomal RNA (rRNA) genes, within specialized nuclear sites – nucleoli – and RNA polymerase III (pol III), a minor nucleoplasmic activity, transcribes transfer RNA (tRNA) and 5S rRNA genes. Small nuclear RNA (snRNA) and small nucleolar RNA (snoRNA) genes encode structural RNAs needed for RNA processing; some are transcribed by pol II and others by pol III.

Proliferating mammalian cells have roughly 30 000, 60 000 and 3000 engaged pol I, II and III complexes/cell, respectively (Jackson *et al.*, 2000). RNA is polymerized at ~ 2–2.5 kbp/min; this rate can be estimated from the time taken to transcribe RNA molecules of known length. The average length of a primary transcript is about 15 kbp in mammalian cells, although the primary transcripts are processed, during splicing, to generate mRNAs that are roughly one-tenth of this size. The mammalian genome (haploid) has an estimated 35 000 genes and $\sim 3 \times 10^9$ bp DNA. Proliferating cells contain roughly 3.5×10^6 copies of the 18 and 28S ribosomal RNAs, 3×10^7 tRNA and 0.4×10^6 mRNA molecules; of these, a few mRNAs may have many thousand copies per cell, although most have less than 10 (Jackson *et al.*, 2000). In human cells, an active ribosomal RNA gene transcribed by pol I gene has ~ 120 engaged synthetic complexes and typical pol II transcribed genes has very few; most active genes have only one to three transcription complexes. With these facts in mind, it is important to remember that in mammalian cells the expression of single copy genes can vary by many orders of magnitude. This will inevitably reflect rates of RNA synthesis, although it is important to note that factors such as mRNA turnover and rates of protein synthesis also play a role. The amplification afforded by protein synthesis ensures that very few templates are required for most proteins and, in corollary, that very few genes are expressed continually, so that most genes in euchromatin remain transcriptionally inert for long periods of time.

This raises obvious questions about the dynamics of gene expression. If transcription factor recruitment to a promoter allows productive association of the transcription machinery, RNA synthesis will ensue. Most transcripts are synthesized within 5–10 min. During this time RNA processing and maturation will begin. After roughly 15 min processing and maturation are complete so that the fully mature mRNA is released from the transcription sites. As a general rule, protein complexes involved in maturation and particularly splicing ensure that only mature mRNAs pass into the RNA export pathway. During export, mRNAs are believed to move by diffusion, so that within a few minutes mRNAs pass to the nuclear periphery and through nuclear pores to the cytoplasm (Femino *et al.*, 1998).

1.2.1 Euchromatin and heterochromatin

During embryonic development, defined cell-type specific patterns of gene expression are established through developmentally controlled expression of relevant transcription factors. The patterns of gene expression that arise are inherited throughout subsequent cell division cycles (Cavalli and Paro, 1999). A fundamental feature of this process involves chromatin status and the ability with which DNA sequence motifs are able to engage the factors needed to activate expression. This process is complex and multifactorial. At one level, modifications in DNA, such as methylation, and changes in nucleosome structure, generated by covalent histone modifications, correlate with the expressional capabilities of different classes of chromatin. Highly specialized nucleosome assembly factors and numerous sophisticated chromatin-remodelling complexes contribute to the maintenance and modulation of chromatin structure. But on top of this, more stable chromatin forms that characterize heterochromatin can have a dramatic impact on gene expression.

In most situations heterochromatin is the dominant stable chromatin form. It has been known for many years that active chromatin translocated adjacent to heterochromatin will generally acquire a heterochromatic structure and lose the ability to support gene expressions. This phenomenon is emphasized if genes are introduced close to the centromeric heterochromatin in transgenic mice. In this case, modulating the concentration of heterochromatin proteins has been shown to influence the efficacy of gene silencing (Festenstein *et al.*, 1999). In *Drosophila*, the suppressive influence of heterochromatin was characterized in the study of position effect variegation, where the position of a gene in the genome can dramatically influence its level of transcription. These early studies characterized chromatin proteins that are known to regulate gene expression, such as HP1 (heterochromatin protein 1) and Pc (polycomb). Mechanistically, a chromatin organization-modifying chromodomain (CD) within both HP1 and Pc is known to play a critical role in heterochromatin formation. Recent experiments in mammalian cells have shown that the chromodomain of HP1 α binds to histone H3 and also to the linker histone H1 (Breiling *et al.*, 1999; Nielsen *et al.*, 2001). The polycomb CD also binds to histone H3. These observations suggest that interactions between HP1 and the histone proteins, together with the self-association properties of HP1, are crucial factors in HP1-mediated heterochromatin assembly. In addition to this, specific modification of chromatin – such as methylation of histone H3 lysine 9 – creates a binding site for HP1 proteins and stabilizes the heterochromatic state (Lachner *et al.*, 2001). Polycomb group proteins and trithorax proteins are thought to act antagonistically to control dynamic transitions between inert and open chromatin states (Muyrers-Chen and Paro, 2001).

1.2.2 Chromatin status and gene activation

In heterochromatin, specific chromatin structures restrict the ability of transcription factors and the transcription machinery to access binding sites within gene promoters and so prevent gene expression. Before gene expression can be activated, the appropriate chromatin structure must be established. Changes that occur in chromatin in response to gene activation and transcription have been reviewed in detail (Workman and Kingston, 1998; Jenuwein and Allis, 2001; Becker 2002). For the purposes of the present discussion, it will be sufficient to recognize that a combination of ATP-dependent chromatin remodelling machines, together with histone acetylase and deacetylase complexes, cooperate to generate chromatin states that are permissive for gene expression. For genes located in inert chromatin, the activation process is stochastic. The initial process will involve activating factors that recruit appropriate ATP-dependent chromatin remodelling machinery and open chromatin sufficiently for the promoter to now be accessibe to further factors (Becker, 2002). Remodelling complexes of the ISWI and SWI/SNF families are recruited by a variety of activating factors and induce nucleosme sliding to reveal binding sites that were previously masked by chromatin structure. These early changes then allow the binding of other transcription factors, other remodelling machines and histone acetylases that stabilize the open chromatin status further to facilitate assembly of the transcription machinery.

Finally, as transcription proceeds, proteins associated with the RNA polymerase holoenzyme generate an extended region of modified, transcriptionally competent, chromatin. This is a complex process, but it is also worth remembering that the process is dynamic, so that these different chromatin modifications will be in a constant state of flux.

This pathway has many interesting features. The transition from inert to active chromatin involves structural changes within chromatin at promoters and the spatial relocation of the relevant gene (away from heterochromatin) prior to the onset of RNA synthesis (Lundgren *et al.*, 2000). Decondensation of compact chromatin is seen during the activation of gene expression, although the process of RNA synthesis appears to have little further impact on the dynamic properties of the active domains (Tsukamoto *et al.*, 2000). Many proteins – perhaps 100 or more – are subsequently required to activate gene expression and many of these are dynamic in nature. The dynamic properties of glucocorticoid receptor coupled to green fluorescent protein (GR-GFP) emphasize this point (McNally *et al.*, 2000). A mouse cell line, 3617, containing a single locus with a 200-copy MMTV LTR array (and some 1000 GR-GFP binding sites), was used. Dexamethasone treatment of GR-GFP cells induces the nuclear accumulation of the activated receptor, so that in most cells a single region of intense staining correlates with the location of the MMTV array. Under these conditions, a FRAP (fluorescence reactivation after photobleaching) analysis showed that bound receptor molecules were only transiently associated with the promoter, with almost all bound factors exchanging in ~10 s.

1.2.3 Gene location and expressional competence

Gene expression from ectopic chromosomal sites is extremely unpredictable. Predictably, genes introduced into heterochroamtin are generally inactive, although expression can be modulated by the availability of hetcrochromatin proteins (Festenstein *et al.*, 1999). Ectopic genes in active chromatin also behave unpredictably in cultured cells or transgenic animals (Wilson *et al.*, 1990). For example, it is possible to analyse gene expression from artificial gene constructs introduced in predetermined target sites by homologous recombination. In one case mouse ES cells were used to express an Oct4/*lacZ* transgene (Wallace *et al.*, 2000). Using homologous recombination, four developmentally controlled promoters were introduced into the target locus of two selected clones and their ability to drive *lacZ* expression in transgenic animals determined. Animals derived from one clone showed appropriate regulation of the four different promoters. However, animals derived from the second showed additional, unexpected patterns of ectopic expression, confirming that gene location can have profound effects on expression throughout development.

While chromosome location can influence gene expression, it transpires that nuclear location might also play a role. Studies on the lymphoid lineage of mammalian cells have demonstrated that the inactivation of gene expression correlates with relocation of a gene to heterochromatic nuclear sites, while the reverse process occurs during gene activation. During this process, the sequence-specific transcription factor Ikaros becomes associated both with the silenced gene and local centromeric heterochromatin. Ikaros is able to bind both a target promoter and sites within the centromeric satellite repeats, so providing a means of driving appropriate genes into nuclear sites that are not permissive for gene expression (Brown *et al.*, 1997). Other examples are known where nuclear location influences gene function. During B lymphocyte development, the immunoglobulin H (IgH) and IgK loci are located at the nuclear periphery in haematopoietic progenitors and pro-T cells and in the nuclear interior in pro-B nuclei (Kosak *et al.*, 2002). The inactive loci are associated with the nuclear lamina and must move to active sites within the nuclear interior before recombination and transcription of the IgH and IgK loci can occur.

1.2.4   Chromatin domains as units of gene expression

The unpredictable behaviour of gene expression from ectopic sites implies that natural gene loci have evolved mechanisms to protect the required levels and patterns of gene expression. One way of achieving this is to place genes in chromatin domains that are spatially isolated for local epigenetic factors. Clear evidence for the existence of domains of chromatin function comes from the ease with which chromatin is cut by nucleases (Workman and Kingston, 1998); transcriptional status correlates with a generalized nuclease 'sensitivity' that results from an open (10 nm) chromatin fibre. Moreover, chromatin modifications that correlate with gene activity, such as histone acetylation, play a role in maintaining chromatin status that can be perpetuated once the initial developmental transactivators are removed (Cavalli and Paro, 1999). Hence, chromatin modifications are capable of generating epigentic signals that could play a role in the control gene expression, provided that this information can be reproduced when chromatin is duplicated (Taddei *et al.*, 1999).

As chromatin remodelling accompanies transcription, RNA synthesis will play a role in maintaining chromatin status. However, it is well known that the general nuclease sensitivity at active loci is not restricted to genes. Chromatin immunoprecipitation experiments (using antibodies to histone variants found in active chromatin) confirm that zones of active chromatin might spread well beyond the coding region of a particular gene. Moreover, regions of transition from active to inactive chromatin structure imply that chromatin domains are demarcated by elements that determine the boundaries of functional genetic units (Elefant *et al.*, 2000; Litt *et al.*, 2001). A high-resolution analysis of acetylation across three gene loci in the vicinity of the chicken β-globin locus has emphasized this point (Litt *et al.*, 2001). Across this region, three levels of chromatin acetylation can be defined. The condensed chromatin maintains very low levels of histone acetylation at all developmental stages, with similar levels maintained in inactive genes. Much higher levels of acetylation are seen throughout transcribed gene domains, while chromatin in the vicinity of upstream regulatory sites maintains the highest levels of acetylation. Most significantly, a very strong constitutive focus of hyperacetylation corresponds with an insulator element that appears to demarcate the globin and adjacent folate receptor domains. This demonstrates how epigenetic modifications might influence chromatin domain structure and also implies that different classes of histone acetyl transferase, with different chromatin targets, control gene expression.

Interestingly, experiments analysing expression of human β-globin gene constructs in transgenic mice have shown that the locus is composed of three functional sub-domains, which acquire an active chromatin status at the time of their expression, at the appropriate stage of development (Gribnau *et al.*, 2000). Intriguingly, the appearance of these nuclease-sensitive domains correlates with the activity of intergenic, non-globin transcripts that extend throughout the corresponding sub-domains. However, the maintenance of chromatin status does not demand continual transcription of the non-genic sequences, as most synthesis is seen in the G_1 cell cycle phase (see Chapter 2 for description of the cell cycle) of appropriate cells. With this in mind, it is notable that a substantial fraction of primary transcripts synthesized in mammalian cells by RNA pol II are not destined to produce mature mRNA (Jackson *et al.*, 2000). Perhaps non-productive RNA synthesis acts to keep chromatin domains 'open', so preventing the formation of inactive chromatin states.

1.2.5 Domains of gene expression

We have seen that in many eukaryotes, individual genes are expressed to quite different extents and many are expressed in specific cell types and at precise times during development. This begs the question: how are different levels of gene expression maintained? For a specific gene, chromatin status and the availability of activating transcription factors will combine to establish an engaged transcription complex that drives RNA synthesis. Following activation, certain combinations of promoter-bound proteins must drive many cycles of synthesis, while others must decay and so generate many fewer transcripts. Gene promoters define the location of the transcription start site, but in the chromosomal context, promoters are rarely sufficient to define natural levels of expression. Operationally, expression is augmented and modulated by genetic elements such as enhancers, locus control regions (LCRs), scaffold/matrix attachment regions (S/MARs) and insulators. Together these elements help us to define a unit or domain of gene expression.

Enhancers generally contain arrays of transcription factor recognition motifs that bind appropriate factors to augment levels of gene expression by directly upregulating rates of transcription (Blackwood and Kadonaga, 1998). Enhancer function is dependent on factor binding and is likely to contribute to the protein complex that drives assembly of the transcription machinery. Enhancers may be responsible for establishing a productive spatial configuration in chromatin or for delivering chromatin to specific nuclear sites. Related elements called locus control regions (LCRs) are also complex arrays of factor binding motifs that make critical contributions to levels of gene expression at the natural chromosomal locus and allow position-independent expression from ectopic sites (Engel and Tanimoto, 2000; Bulger *et al.*, 2002). LCRs contribute to the maintenance of expressional status and are believed to achieve this by maintaining adjacent chromatin in an active configuration; probably by recruiting chromatin remodelling machinery. However, the view that LCRs directly influence chromatin status has been challenged, using recombination strategies to investigate LCR function by deletion of elements at the natural locus (Bender *et al.*, 2000). This approach appears to suggest that LCR elements operate as very efficient enhancers at the natural locus that do not themselves directly influence chromatin status across the locus. An explanation for this apparent controversy is that LCR function is dependent on chromosome context.

1.2.6   The nuclear matrix and gene domains

The nuclear matrix and related scaffold are believed to play important roles in different aspects of chromatin function (Bode *et al.*, 2000). Scaffold/matrix attachment regions (S/MARs) clearly augment transcription and may achieve this using AT-rich sequences to modulate superhelical stress that arises during transcription. S/MARs may also play a role in targeting gene domains to matrix-associated transcription centres prior to gene expression. As an example, MAR elements play a critical role in orchestrating the temporal and spatial expression of many genes during T cell development (Alvarez *et al.*, 2000). In a comprehensive comparison of different classes of potential boundary elements, a strong MAR was shown to support efficient and long-term gene expression from ectopic chromosomal sites (Zahn-Zabal, 2001).

Insulator or boundary elements are believed to function as chromatin domain boundaries (Bell and Felsenfeld, 1999; Udvardy, 1999). A possible role for boundary elements was first suggested by their ability to uncouple linked promoters and enhancers, presumably by 'blocking' enhancer function. This property supports their role as barriers to the influence of neighbouring DNA elements. Many elements that are capable of providing insulator activity and proteins that bind to them have now been described. Characterized boundary element attachment factors (BEAF-32A and -32B) and their recognition motifs within the specialized chromatin structures of *Drosophila* may be important determinants of chromosome structure (Zhao *et al.*, 1995). In vertebrates, recent studies have confirmed that sequence elements at the upstream border of the chicken β-globin LCR act as classical insulators (Bell *et al.*, 1999). A protein, CTCF, has been shown to bind to critical sequence motifs within this region and is thought to play a role in determining insulator function (Farrell *et al.*, 2002). Intriguingly, this same protein appears to play a role in the process of 'genomic imprinting' to control the differential expression of maternally and paternally inherited genes (Ferguson-Smith and Surani, 2001).

1.3 Chromosome structure and DNA loops

Gene distribution within the chromosomes of mammalian genomes is not uniform. Even within individual chromosomes, the majority of genes are clustered within chromatin that can be defined using cytological criteria. These chromosomal R-bands are interspersed with G-bands that contain very few genes and are rich in inert, repetitive sequences. The arrangement of these elements will clearly impact on both local (Gasser, 2002) and long-range (Cremer and Cremer, 2001) chromatin organization. It is important to realise that mammalian chromosomes occupy discrete regions within interphase nuclei. These chromosome 'territories' do not occupy specific nuclear positions, although gene activity can influence interphase chromosome location, so that chromosomes with a high density of active genes tend to be located towards the nuclear centre (Bridger *et al.*, 2000). Interestingly, there is some evidence that different classes of chromatin within individual territories may be spatially separated – perhaps through aggregation of heterochromatin – to generate functional 'polarity' within territories (Sadoni *et al.*, 1999). This potential to develop local compartments of active and inert chromatin must reflect the path of the chromosome axis within each territory and the organization of chromosomal sub-domains represented by DNA foci. These structures are thought to represent structurally related gene clusters that are stable over many cell generations (Jackson and Pombo, 1998); they are derived for individual chromosome bands and typically contain ~ 1 Mbp DNA.

In mammals, a typical interphase chromosome territory might be represented as a sphere that measures roughly 2.5 μm in diameter. Within this volume, chromatin is only locally dynamic; for most chromatin relative movements of > 500 nm are seen only rarely (Abney *et al.*, 1997). This implies that long-range movements are constrained by chromosome structure. The borders of individual territories appear distinct, implying that interaction of chromatin for adjacent chromosomes is spatially restricted. However, in some instances, chromatin can be seen to escape from the confines of an individual territory to form loops that spread ~ 1–2 μm into the adjacent nucleoplasm. Chromatin loops of this sort probably reflect the transcriptional status of highly expressed gene-rich chromosomal regions, and their extruded appearance supports the impression that they expand to occupy local chromatin channels. The functional implications of this type of chromatin movement are unclear.

1.4 Duplicating the genetic code

In multicellular organisms it is self-evident that many cells must arise by duplication of a single precursor, the fertilized egg; the adult human has roughly 5×10^{13} somatic cells. Moreover, it is critical that both the genetic code and epigenetic information are faithfully reproduced during each cell division cycle (see Chapter 2 for further details of the cell cycle). In fact it is reasonable to suggest that the precision with which DNA replication occurs is one of the greatest feats performed by higher eukaryotic cells (Berezney *et al.*, 2000; Gilbert, 2002). Control of DNA synthesis relies on the complex interplay of chromosome structure, chromatin organization and perhaps even global nuclear architecture. This complex process has a number of steps that occur at different points of the cell cycle. Initially, replicons that are activated at the onset of S phase and most probably the downstream replication programme are established during the first 2 h of the G_1 phase of the cell cycle. This 'temporal decision point' coincides with the period when chromosomal territories re-establish their interphase structure and full transcriptional potential. The process of engaging an S phase programme that duplicates the genome in a predetermined order is distinct from the process that establishes active origins of DNA replication. This second process is thought to involve the binding of the origin recognition complex (orc) to particular chromosomal sites. This process occurs during the mid-G_1 phase and is designated the 'origin decision point'. About 2 h after this, mammalian cells pass a restriction point, at which time they are committed to proceed through S phase and on to cell division.

During S phase, replication takes place in dedicated replication factories. These sites contain all the machinery need to duplicate the DNA of small replicon (replication unit) clusters with about 1 Mbp DNA. Like transcription factories, these structures appear to be assembled on the nucleoskeleton; many components of the replication machinery have defined nucleoskeleton/replication factory-targeting sequences. At the onset of S phase, a specific sub-set of DNA foci is selected for replication. These foci are from chromosomal R bands and are rich in transcribed genes. Typical replicons within these early replicon clusters are ~150 kbp in length, so that most replicons activated at the onset of S phase will be completed 30–60 min later. As replication of these replicons begins to terminate, DNA synthesis is activated at new sites. This process continues until after about 10 h the entire genome has been duplicated. Although the precise mechanisms are unknown, chromosome architecture is believed to play a pivotal role in defining this S phase programme.

1.5   Nuclear compartmentalization

The cytoplasm of a mammalian cell contains various membranated organelles, which provide compartments that are dedicated to specific roles. The nucleus contains no equivalent structures and appears ill-organized in comparison. A classical electron micrograph of a typical human cell emphasizes this impression (Figure 1.1). The nuclear membrane, nucleolus and regions rich in condensed heterochromatin stand out in contrast to the relatively amorphous nuclear interior. Detailed inspection of this interior reveals an interchromatin space, which is rich in hnRNP (an RNA-binding protein) and contains characteristic structures, such as perichromatin fibrils, perichromatin granules, interchromatin granules and interchromatin granule clusters. These are known to be related to different aspects of RNA metabolism (Fakan, 1994). The major message from this type of image is that chromatin-rich regions are generally distinct

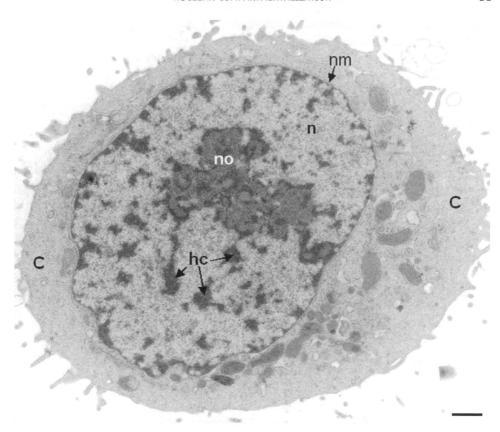

Figure 1.1 Morphology of a mammalian cell. Proliferating HeLa cells (this is a transformed human cell line) were fixed, stained and processed using standard electron microscopy techniques. Thin sections were then prepared and inspected. This typical cell section shows a central nucleus (n), typically of about 10 μm diameter in an equatorial section, which is separated from the cytoplasm (c) by a double nuclear membrane (nm). The nuclear membrane is studded with a few thousand nuclear pores that allow molecules to pass between the two major cell compartments. Note that the cytoplasm contains a large number of membrane-bound organelles. The most conspicuous of these, the mitochondria, are dark staining and typically about 1 μm in diameter in this example. The nucleus does not contain any membrane-bound sub-compartments. Even so it is highly organized. In this type of image, the most obvious substructure is the nucleolus (no). In this example a single large nucleolus occupies the centre of the nucleus. The nucleolus is densely staining and has three distinct subcompartments. The most densely-staining regions are called the dense fibrillar component; these structures are generally crescent-shaped and a few hundred nm in size. The dense fibrillar components represent the nascent transcripts that surround the much paler fibrillar centres. The processed ribosomal RNAs are assembled into ribosomal subunits in the granular component, which comprises the bulk of the nucleolus. Dark-staining clumps of material that lie against the surface of the nucleolus, the nuclear face of the membrane and are scattered throughout the centre of the nucleus are dense heterochromatin (hc). The remainder of the nucleus has an amorphous appearance and little discernable structure. This nucleoplasmic region contains the transcribed genes in euchromatin and is composed of a mixture of this open chromatin and the interchromatin material, which is predominantly RNA and protein. Bar = 1 μm

from the RNA-rich interchromatin regions and that gene transcription occurs predominantly at the interface between these two zones.

The nucleolus provides the best example of a nuclear compartment with a well-characterized role (see Chapter 7 for details of the nucleolus). The anatomy of a mammalian cell nucleolus is shown in Figure 1.1. Fibrillar centres are rich in the synthetic machinery – RNA polymerase I – and are surrounded by a dense fibrillar component, which contains the nascent RNAs. An individual fibrillar centre is likely to be associated with the genes and nascent products from a single active rDNA cluster, typically three to five genes. Following transcription, the nascent transcripts are processed and assembled into the preribsomal subunits that form the granular component of the nucleolus. The fibrillar centres of nucleoli are the most active transcription factories in mammalian cells. A proliferating cell produces at least 2.5×10^6 ribosomes in each cell cycle. Each ribosome contains the 28S, 18S and 5.8S RNA molecules that are derived from the 45S primary transcript of a ribosomal RNA gene. A proliferating cell has roughly 25 active fibrillar centres with roughly 100 active genes. Each of these has about 120 engaged polymerase complexes at any time. A recent study has used many nucleolar proteins coupled to a green fluorescent protein (GFP) to provide a detailed insight into the dynamic features of ribosomal RNA gene transcription (Dundr *et al.*, 2002).

The complexity of nuclear organization is emphasized when the nuclear distribution of different proteins is analysed by immunostaining. This approach shows that many proteins involved in different aspects of nuclear function are found to be concentrated at discrete nuclear sites, often with a punctate distribution (van Driel *et al.*, 1996; Hendzel *et al.*, 2001). This general impression implies that different proteins accumulate at sites where specific functions are performed but, perhaps more importantly, suggests that many proteins have specific mechanisms that ensure that their distribution throughout nuclear space is not uniform. The nuclear speckles provide an excellent example of this. The term 'nuclear speckle' was used to describe the appearance of the structures that immunostained with a human autoimmune Sm serum. The structures were later shown to be equivalent to interchromatin granule clusters that can be visualized by electron microscopy. The reacting protein and many others that stain the same structures are now known to be involved in RNA metabolism and particularly RNA processing. However, the structures themselves are not active centres; instead they appear to serve as storage depots. Interestingly, while speckles form a distinct nuclear compartment, which can be purified, the proteins within them are very dynamic. Hence, speckles represent a steady-state snapshot of the distribution of a protein population that rapidly moves throughout most of the available nuclear space.

The fact that high concentrations of particular proteins exist at defined nuclear sites raises obvious questions about the role of these compartments, particularly with respect to structure–function relationships. With this in mind, careful studies have demonstrated that many very active genes are located at the periphery of nuclear speckles (Smith *et al.*, 1999). However, this is clearly not the case for all transcribed genes, most of which are not located at borders of the speckle domains; at these remote sites, proteins that are found at high concentrations in speckles are supplied from the dynamic nucleoplasmic pools. The fact that highly active genes and speckles undergo function-dependent interactions gives us a valuable insight into the principles of nuclear architecture. Here it seems that the major structures are fundamentally plastic, so that changes in the structure of chromosome territories and local changes in the arrangement of dynamic compartments, like speckles, determine how the different components might interact. In this way, highly active genes might be located to the periphery of a chromosome territory, where the splicing components can accumulate in close proximity to sites of greatest demand (Misteli and Spector, 1998). Mechanistically this is not outrageous, as the... carboxy-terminal domain (CTD) of pol II is known to couple RNA synthesis and processing (McCraken *et al.*, 1997). If the cross-talk between RNA polymerase and splicing is eliminated by removing most of the CTD, the association between highly active genes and nuclear speckles is lost (Misteli and Spector, 1999). More generally, disrupting global RNA synthesis with inhibitors causes a dramatic alteration in the appearance of speckles, which are seen to lose surface texture, become more spherical and aggregate. This confirms the dynamic nature of this compartment and the fact that its appearance is related to function. In addition to this, recent studies have indicated that the efficiency of RNA processing is disrupted if the components normally found within speckles are artificially dispersed (Sacco-Bubulya and Spector, 2002). This observation suggests that the speckle compartment plays an important role in the cycle of events that is required to generate active spliccosomes and then splice the nascent RNA; if the speckles are not essential, they certainly influence the efficiency of the splicing process.

Speckles are large subnuclear structures, many of which will measure hundreds of nanometres across. However, it is clear that the same principles apply to other aspects of nuclear organization. For example, nascent RNA can be labelled in mammalian cells using Br-UTP (Jackson *et al.*, 1993; Wansink *et al.*, 1993) and Biotin-CTP (Jackson *et al.*, 1998; Figure 1.2) and sites of incorporation detected by immunolabelling to reveal transcripts associated with the dense fibrillar component of nucleoli and many (~5000) sites, dispersed throughout the nucleoplasm. High-resolution immunostaining techniques show that the active transcription sites contain transcript-rich zones measuring ~50 nm with adjacent regions rich in chromatin, transcription factors, the active form of RNA polymerase II or proteins required for RNA processing (Pombo *et al.*, 1999; Iborra *et al.*, 1996). As these sites perform all steps needed to generate mature mRNA, they have been called transcription 'factories'. Perhaps surprisingly, detailed studies have shown a typical factory contains many active RNA polymerase complexes (typically about five), although the majority of transcribed genes contain fewer engaged polymerase complexes. This raises the possibility that groups of genes are transcribed together within specialized nuclear compartments that are dedicated to performing all the steps needed to generate mature mRNA–protein complexes. Protein–nucleic acid interactions within these compartments are thought to account for the majority of DNA loops found within mammalian cells (Cook, 1999).

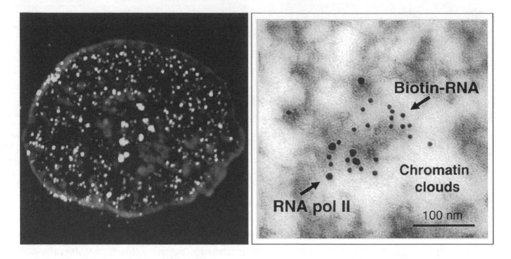

Figure 1.2 Transcription centres in human cells. Patterns of transcription in human cells can be visualized by either light (left) or electron (right) microscopy. In these examples, the light microscopy is used to demonstrate the global distribution of the transcription centres and the electron microscopy a more detailed image of the anatomy of a single transcription factory. For the first image, Hela cells were permeabilized, nascent transcripts extended in Br-UTP and cryosections (~ 100 nm) prepared. Br-RNA was indirectly immunolabelled with Cy3 and nucleic acids were counterstained with TOTO-3. Red and far-red images were collected using a 'confocal' microscope. The transcription sites are bright (shown green) foci. Note the distribution of very bright nucleolar foci in the centre and many smaller sites – RNA polymerase II/III transcription centres – throughout the nucleoplasm. The very bright foci will typically contain ~ 4 ribosomal RNA genes and ~ 500 nascent transcripts. The nucleoplasmic sites have ~ 5–10 active transcription complexes. TOTO-stained heterochomatin is clearly visibly around the nuceolus and along the nuclear periphery. Reproduced from Pombo *et al.* (1999) with *permission* of Oxford University Press. The second image is an electron microscopy section that has been prepared in LR-white, treated using the EDTA regressive staining technique and immunolabelled using antibodies to RNA polymerase II and Biotin-RNA (labelled *in vitro* using Biotin-CTP). Note the relative distributions of RNA polymerase II (large, 10 nm, gold particles), Biotin-RNA (small, 5 nm, gold particles), chromatin clouds (pale areas) and RNA rich interchromatin channels (grey areas). The transcription centre of ~ 100 nm in this example corresponds to 1 nucleoplasmic focus in the light microscopy image. Image courtesy Francisco Iborra

The idea that active RNA polymerase complexes might be localized within discrete and dedicated nuclear sites can have profound implications for gene expression. In particular, the fact that groups of engaged polymerases are found at each site means that the enzymes must be 'fixed' within the sites and the chromatin mobile. How chromatin interacts with the active centre would then dictate levels of gene expression. Moreover, it is also interesting to consider how chromatin structure and genetic elements outside promoters might influence gene activity. *A priori*, if chromatin is dynamic the fluid properties inherent in different chromatin states might have a profound influence on the potential of a promoter to engage any local active centres – genes in or close to heterochromatin would be more inert and their movement locally constrained, while sequences such as enhancers, LCRs and S/MARs might contribute to the overall efficiency of gene expression by targeting genes to the active centres. All of these components are likely to be in a finely balanced dynamic equilibrium that reflects both local and long-range chromatin organization and contributes in some way to levels of gene expression.

While many components involved in chromatin function are extremely dynamic or mobile, others appear to be rather static or fixed. We have seen that many RNA-binding proteins have a preferred location within the speckle compartment, even though they are normally in a continuous state of flux (Misteli and Spector, 1998). The same applies for many transcription factors and the way they interact with both promoters and the nuclear substructure. Intriguingly, many nuclear proteins appear to be stably associated with the nucleoskeleton. The most comprehensive study of this question has been performed on the DNA replication protein PCNA (proliferating cell nuclear antigen). PCNA is a major component of the replication factories and in living cells has been shown to be associated with sites that behave as though they are associated with a much larger nuclear structure (Sporbert *et al.*, 2002).

1.5.1   The nucleoskeleton, nuclear matrix and lamina

We have seen that nuclei can be plastic, so that they appear to have a complex structure of ill-defined nuclear compartments, which generally represent centres of nuclear function. These centres have components that are inherently dynamic and in general are not stable entities. But while these compartments might be plastic, it is also interesting to note that the spatial organization of sites such as transcription factories and speckles is unaffected if nuclei are depleted of chromatin (Figure 1.3). This suggests that the compartments are spatially constrained and raises the possibility that a nuclear 'solid phase' exists to provide a platform upon which the active sites might assemble. A 'nucleoskeleton' has been proposed to fulfil this role.

As the molecular structure of a filament-based network equivalent to the cytoskeleton has not been unequivocally demonstrated in nuclei, the nature of the nucleoskeleton remains a matter for debate. Core filaments that could provide a form of continuity throughout nuclei have been described (Nickerson, 2001). These filaments have characteristic morphological features of the intermediate filaments seen throughout the cytoplasm. The best-characterized intermediate filaments within nuclei are undoubtedly the nuclear lamins, which form a mesh-like structure at the nuclear periphery, where they play important roles in maintaining nuclear shape and controlling chromatin organization. Recently, it has been suggested that the lamin proteins spread further into the nucleus as a veil of lamin filaments that pervades the nucleoplasm. Interestingly, the integrity of nuclear lamin filaments has been shown to be a major determinant of nuclear function. For example, using dominant negative mutant lamin proteins to disrupt lamin assembly, it has been demonstrated that intact lamin filaments are

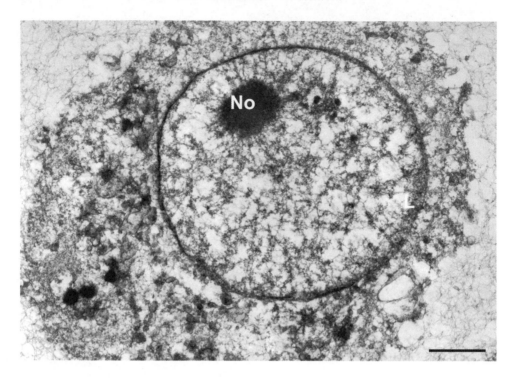

Figure 1.3   The nucleoskeleton in mammalian cells. To reveal the nucleoskeleton, HeLa cells encapsulated in agarose microbeads were permeabilized and chromatin cut with nucleases, so that ~90% DNA could be removed before preparing a resinless electron micrograph. In this ~250 nm thick section, note that a diffuse agarose mesh surrounds the densely stained cell. The spherical, central nucleus is demarcated by a dense nuclear lamina (L). It is important to recognize that nuclear architecture is preserved even though most chromatin – half of the nuclear mass – is removed. The residual nucleoskeleton can be visualized as a diffuse network of coated filaments that pervade the nuclear volume and provides a 'solid phase' upon which nuclear compartments such as nucleoli (No), replication and transcription factories and interchromatin granule clusters are assembled. The distribution of transcription and replication centres is preserved in preparations such as this, even though almost all chromatin has been removed. Bar = 2 µm. Reproduced from Hozak *et al.* (1994) with permission of the Company of Biologists

required for both DNA and RNA synthesis. The disruption of RNA synthesis (Spann *et al.*, 2002) is especially compelling, as experiments have demonstrated that destroying the lamin filaments can ablate synthesis by RNA polymerase II in the nucleoplasm, while synthesis by other classes of polymerase is unaffected. This approach supports the existence of a network that is required for expression of protein coding genes but which is either related to or dependent on the nuclear lamin filaments for its function. With this in mind, it is worth reflecting on the many examples where transcription factors are recruited to the nuclear matrix prior to activating differential gene expression in cell lineages where expression is maintained (Zaidi *et al.*, 2001).

The nuclear matrix (Berezney *et al.*, 1996) is classically described as an amorphous fibro-granular structure that can be isolated from nuclei by hypertonic treatment following nuclease digestion. The matrix typically contains many hundreds of different proteins, many of which have been studied in detail. Two of the best-characterized, SAF-A (scaffold attachment factor A) and ARBP (attachment region binding protein), were discovered because of their association with matrix-attached DNA elements. SAF-A (Romig *et al.*, 1992) turns out to be a major RNA binding protein (hnRNP-U) and ARBP (Stratling and Yu, 1999) a protein that binds methylated DNA (MeCP2). The behaviour of this protein is undoubtedly complex. SAF-A, for example, has been shown to bind p300, a major transcriptional co-activator, and in this way can recruit active genes to the nuclear matrix. ARBP, in contrast, interacts with methylated DNA in MAR elements and, through an intermediary that contains Sin3A protein, recruits a histone deacetylase to generate a silenced chromatin state (Stratling and Yu, 1999). Another protein, SAF-B (scaffold attachment factor B) specifically binds to S/MAR regions, interacts with RNA polymerase II (RNA pol II) and a subset of serine/arginine-rich RNA processing factors (SR proteins). It was proposed that these interactions allow SAF-B to provide a surface for the assenbly of the transcription apparatus (Nayler *et al.*, 1998). As with the vast majority of characterized matrix proteins, these are involved in different aspects of chromatin function. Extending this idea implies that the nucleoskeleton and nuclear matrix (which forms during salt extraction) are an expression of different processes that are performed within the interchromatin space. In this way, the complexity of the matrix simply reflects that fact that many of the proteins involved in DNA and RNA metabolism are engaged in multiple interactions within protein megacomplexes that, through direct or indirect associations (via a filament network), are able to form a stable structure in extracted cells. This implies that interaction and associations between these structures and chromatin will be the major determinants of nuclear structure in mammalian cells.

1.5.2 Other nuclear compartments

Many other discrete nuclear compartments have been described (Spector, 2001; Wang *et al.*, 2002). A prominent example is the coiled or Cajal bodies, which are generally spherical, measure ~1 µm across and in cross section appear as a network of coiled fibres (Matera, 1999; see also Chapter 3). Coiled bodies are found in both animals and plants, and contain many different proteins, including p80 coilin (the diagnostic marker), various transcription factors, as well as small ribonucleoproteins. While these bodies are not transcriptionally active, nascent snRNAs are found immediately next to them in mammalian cells. PML bodies are defined by their content of promye-locytic leukemia protein (Matera, 1999; see also Chapter 8). They are roughly similar in size to coiled bodies but are usually more abundant, so a mammalian cell in tissue culture may have one to three coiled bodies and 10–20 PML-bodies, and the PML bodies split up when cells experience the stresses associated with heat-shock and inflammation. PML bodies have also been referred to as the ND10 compartment. One report suggests that they are transcriptionally active and another that their surface can provide a site of efficient viral transcription and replication in infected cells (Maul, 1998).

Other sites have been described that might be classified as 'transcription super-factories'. The OPT (Oct1/PTF/transcription) domain provides the best example (Pombo *et al.*, 1998). Immunofluorescence shows that in HeLa cells the transcription factors Oct1 and PTF are found in one to three large domains (diameter ~ 1.3 μm) that appear during G_1 phase and disappear in S phase. Each OPT domain typically contains two or three factories where Br-UTP is incorporated into nascent transcripts, as well as RNA polymerase II, TBP and Sp1. Some chromosomes, notably chromosome 6, are found to be closely associated with the OPT domains. A variety of other structures, such as the perinucleolar compartment and gemini bodies, have been described (Spector, 2001). However, none of the extranucleolar domains discussed above yet have well-defined roles. They all seem to be dynamic structures. Most importantly, these domains are all defined by a local concentration of a particular marker. However, these markers are also found elsewhere in the nucleus, usually in many thousands of smaller sites. Therefore, the major domains discussed above may simply be uninteresting depots where components are stored. Thus, a component like p80 coilin may be stored in a coiled body, but function elsewhere. Note in this context that microinjecting antibodies directed against p80 coilin disrupts coiled bodies, without obvious detriment to the cell (Almeida *et al.*, 1998).

1.6   Nuclear architecture – implications for viral infection

The preceding discussion gives an overview of nuclear architecture in higher eukaryotes and emphasizes how chromatin domains and specialized nuclear compartments play a vital role in defining programmes of gene expression. With these features in mind, we now need to ask how these aspects of nuclear architecture influence the behaviour of a viral genome as it enters the nucleus soon after infecting the host cell (for the sake of clarity, I will not discuss viruses that only occupy the cytoplasm). As a generalization, it is safe to assume that the major functions of a viral genome are performed predominantly by host cell components. This is not surprising, as most viral genomes are too small to code for each of the proteins that are needed to generate the very sophisticated machines required for DNA replication and RNA synthesis and processing. Following infection, the virus must enter the nucleus. Many different mechanisms for nuclear targeting have been described (see Chapter 3). Once inside the nucleus, the viral genome must engage the relevant host cell apparatus. In most cases, this early phase of the infection cycle requires that the viral genome accesses the nuclear compartment where gene expression occurs. Hence, viruses target transcription sites and the nucleoskeleton. In reality, nuclear structure ensures that this is almost inevitable. Hence, the location of transcription sites at the interface between chromatin domains and the interchromatin channels ensures that mature mRNPs can pass efficiently to the cytoplasm just as the same organization ensures that viruses entering the nucleus will pass efficiently to transcription sites via the same channels. Some viruses might seek preferred nuclear target sites, such as PML bodies, but these too are established in the highly accessible interchromatin domains. Once the early phase of transcription is established, viruses commonly engage other mechanisms to ensure that expression of viral genes will proceed at the expense of host expression. In some instances host synthesis shuts off and the host chromatin architecture is destroyed. Clearly, different viruses use different mechanisms, which it is not appropriate to catalogue here. At a later stage of infection the focus of activities might switch to viral genome synthesis. Once again, those viruses with DNA in their infectious particles will subvert the host replication machinery to reproduce large quantities of the viral genome. During this process, some features of host cell replication are retained while others are discarded.

1.7   Conclusion

In higher eukaryotes, many layers of organization contribute to the complex processes that control chromatin function. We have seen how changes in chromatin structure modulate the access of transcription factors that activate pathways of gene expression. For genes in chromatin that are competent for gene expression, simple recognition motifs in DNA can interact with transcription factors that eventually recruit the transcription machinery so that RNA synthesis can begin. During this process, the constellation of factors assembled on promoters is likely to be in a continual state of flux until some dynamic equilibrium is established that allows transcription to proceed. This dynamic process explains some of the complexity of transcriptional regulation, as individual factors will usually interact weakly with DNA but operate cooperatively with other factors in the promoter complex. Many other features of genome architecture and nuclear structure can have profound effects on gene expression. This is clear from many experiments that demonstrate that different chromosomal locations have quite different expressional capabilities. *A priori*, we might assume that genes expressed from their natural chromosomal sites would operate with the required efficiency. It is now obvious that the combination of genetic factors determined by the organization of DNA sequence elements and different epigenetic factors ensure that the expression of any particular gene is a very complex process.

   In higher eukaryotes, global features of nuclear organization also appear to influence gene expression. Over recent years a significant body of literature has described the compartmentalization of different nuclear functions within dedicated sites. In mammalian cells the density of transcription sites is far lower than expected from estimates of active transcripts and transcription unit complexity, implying that multiple transcription units are active in each site. Morphological analyses, together with immunostaining, indicate that these sites have a zonal organization, with different regions performing specialized roles. Furthermore, as the transcript-containing regions occupy only 0.2% of a HeLa nucleus (Iborra *et al.*, 1996), this arrangement is inconsistent with the view that transcripts will be uniformly dispersed throughout euchromatin (occupying ~ 10% of the nucleus) bound to tracking RNA polymerases. Instead it appears that many chromatin domains may be served by a single, active transcription compartment.

Although many ideas have been proposed to explain the observed arrangement of active sites in mammalian cells, it is still unclear whether this level of organization is an essential component of nuclear structure. Nevertheless, it does at least appear reasonable to argue that by concentrating components required to perform coupled functions in a limited number of sites, it will be possible to execute these functions with optimal efficiency. It is clear that active chromatin domains must maintain their active configuration under circumstances where the majority of chromatin is inert. Although different mechanisms will tend to stabilize these forms, it is important that they are both sufficiently stable to maintain expression and sufficiently unstable to allow reprogramming. Interestingly, proteins capable of remodelling chromatin have been shown to form an integral part of the transcription machinery. Restricting the spatial distribution of the complex in this way will tend to preserve existing chromatin states.

How, then, does this impact on the events that follow virus infection? For the point of view of a virus, the critical requirement is to ensure that the viral genome has an efficient mechanism to engage the active nuclear domains and evade the inert chromatin compartments – where they too might acquire a condensed chromatin status. I have described how nuclear architecture ensures that the active sites are linked to the cytoplasm through a network of interchromatin channels. It is proposed that this interchromatin compartment will represent a 'route of least resistance', so that mRNP complexes passing from transcription sites into the channel network will move quickly to the nuclear periphery. Inevitably, the same route will be used in the reverse direction – from cytoplasm to active sites – by the substrates for RNA synthesis and other dynamic components of the active centres. By engaging this very pathway, a virus will ensure that the active compartments are encountered with optimal efficiency.

References

Abney, J. R., Cutler, M. L., Axelrod, D. and Scalettar, B. A. (1997). Chromatin dynamics in interphase nuclei and its implications for nuclear structure. *J Cell Biol* **137**, 1459–1468.

Alvarez, J. D., Yasui, D. H., Niida, H., Joh, T., Loh, D. Y. and Kohwi-Shigematsu, T. (2000). The MAR-binding protein SATB1 orchestrates temporal and spatial expression of multiple genes during T-cell development. *Genes Dev* **14**, 521–535.

Almeida, F., Saffrich, R., Ansorge, W. and Carmo-Fonseca, M. (1998). Microinjection of anti-coilin antibodies affects the structure of coiled bodies. *J Cell Biol* **14**, 2899–2912.

Becker, P. B. (2002). Nucleosome sliding: facts and fiction. *EMBO J* **21**, 4749–4753.

Bell, A. C. and Felsenfeld, G. (1999). Stopped at the border: boundaries and insulators. *Curr Opin Genet Dev* **9**, 191–198.

Bell, A. C., West, A. G. and Felsenfeld, G. (1999). The protein CTCF is required for the enhancer blocking activity of vertebrate insulators. *Cell* **98**, 387–396.

Bender, M. A., Bulger, M., Close, J. and Groudine, M. (2000). β-globin gene switching and Dnase I sensitivity of the endogenous β-globin locus in mice do not require the locus control region. *Mol Cell* **5**, 387–393.

Berezney, R., Dudey, D. D. and Huberman, J. A. (2000). *Chromosoma* **108**, 471–484.

Berezney, R., Mortillaro, M. J., Ma, H., Wei, X. and Samarabandu, J. (1996). The nuclear matrix: a structural milieu for genomic function. *Int Rev Cytol* **162A**, 1–65.

Blackwood, E. M. and Kadonaga, J. T. (1998). Going the distance: a current view of enhancer action. *Science* **281**, 60–63.

Bode, J., Benham, C., Knopp, A. and Mielke, C. (2000). Transcriptional augmentation: modulation of gene expression by scaffold/matrix attached regions (S/MAR elements). *Crit Rev Eukaryotic Gene Expr* **10**, 73–90.

Breiling, A., Bonte, E., Ferrari, S., Becker, P. B. and Paro, R. (1999). The Drosophila polycomb protein interacts with nucleosomal core particles *in vitro* via its repression domain. *Mol Cell Biol* **19**, 8451–8460.

Bridger, J. M., Boyle, S., Kill, I. R. and Bickmore, W. A. (2000). Re-modelling of nuclear architecture in quiescent and senescent human fibroblasts. *Curr Biol* **10**, 149–152.

Brown, K. E., Guest, S. S., Smale, S. T., Hahm, K., Merkenschlager, M. and Fisher, A. G. (1997). Association of transcriptionally silent genes with Ikaros complexes at centromeric heterochromatin. *Cell* **9**, 845–854.

Bulger, M., Sawado, T., Schubeler, D. and Groudine, M. (2002). ChIPs of the β-globin locus: unraveling gene regulation within an active domain. *Curr Opin Genet Dev* **12**, 170–177.

Cavalli, G. and Paro, R. (1999). Epigenetic inheritance of active chromatin after removal of main transactivator. *Science* **286**, 955–958.

Cook, P. R. (1999). The organization of replication and transcription. *Science* **284**, 1790–1795.

Cremer, T. and Cremer, C. (2001). Chromosome territories, nuclear architecture and gene regulation in mammalian cells. *Nat Rev Genet* **2**, 292–301.

Dundr, M., Hoffmann-Rohrer, U., Hu, Q. Y., Grummt, I., Rothblum, L. I., Phair, R. D. and Misteli, T. (2002). A kinetic framework for a mammalian RNA polymerase *in vivo*. *Science* **298**, 1623–1626.

Elefant, F., Cooke, N. E. and Liebhaber, S. K. (2000). Targeted recruitment of histone acetyltransferase activity to a locus control region. *J Biol Chem* **275**, 13827–13834.

Engel, J. D. and Tanimoto, K. (2000). Looping, linking and chromatin activity: new insights into β-globin locus regulation. *Cell* **100**, 499–502.

Fakan, S. (1994). Perichromatin fibrils are *in situ* forms of nascent transcripts. *Trends Cell Biol* **4**, 86–90.

Farrell, C. M., West, A. G. and Felsenfeld, G. (2002). Conserved CTCF insulator elements flank the mouse and human β-globin loci. *Mol Cell Biol* **22**, 3820–3831.

Femino, A. M., Fay, F. S., Fogarty, K. and Singer, R. H. (1998). Visualization of single RNA transcripts *in situ*. *Science 280*, 585–590.

Ferguson-Smith, A. C. and Surani, M. A. (2001). Imprinting and the epigenetic asymmetry between parental genomes. *Science* **293**, 1086–1089.

Festenstein, R., Sharghi-Namini, S., Fox, M., Roderick, K., Tolaini, M., Norton, T., Saveliev, A., Kioussis, D. and Singh, P. (1999). Heterochromatin protein 1 modifies mammalian PEV in a dose- and chromosomal context-dependent manner. *Nat Genet* **23**, 457–461.

Gasser, S. M. (2002). Visualizing chromatin dynamics in interphase nuclei. *Science* **296**, 1412–1416.

Gilbert, D. M. (2002). Replication timing and transcriptional control: beyond cause and effect. *Curr Opin Cell Biol* **14**, 377–383.

Gribnau, J., Diderich, K., Pruzina, S., Calzolari, R. and Fraser, P. (2000). Intergenic transcription and developmental remodeling of chromatin subdomains in the human β-globin locus. *Mol Cell* **5**, 377–386.

Hendzel, M. J., Kruhlak, M. J., MacLean, N. A. B., Boisvert, F. M., Lever, M. A. and Bazett-Jones, D. P. (2001). Compartmentalization of regulatory proteins in the cell nucleus. *J Steroid Biochem* **76**, 9–21.

Hozak, P., Jackson, D. A. and Cook, P. R. (1994). Replication factories and nuclear bodies: the ultrastructural characterization of replication sites during the cell cycle. *J Cell Sci* **107**, 2191–2202.

Iborra, F. J., Pombo, A., Jackson, D. A. and Cook, P. R. (1996). Active RNA polymerases are localized within discrete transcription 'factories' in human nuclei. *J Cell Sci* **109**, 1427–1436.

Jackson, D. A. and Pombo, A. (1998). Replicon clusters are stable units of chromosome structure: evidence that nuclear organization contributes to the efficient activation and propagation of S-phase in human cells. *J Cell Biol* **140**, 1285–1295.

Jackson, D. A., Hassan, A. B., Errington, R. J. and Cook, P. R. (1993). Visualization of focal sites of transcription within human nuclei. *EMBO J* **12**, 1059–1065.

Jackson, D. A., Iborra, F. J., Manders, E. M. M. and Cook, P. R. (1998). Numbers and organization of RNA polymerases, nascent transcripts, and transcription units in HeLa nuclei. *Mol Biol Cell* **9**, 1523–1536.

Jackson, D. A., Pombo, A. and Iborra, F. (2000). The balance sheet for transcription: an analysis of nuclear RNA metabolism in mammalian cells. *FASEB J* **14**, 242–254.

Jenuwein, T. and Allis, C. D. (2001). Translating the histone code. *Science* **293**, 1074–1080.

Kosak, S. T., Skok, J. A., Medina, K. L., Riblet, R., Le Beau, M. M., Fisher, A. G. and Singh, H. (2002). Subnuclear compartmentalization of immunoglobulin loci during lymphocyte development. *Science* **296**, 158–162.

Lachner, M., O'Carroll, N., Rea, S., Mechtler, K. and Jenuwein, T. (2001). Methylation of histone H3 lysine 9 creates a binding site for HP1 proteins. *Nature* **410**, 116–120.

Litt, M. D., Simpson, M., Recillas-Targa, F., Prioleau, M. N. and Felsenfeld, G. (2001). Transitions in histone acetylation reveal boundaries of three separately regulated neighboring loci. *EMBO J* **20**, 2224–2235.

Lundgren, M., Chow, C. M., Sabbattini, P., Georgiou, A., Minaee, S. and Dillon, N. (2000). Transcription factor dosage affects changes in higher order chromatin structure associated with activation of a heterochromatic gene. *Cell* **103**, 733–743.

Matera, A. G. (1999). Nuclear bodies: multifaceted subdomains of the interchromatin space. *Trends Cell Biol* **9**, 302–309.

Maul, G. G. (1998). Nuclear domain 10, the site of DNA virus transcription and replication. *Bioessays* **20**, 660–667.

McCracken, S., Fong, N., Yankulov, K., Ballantyne, S., Pan, G., Greenblatt, J., Patterson, S. D., Wickens, M. and Bentley, D. L. (1997). The C-terminal domain of RNA polymerase II couples mRNA processing to transcription. *Nature* **385**, 357–361.

McNally, J. G., Muller, W. G., Walker, D., Wolford, R. and Hager, G. L. (2000). The glucocorticoid receptor: rapid exchange with regulatory sites in living cells. *Science* **287**, 1262–1265.

Misteli, T. and Spector, D. L. (1998). The cellular organization of gene expression. *Curr Opin Cell Biol* **10**, 323–331.

Misteli, T. and Spector, D. L. (1999). RNA polymerase II targets pre-mRNA splicing factors to transcription sites *in vivo*. *Mol Cell* **3**, 697–705.

Muyrers-Chen, I. and Paro, R. (2001). Epigenetics: unforeseen regulators in cancer. *Biochim Biophys Acta* **1552**, 15–26.

Nayler, O., Stratling, W., Bourquin, J. P., Stagljar, I., Lindemann, L., Jasper, H., Hartmann, A. M., Fackelmayer, F. O., Ullrich, A. and Stamm, S. (1998). SAF-B protein couples transcription and pre-mRNA splicing to SAR/MAR elements. *Nucleic Acids Res* **26**, 3542–3549.

Nickerson, J. A. (2001). Experimental observations of a nuclear matrix. *J Cell Sci* **114**, 463–474.

Nielsen, A. L., Oulad-Abdelghani, M., Ortiz, J. A., Remboutsika, E., Chambon, P. and Losson, R. (2001). Heterochromatin formation in mammalian cells: interaction between histones and HP1 proteins. *Mol Cell* **7**, 729–739.

Pombo, A., Cuello, P., Schul, W., Yoon, J.-B., Roeder, R. G., Cook, P. R. and Murphy, S. (1998). Regional and temporal specialization in the nucleus: a transcriptionally-active nuclear domain rich in PTF, Oct1 and PIKA antigens associates with specific chromosomes early in the cell cycle. *EMBO J* **17**, 1768–1778.

Pombo, A., Jackson, D. A., Hollinshead, M., Wang, Z., Roeder, R. G. and Cook, P. R. (1999). Regional specialization in human nuclei: visualization of discrete sites of transcription by RNA polymerase III. *EMBO J* **18**, 2241–2253.

Romig, H., Fackelmayer, F. O., Renz, A., Ramsperger, U. and Richter, A. (1992). Characterization of SAF-A, a novel nuclear DNA binding protein from HeLa cells with high affinity for nuclear matrix scaffold attachment DNA elements. *EMBO J* **11**, 3431–3440.

Sacco-Bubulya, P. and Spector, D. L. (2002). Disassembly of interchromatin granule clusters alters the coordination of transcription and pre-mRNA splicing. *J Cell Biol* **156**, 425–436.

Sadoni, N., Langer, S., Fauth, C., Bernardi, G., Cremer, T., Turner, B. M. and Zink, D. (1999). Nuclear organization of mammalian genomes: polar chromosome territories build up functionally distinct higher order compartments. *J Cell Biol* **146**, 1211–1226.

Smith, K. P., Moen, P. T. Jr, Wynder, K. L., Coleman, J. R. and Lawrence, J. B. (1999). Processing of endogenous pre-mRNAs in association with SC-35 domains is gene specific. *J Cell Biol* **144**, 617–629.

Spann, T. E., Goldman, A. E., Wang, C., Huang, S. and Goldman, R. D. (2002). Alteration of nuclear lamin organization inhibis RNA polymerase II-dependent transcription. *J Cell Biol* **156**, 603–608.

Spector, D. L. (2001). Nuclear domains. *J Cell Sci* **114**, 2891–2893.

Sporbert, A., Gahl, A., Ankerhold, R., Leonhardt, H. and Cardoso, M. C. (2002). DNA polymerase clamp shows little turnover at established replication sites but sequential *de novo* assembly at adjacent origin clusters. *Mol Cell* **10**, 1355–1365.

Stratling, W. H. and Yu, F. (1999). Origin and roles of nuclear matrix proteins. Specific functions of the MAR-binding protein MeCP2/ARBP. *Crit Rev Eukaryot Gene Expr* **9**, 311–318.

Taddei, A., Roche, D., Sibarita, J.-B., Turner, B. M. and Almouzni, G. (1999). Duplication and maintenance of heterochromatin domains. *J Cell Biol* **147**, 1153–1166.

Tsukamoto, T., Hashiguchi, N., Janicki, S. M., Tumbar, T., Belmont, A. S. and Spector, D. L. (2000). Visualization of gene activity in living cells. *Nat Cell Biol* **2**, 871–878.

Udvardy, A. (1999). Dividing the empire: boundary chromatin elements delimit the territory of enhancers. *EMBO J* **18**, 1–8.

van Driel, R., Wansink, D. G., van Steensel, B., Grande, M. A., Schul, W. and de Jong, L. (1996). Nuclear domains and the nuclear matrix. *Int Rev Cytol* **162A**, 151–189.

Wang, I. F., Reddy, N. M. and Shen, C. K. J. (2002). Higher order arrangement of the eukaryotic nuclear bodies. *Proc Natl Acad Sci USA* **99**, 13583–13588.

Wansink, D. G., Schul, W., van der Kraan, I., van Steensel, B., van Driel, R. and de Jong, L. (1993). Fluorescent labeling of nascent RNA reveals transcription by RNA polymerase II in domains scattered throughout the nucleus. *J Cell Biol* **122**, 283–293.

Wallace, H., Ansell, R., Clark, J. and McWhir, J. (2000). Pre-selection of integration sites imparts repeatable transgene expression. *Nucleic Acids Res* **28**, 1455–1464.

Workman, J. L. and Kingston, R. E. (1998). Alterations of nucleosome structure as a mechanism of transcriptional regulation. *Annu Rev Biochem* **67**, 545–579.

Wilson, C., Bellen, H. J. and Gehring, W. J. (1990). Position effects on eukaryotic gene expression. *Annu Rev Cell Biol* **6**, 679–714.

Zahn-Zabal, M., Kobr, M., Girod, P. A., Imhof, M., Chatellard, P., de Jesus, M., Wurm, F. and Mermod, N. (2001). Development of stable cell lines for production or regulated expression using matrix attachment regions. *J Biotechnol* **87**, 29–42.

Zaidi, S. K., Javed, A., Choi, J. Y., van Wijnen, A. J., Stein, J. L., Lian, J. B. and Stein, G. S. (2001). A specific targeting signal directs Runx2/Cbfa1 to subnuclear domains and contributes to transactivation of the osteocalcin gene. *J Cell Sci* **114**, 3093–3102.

Zhao, K., Hart, C. M. and Laemmli, U. K. (1995). Visualization of chromosomal domains with boundary element-associated factor BEAF-32. *Cell* **81**, 879–889.

2 The Eukaryotic Cell Cycle

Jane V. Harper and Gavin Brooks

Cardiovascular Research Group, School of Animal and
Microbial Sciences, University of Reading,
Whiteknights, Reading, Berkshire, UK

2.1 Introduction

Cell division in eukaryotes is an evolutionarily conserved process that involves an ordered and tightly controlled series of events. The cell cycle consists of five distinct phases: three gap phases, G_0, during which cells remain in a quiescent or resting state, and G_1 and G_2, during which RNA and protein synthesis occur; S-phase, during which DNA is replicated; and M phase, during which cells undergo mitosis and cytokinesis (Figure 2.1). G_0, G_1, S and G_2 are collectively referred to as interphase (between mitoses). Some cells in the body remain quiescent for their whole lifetime and do not undergo cell division; however, stimulation of the cell by external factors such as mitogens (e.g. growth factors) causes these quiescent cells to re-enter the cell cycle and undergo division. Binding of a growth factor molecule to its cell surface receptor can stimulate a number of signalling pathways, an example of which is the Ras-dependent mitogen-activated protein kinase (MAPK) pathway, which plays a major role in entry into G_1, as discussed in more detail later in this chapter (Section 2.2.1). Once cells enter G_1, synthesis of mRNAs and proteins necessary for DNA synthesis occur to allow cells to enter S phase.

As discussed in more detail later in this book, viruses can manipulate cellular function in order to promote viral replication. Evidence suggests that viruses and viral proteins may target subnuclear structures (e.g. the nucleolus, Cajal bodies) to increase transcription and translation and this may alter the host cell cycle (reviewed in Hiscox, 2002); indeed, the nucleolus and associated proteins have been implicated in influencing host cell cycle progression and these structures are, in turn, regulated by the cell cycle (Carmo-Fonseca *et al.*, 2000). Furthermore, a number of viral

Viruses and the Nucleus   Edited by Julian A. Hiscox

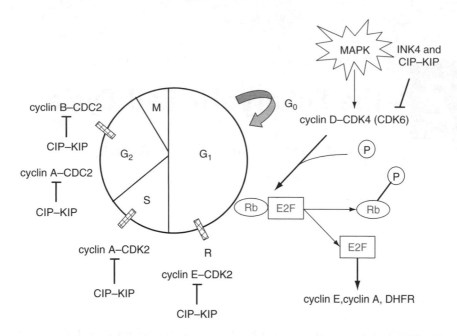

Figure 2.1 The five distinct phases of the cell cycle are each controlled by specific cyclin–CDK complexes, which are in turn negatively regulated by CIP/KIP and INK4 CDKI family members. E2F transcription factors function at the restriction point (R), leading to the activation of genes essential for DNA synthesis and cell cycle progression. E2F complexed with hypophosphorylated Rb cannot activate transcription. Hyperphosphorylation of Rb causes dissociation from E2F. Cell cycle checkpoints are shown as hashed bars. ⊤, inhibition step; ↑, activation step

proteins have been shown to localize to the nucleolus (reviewed in Hiscox, 2002). Thus, by targeting the nucleolus and interacting directly with nucleolar proteins, viruses can alter host cell transcription and translation and possibly cell cycle machinery to ensure the successful propagation of new virus.

The cell cycle consists of a number of checkpoints that exist to ensure normal cell cycle progression. The primary checkpoint acts late in G_1 and is known as the restriction (R) point (Figure 2.1). Once cells have passed this point they normally are committed to a round of cell division. Other checkpoints exist in S phase to activate DNA repair mechanisms if necessary and at the G_2–M transition to ensure cells have fully replicated their DNA and that it is undamaged before they enter mitosis.

The average length of time for a mammalian cell to progress around the cell cycle and undergo division is approximately 24 h. Cell cycle time varies depending on cell type and this variation is due to differences in the time spent between cytokinesis and the restriction point (i.e. G_1). The time taken for a cell to pass from S phase into M is extremely constant between cells and typically is in the region of ~ 6 h for S phase, 4 h for G_2 and 1–2 h for mitosis and cytokinesis (Alberts *et al.*, 2002).

Progression through the cell cycle is under the strict control of various cell cycle molecules that themselves are regulated by phosphorylation and dephosphorylation events (Alberts *et al.*, 2002). The cyclin-dependent kinases (CDKs) play a crucial role in regulating cell cycle events along with their regulatory subunits, the cyclins. Cyclin levels vary dramatically through the cell cycle as a consequence of changes in transcription and ubiquitin-mediated degradation. CDK activity is negatively regulated by the cyclin-dependent kinase inhibitors (CDKIs). Each cyclin–CDK complex has a defined role at a distinct phase(s) of the cell cycle (Figure 2.1). For example, cyclin D–CDK4(CDK6) complexes, which are one of the major targets for the Ras/MAPK pathway, initiate progression through G_1 by phosphorylating substrates that eventually lead to the activation of transcription of genes necessary for DNA synthesis and subsequent cell cycle progression; the cyclin E–CDK2 complex is important in the G_1–S transition, where levels peak at the restriction point (Dulic *et al.*, 1992; Koff *et al.*, 1992); cyclin A–CDK2 is important during S phase progression; and cyclin A–CDC2 (also known as CDK1) and cyclin B–CDC2 are important for progression through G_2 and M. The regulation of cyclin synthesis and degradation in addition to CDK activity are tightly controlled and are key to the control of ordered progression through the cell cycle. In the following sections we will overview the various stages of the cell cycle and the molecules that regulate progression through each stage of the cycle.

2.1.1 Cyclins and cyclin-dependent kinases (CDKs)

The eukaryotic cell cycle is regulated by the sequential formation, activation and inactivation of a series of cell cycle regulatory molecules that include the cyclins (regulatory subunits) and the CDKs (catalytic kinase subunits) (Brooks *et al.*, 1998; Tyson *et al.*, 1996). Different cyclins bind specifically to different CDKs to form distinct complexes at specific phases of the cell cycle and thereby drive the cell from one phase to another (Table 2.1). The cyclins are a family of proteins which, as their name suggests, are synthesized and destroyed during each cell cycle. To date, eight cyclins have been described that directly affect cell cycle progression, these being: cyclins A_1 and A_2, $B_{1,2,3}$, C, $D_{1,2,3}$, $E_{1,2}$, F, G_1 and G_2, and H, that all share a ~ 150 amino acid region of homology called the 'cyclin box' that binds to the N-terminal end of specific CDKs (McGill and Brooks, 1995). Cyclins C, D and E are short-lived proteins that function mainly during the G1 phase and at the G1–S transition before being destroyed via the ubiquitin pathway (Tyson *et al.*, 1996). Cyclins A and B, on the other hand, are mitotic cyclins that remain stable throughout interphase, but are rapidly proteolysed during mitosis also by an ubiquitin-dependent pathway. Little information currently is available regarding the recently described cyclins F and G, whereas cyclin H has been shown to form complexes specifically with CDK7 to produce an enzyme known as cyclin-dependent kinase-activating kinase (CAK) that is involved in the activation of CDC2 and CDK2 kinases by phosphorylating Thr^{160} and Thr^{161}, respectively (Martinez *et al.*, 1997). Another cyclin, cyclin T, has also been reported in the literature. This protein is known to pair with CDK 9 and is involved in various processes, e.g. basal

Table 2.1 Summary of CDK–cyclin complexes that are activated during cell cycle progression and the CDKI molecules that inhibit their function

CDK	Cyclin partner	Stage of cell cycle where complex is activated	CDKIs that inhibit cyclin–CDK complex	References
1 (CDC2)	A	G_2/M	p21, p27, p57	Bicknell *et al.*, 2003
	B	M (prophase)		
2	A	Late S/G_2	p21, p27, p57	Bicknell *et al.*, 2003
	E	G_1/S		
3	NK	G_1	NK	Braun *et al.*, 1998; Keezer and Gilbert, 2002
4	D1, D2, D3	G_1	p15, p16, p21, p27	Bicknell *et al.*, 2003
5	NK	G_0	NK	Dhavan and Tsai, 2001
6	D1, D2, D3	G_1	p15, p16, p21, p27	Bicknell *et al.*, 2003
7	H	Ubiquitous	NK	Nigg, 1996
8	C	Possibly G_1	NK	Leclerc and Leopold, 1996
9	T	G_1	NK	Sano and Schneider, 2003

NK, not known.

transcription, signal transduction and differentiation (reviewed in Simone and Giordano, 2001; Napolitano *et al.*, 2002).

The CDKs are a family of serine/threonine protein kinases which bind to, and are activated by, specific cyclins. To date, at least nine CDKs have been described, viz. CDC2 (CDK1), CDK2, CDK3, CDK4, CDK5, CDK6, CDK7, CDK8 and CDK9. CDKs 4, 5 and 6 complex mainly with the cyclin D family and function during the G0/G1 phases of the cycle; CDK2 also can bind with members of the cyclin D family, but more commonly associates with cyclins A and E and functions during the G1 phase and during the G1–S transition. As mentioned above, CDK7 is found in association with cyclin H and is able to phosphorylate either CDC2, CDK2 or the C-terminal domain of the largest subunit of RNA polymerase II, in addition to the TATA box-binding protein or TFIIE (Martinez *et al.*, 1997). CDC2 is the mitotic CDK and forms complexes with cyclins A and B and functions in the S, G2 and M phases of the cell cycle. CDK8 pairs with cyclin C and is found in a large multiprotein complex with RNA polymerase II. CDK 8 and cyclin C may control RNA polymerase function (reviewed in Leclerc and Leopold, 1996). Finally, CDK 9 is a serine-threonine CDC2-related kinase and pairs with T-type cyclins. Its activity is not cell cycle-regulated and it is involved in many processes, such as differentiation and basal transcription (reviewed in de Falco and Giordano, 1998; Simone and Giordano, 2001; Napolitano *et al.*, 2002).

As stated above, specific CDKs bind to specific cyclins to form an active complex that integrates signals from extracellular molecules and controls progression through the cell cycle. The CDK subunit on its own has no detectable kinase activity and requires sequential activation by cyclin binding, subsequent phosphorylation by CAK and dephosphorylation by CDC25 protein phosphatase (see Section 2.6.1). This activation process occurs in a two-step manner as follows.

1. Binding of the cyclin to the CDK confers partial activity to the kinase. Cyclin binding causes a conformational change in the CDK which brings together specific residues involved in orientating ATP phosphate atoms ready for catalysis within the catalytic cleft. These conformational changes also set the stage for subsequent phosphorylation and full activation.

2. Phosphorylation of the cyclin–CDK complex is performed by CAK which increases CDK activity approximately 100-fold (Russo *et al.*, 1996). Phosphorylation occurs on a conserved threonine residue within the T-loop region of the CDK (Thr^{160} in CDC2 and Thr^{161} in other CDKs). Cyclin binding moves the T-loop to expose the phosphorylation site, allowing full activation of the CDK.

Once activated, the various cyclin–CDK complexes phosphorylate a number of specific substrates involved in cell cycle progression. Such substrates include the retinoblastoma (Rb) pocket proteins (reviewed in Grana *et al.*, 1998; Vidal and Koff, 2000) and histones. Evidence exists to suggest that cyclins may be involved in determining the substrate specificity of CDKs (reviewed in Miller and Cross, 2001). For example, cyclin A–CDK2 and cyclin A–CDC2, but not cyclin B–CDC2, can phosphorylate p107, showing regulation of substrate specificity between kinases complexed with cyclins A and B (Peeper *et al.*, 1993). The E2F-1/DP-1 heterodimer is not a substrate for the active cyclin D-dependent kinases but is efficiently phosphorylated by cyclin B-dependent kinases (Dynlacht *et al.*, 1997). Interestingly, whereas phosphorylation of E2F-1/DP-1 by cyclin B-dependent kinases does not result in a loss of DNA binding activity, phosphorylation of this same heterodimer by cyclin A-dependent kinases does lead to loss of DNA binding (Dynlacht *et al.*, 1997). Thus, different CDK complexes can exert contrasting effects on a common substrate, depending upon the complexed cyclin. The regulation of CDKs themselves by other molecules may also differ depending on the bound cyclin. Thus, cyclin A–CDC2 complexes do not require activation by CDC25, whereas cyclin B–CDC2 complexes do (Devault *et al.*, 1992).

2.1.2 Cyclin-dependent kinase inhibitors (CDKIs)

The cyclins and CDKs often are referred to as positive regulators of the eukaryotic cell cycle. A family of negative regulators also exists, called the cyclin-dependent kinase inhibitors (CDKIs) (Reed *et al.*, 1994; McGill and Brooks, 1995; Pines, 1997; Brooks *et al.*, 1998) The CDKIs comprise two structurally distinct families, the

INK4 and CIP/KIP families (reviewed in Paveltich, 1999). The INK4 family includes p14, p15 (INK4B), p16 (INK4A), p18 (INK4C) and p19 (INK4D), which inhibit specifically G_1 cyclin–CDK complexes (cyclin D–CDK4 and cyclin D–CDK6) and are involved in G_1 phase control. The CIP/KIP family includes p21 (CIP1/WAF1/SDI1), p27 (KIP1) and p57 (KIP2), which are 38–44% identical in the first 70 amino acid region of their amino-termini – a region that is involved in cyclin binding and kinase inhibitory function (Pines, 1997; Brooks *et al.*, 1998; Gartel *et al.*, 1996). The CIP/KIP family displays a broader specificity than the INK4 family, since members interact with, and inhibit the kinase activities of, cyclin E–CDK2, cyclin D–CDK4, cyclin D–CDK6, cyclin A–CDK2 and cyclin B–CDC2 complexes, and function throughout the cell cycle (Table 2.1) (Pines, 1997). The tumour suppressor protein, p53, also plays an important role in cell cycle arrest at the G1 and G2 checkpoints subsequent to inducing apoptosis (Prives, 1993; Ewen, 1996; Jacks and Weinberg, 1996). Each of these families inhibits CDK activity by distinct mechanisms. The CIP/KIP family of CDKIs bind and inhibit the active cyclin–CDK complex (Sherr and Roberts, 1995), whereas the INK4 family bind isolated CDKs preventing association with cyclins. INK4 members can also bind cyclin–CDK complexes causing inhibition without disrupting the complex (Serrano, 1997). The p53 protein has a central sequence-specific DNA binding domain and a transcriptional activation domain at its amino-terminus and, in response to DNA damage, can induce the transcription of the CDKI p21, which inhibits the activation of various G_1 cyclin–CDK complexes (Ewen, 1996; Gartel *et al.*, 1996).

In the case of p21, this CDKI has been shown to exist in both active and inactive cyclin–CDK complexes and it has been suggested that the stoichiometry of p21 binding to the cyclin–CDK complex controls activation/inhibition of the complex (Zhang *et al*, 1994). In support of this hypothesis, Zhang and colleagues demonstrated that p21 exists both in catalytically active and inactive cyclin–CDK complexes and that the addition of sub-saturating concentrations of p21 to cyclin A–CDK2 complexes resulted in a progressive increase in CDK2 activity, suggesting that low concentrations of p21 might function as a cyclin–CDK assembly factor, whereas the binding of more than one p21 molecule is required to inhibit CDK2 activity.

Anti-proliferative signals, such as contact inhibition, senescence (Alcorta *et al.*, 1996), extracellular anti-mitogenic factors (Reynisdottir *et al.*, 1995), and cell cycle checkpoints, such as p53 (el-Deiry *et al.*, 1993), induce expression of p27, p16, p15 and p21, respectively. The role of cell cycle molecules in regulating proliferation is highlighted by the fact that a number of these molecules are found mutated or deregulated in numerous tumours. For example, p16 is mutated in approximately one-third of all human cancers (Kamb *et al.*, 1994; Nobori *et al.*, 1994; Serrano, 1997) and p53 is the most frequently mutated gene identified in human tumours (Levine, 1997). Also, many types of tumour show low expression levels of p27 that is associated with a poor prognosis (Porter *et al.*, 1997) and cyclin D_1 often is found at increased levels in breast cancer (Hunter and Pines, 1994; Sherr, 1996). CDKs have also been found deregulated in tumours; e.g. CDK4 is mutated in melanoma (Wolfel *et al.*, 1995; Zuo *et al.*, 1996).

2.2 The G$_0$/G$_1$ transition

The mammalian cell cycle is influenced by external signals during the G$_0$ and G$_1$ phases. The MAPK cascade is one of the most ubiquitous signal transduction pathways that regulates several biological processes, including progression of the cell cycle. The MAPK cascade consists of three evolutionary conserved protein kinases that are activated sequentially in a Ras-dependent manner (reviewed in Marshall, 1995).

The MAPK cascade influences cellular proliferation by targeting the cyclin D-dependent kinases (Lavoie *et al*, 1996; Cheng *et al.*, 1998; Balmanno and Cook, 1999). Cyclin D-dependent kinases are essential in early G$_1$, where they initiate passage through the cell cycle, since they are required to phosphorylate the Rb family of pocket proteins, thereby causing activation of genes necessary for cell cycle progression (see Figure 2.1). Evidence for this comes from the fact that cells which proliferate in the absence of mitogens, e.g. during embryogenesis, have very little cyclin D-dependent kinase activity (Meyer *et al.*, 2002).

2.2.1 Role of MAP kinase in G$_1$ cell cycle progression

The activation of cyclin D–CDK4 and cyclin D–CDK6 complexes is essential for passage through the G$_1$ phase and they exert their regulation on cell cycle progression by phosphorylating Rb pocket proteins. The Rb pocket protein family serves to repress the activity of the E2F transcription factors, which are themselves essential for transcription of genes necessary for entry into S phase (discussed in more detail in Section 2.3). The Ras/MAPK pathway has been shown to directly control cyclin D expression. This is mediated primarily by MAPK controlling activation of the AP-1 and ETS transcription factors, which then transactivate the cyclin D promoter that contains specific binding sites for both AP-1 and ETS (Lavoie *et al.*, 1996; Liu *et al.*, 1995). Furthermore, expression studies using direct inhibitors of cyclin D–CDK4(CDK6) complexes (e.g. p21) inhibits Ras-induced proliferation (Serrano *et al.*, 1995). These data demonstrate that MAPK directly regulates cyclin D expression, hence its effects on CDK4 and CDK6 activity.

The Ras–MAPK pathway also has been shown to regulate CDKs post-transcriptionally by affecting their assembly and catalytic activities. Although the primary role of p21 and p27 is to negatively regulate the activity of CDKs, they also are involved in the assembly of cyclin D–CDK4(CDK6) complexes during early G$_1$ (Sherr and Roberts, 1999; Cheng *et al.*, 1999) (see Section 2.1.2). In addition, the Ras/MAPK pathway has been shown to directly regulate the synthesis of the CIP/KIP family of inhibitors and it was demonstrated that growth factor stimulation of quiescent cells causes cell cycle re-entry and transient expression of p21, which is dependent on MAPK activity (Bottazzi *et al.*, 1999).

Entry into S phase is partly dependent on proteolytic degradation of p27 and this has also been shown to be dependent on MAPK activity (Sheaff *et al.*, 1997). Rivard *et al.* (1999)

demonstrated that MAPK activity was required for p27 downregulation and S phase entry. These investigators also observed that expression of Ras resulted in decreased p27 protein levels and an increase in E2F-dependent transcriptional activity.

Taken together, these data provide evidence for a role for the Ras–MAPK pathway in controlling G_1–S progression by a number of mechanisms. The first is by induction of cyclin D expression, which is involved in the synthesis of genes necessary for entry into S phase. This occurs as a result of the release of E2F transcription factors subsequent to phosphorylation of Rb pocket proteins by cyclin D-dependent kinases. This pathway is also involved in the assembly of cyclin A–CDK2 and cyclin E–CDK2 complexes by increasing levels of the CDKIs involved in cyclin–CDK assembly. Finally, the Ras–MAPK pathway has been shown to decrease p27 levels and increase E2F-dependent transcription, which is an event essential for entry into S phase (see Section 2.3.2).

More recently, a role for MAPK in regulating the G_2–M transition has been suggested. Thus, it has been shown that ionizing radiation can activate the MAPK pathway (Kasid *et al.*, 1996; Sklar, 1988) and cells expressing a dominant negative MAPKK are unable to recover from radiation-induced G_2–M arrest (Abbott and Holt, 1999). Additionally, treatment of cells with a MAPK inhibitor induces G_2–M arrest, concomitant with a reduction in cyclin B/CDC2 activity (Wright *et al.*, 1999). This data suggests that the Ras–MAPK pathway plays a regulatory role at many points during the cell cycle.

The data discussed above demonstrates the regulation of the cell cycle by the MAPK extracellular mitogenic signalling pathway. If the activity of the MAPK pathway were maintained at an abnormally high level, this could lead to cellular transformation and tumourigenesis. Therefore, cells have developed a safety mechanism in order to counteract this possibility. It has been shown that expression of oncogenic Ras or constitutively active MAPKK causes cell cycle arrest with high levels of p21, which is expressed in a p53-dependent manner (Lin *et al.*, 1998; Serrano *et al.*, 1997).

2.3   The G_1–S transition

As stated above, progression through the cell cycle is tightly regulated by the activation status of various cyclin–CDK complexes. One of the most extensively studied substrates of the cyclin–CDKs is the Rb family of pocket proteins. The Rb pocket proteins play a major role at the restriction point in late G_1 and regulate entry into S phase. The intracellular targets of the Rb family are the E2F family of transcription factors, which are themselves required for the expression of genes necessary for DNA synthesis. In their hypophosphorylated state, the Rb pocket proteins bind to, and inhibit the transcriptional activity of, E2F. Phosphorylation of Rb by cyclin D–CDK4(CDK6) complexes causes disassociation of the pocket protein from E2F and subsequent activation of E2F-dependent genes (see Figure 2.1). The precise relationship between the various E2Fs and Rb family members is extremely complex and is discussed in more detail in the following sections.

2.3.1 The retinoblastoma (Rb) pocket protein family

The Rb family of pocket proteins comprises a group of tumour suppressor proteins consisting of three members; pRb, p107 and p130. As their name suggests, these proteins contain a pocket region that binds cellular targets. This region also is capable of binding a number of viral oncoproteins, such as the adenovirus E1A protein, SV40 large T antigen and the human papillomavirus 16 E7 protein (Vousden, 1995), demonstrating one mechanism by which tumour viruses can interfere with cell cycle progression. The functions of different members of the Rb family are regulated during the cell cycle by phosphorylation events and also by changes in expression. During G$_1$ the pocket proteins are found in a hypophosphorylated state, where they bind to members of the E2F transcription factor family (see Section 2.3.2, below). As cells progress through the cell cycle, these proteins become hyperphosphorylated as a result of phosphorylation by cyclin D–CDK4(CDK6) and cyclin E–CDK2 complexes. Each family member also displays differential expression throughout the cell cycle (Figure 2.2). Thus, pRb is expressed throughout the cell cycle but is hyperphosphorylated and therefore inactivated in late G$_1$, although by mitosis it becomes dephosphorylated; p130 is highly expressed in G$_0$, whereas levels diminish as cells progress into S phase, consistent with a role for p130 in maintaining quiescence (reviewed in Grana *et al.*, 1998); also, p107 shows an opposite expression pattern to p130, such that low levels are found in G$_0$ which then increase as cells progress through G$_1$ into S.

The importance of the Rb family of tumour suppressor proteins in controlling the restriction point is demonstrated by the fact that the Rb proteins are targets of deregulation in most types of human cancer (Weinberg, 1995; Harlow, 1996; Sherr, 1996). Indeed, pRb has been reported to be mutated in ~30% of all human cancers (reviewed in Fearon, 1997).

The different actions of Rb pocket proteins with respect to E2F regulation was demonstrated in a study by Hurford *et al.* (1997). These authors showed that pRb has different functions from p107 and p130. They also demonstrated that p107 and p130 functions overlap, since in cells lacking p107 or p130 there were no changes in E2F-regulated transcription. However, in cells lacking both p107 and p130, or lacking pRb alone, an increase in E2F-regulated transcription was observed (Hurford *et al.*, 1997).

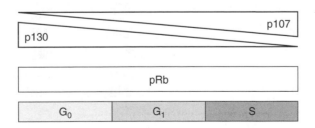

Figure 2.2   Expression patterns of Rb pocket protein family members through G$_0$, G$_1$ and S phase of the cell cycle. Note that phosphorylation of these proteins alters, depending upon the phase of the cell cycle. Thus, in G$_0$–G$_1$ pRb is found in a hypophosphorylated state bound to E2F, whereas it becomes phosphorylated and dissociates from E2F at the G$_1$–S border (see text for details)

2.3.2   The E2F transcription factors

Another family of molecules that regulates the G_1–S transition is that comprising the E2F transcription factors. To date, six E2F members have been described (E2Fs 1–6) and these molecules exist as heterodimers paired with a DP subunit (Figure 2.3). Two mammalian DP genes have so far been identified (DP 1 and 2) (Nevins, 1992). E2F and DP proteins contain highly conserved DNA-binding and dimerization domains (see Figure 2.3). The E2F and DP proteins activate transcription in a synergistic manner and DP proteins appear to act indirectly by enhancing the activity of E2F (Helin *et al.*, 1993).

The Rb pocket proteins bind to, and sterically hinder transcriptional activity of, the E2F–DP complex, thereby enabling the E2F transcription factors to act as repressors of gene transcription (reviewed in Dyson, 1998). Phosphorylation of the pocket protein component of the E2F–pocket protein complex by cyclin D–CDK4(CDK6) complexes in the G_1 phase of the cycle leads to dissociation of the phosphorylated pocket protein and E2F, followed by E2F-mediated transactivation of promoters of genes necessary for S phase progression, e.g. dihydrofolate reductase (DHFR), cyclin E and cyclin A (Figure 2.1) (Waga *et al.*, 1994; Luo *et al.*, 1995; Knibiehler *et al.*, 1996). Structurally and functionally, the E2F transcription factors can be divided into three main categories: E2Fs 1–3, which bind to retinoblastoma protein (pRb) and are believed to be involved in proliferation; E2F-4 and E2F-5, which bind preferentially to p130 and/or p107 and are believed to play a role in differentiation; and E2F-6, which has been described as a transcriptional repressor (Weinberg, 1995). Transcriptional activity of the various E2F members is regulated at different phases of the cell cycle by the Rb pocket proteins and the formation of such complexes occurs at distinct phases of the cell cycle, suggesting a role for E2F in regulating events beyond G1–S phase progression. Thus, in quiescent cells, E2F–p130 complexes predominate, whereas E2F–p107 and E2F–pRb complexes persist through the G_1–S transition in proliferating cells (Woo *et al.*, 1997).

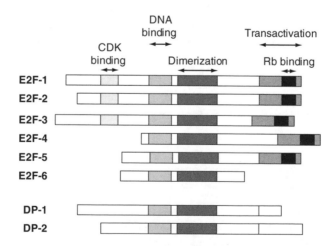

Figure 2.3   E2F and DP conserved domain structures. The Rb binding domain is located in the transactivation domain of the E2F proteins

E2F6 is the most recently identified family member and differs from the other E2Fs in that it lacks the transactivation domain common to all other E2Fs; for this reason E2F6 is believed to be a transcriptional repressor (Morkel *et al.*, 1997; Cartwright *et al.*, 1998; Trimarchi *et al.*, 1998). Indeed, it has been shown that overexpression of E2F6 had no specific effect on the cell cycle of asynchronous cells. However, it did prevent G$_0$-arrested cells from entering S phase when stimulated (Gaubatz *et al.*, 1998), suggesting that E2F6 may play a role in activating genes necessary for exit from G$_0$ but that it is not required for normal cell cycle progression.

Activation of transcription by E2F

The precise mechanism by which E2Fs activate transcription is unclear, although studies have shown that the transactivation domain of E2F1 can interact with cAMP response element binding protein (CBP) (Trouche *et al.*, 1996). CBP is a transcriptional co-activator and possesses intrinsic histone acetyl transferase activity, which can modulate chromatin structure and hence gene transcription (Wang *et al.*, 1998). Acetylation of histones causes weakening of the interaction between DNA and the nucleosome, thereby making the DNA more accessible for transcription (Trouche *et al.*, 1996). E2F complexes have also been shown to 'bend' DNA and this could be important for activation in certain instances (Cress and Nevins, 1996).

Subcellular localization of E2F transcription factors

One level of regulation of E2F function occurs through changes in the subcellular localization of individual E2F transcription factors. For example, it has been demonstrated that E2F4 is expressed in the nucleus and cytoplasm of quiescent cells, but as cells reach S phase this molecule is found almost exclusively in the cytoplasm (Lindeman *et al.*, 1997; Verona *et al.*, 1997). This relocation ensures that repressive E2F4–p107 complexes cannot bind E2F-responsive genes. E2F4 lacks a nuclear localization sequence (NLS) and might therefore gain entry to the nucleus by association with DPs or Rb pocket proteins, both of which contain an NLS. Indeed, studies have shown that when overexpressed in cells, E2F4 is only transported to the nucleus when co-expressed with DP2, p107 or p130 (Magae *et al.*, 1996; Lindeman *et al.*, 1997). E2F5 also lacks an NLS, although nuclear localization has been shown to occur in a DP- and Rb-independent manner, such that transport of this E2F to the nucleus is mediated via formation of nuclear pore complexes (Apostolova *et al.*, 2002).

Inactivation of E2F transcription factors

Inactivation of E2F is as important as E2F activation for continued progression through the cell cycle. It has been shown that expression of a constitutively active mutant of E2F1 or DP1 causes accumulation of cells in S phase that leads eventually to apoptosis (Krek and Livingston, 1995). These results imply that inactivation of E2F could be required for exit from S phase.

Inactivation of E2F may be controlled by phosphorylation of E2F and DP subunits leading to an inhibition of DNA-binding activity (Dynlacht *et al.*, 1994; Krek *et al.*, 1994; Krek and Livingston, 1995). E2Fs 1–3 have been shown to contain a conserved region which enables interaction with cyclin A–CDK2 or cyclin E–CDK2, and these interactions lead to inhibitory phosphorylations on these transcription factors (Krek *et al.*, 1994). There also is evidence for ubiquitin-directed degradation of E2Fs 1–4 (Hateboer *et al.*, 1996; Hofmann *et al.*, 1996) that would lead to regulation of DNA-binding activity.

Mechanism of pRb-dependent repression of E2F transcriptional activity

The exact mechanism of pRb-mediated repression has only recently become understood, following the discovery that histone deacetylase-1 (HDAC-1) is involved (Brehm *et al.*, 1998; Luo *et al.*, 1998; Magnaghi-Jaulin *et al.*, 1998). Recruitment of HDAC-1 to the DNA is thought to repress gene activation by altering chromatin structure. Nucleosomal histones have a high proportion of positively charged amino acids that facilitate interaction with negatively charged DNA. Deacetylation is thought to occur on histone tails protruding from the nucleosome (Loyola *et al.*, 2001) and this increases their positive charge, causing a tighter interaction with DNA, thereby making the DNA less accessible for transcription. Takahashi *et al.* (2000) observed that high levels of acetylation correlated with activation of E2F-responsive genes in late G_1 and at the G_1–S border. However, during quiescence, when transcriptional activity is low, histones showed reduced levels of acetylation (Takahashi *et al.*, 2000). They also showed that acetylation of genes occurred in a cell cycle-dependent manner. Thus, during G_0, when transcription levels are low, histones display reduced acetylation levels due to the recruitment of HDAC-1. However, as cells progress through G_1 into the S phase this repression is relieved by HAT (Figure 2.4). As mentioned earlier, it has been shown that E2F is able to interact with both CBP and HAT *in vitro* and also in transiently transfected cells (Trouche *et al.*, 1996).

The role of HDAC-1 in repressing gene transcription has been demonstrated further, such that HDAC-1 physically interacts with the DHFR promoter to affect cell growth. Thus, an association of HDAC-1 with the DHFR promoter was detected in G_0 and early G_1, when the gene was silent, and also histone H4 showed low acetylation levels. This association then decreased as cells entered S phase, consistent with an increase in DHFR mRNA levels (Ferreira *et al*, 2001).

It has been suggested that chromatin modifying factors may form multienzyme co-repressor complexes at promoter regions. Thus, the modifying factor, mSin3, has been shown to form a co-repressor complex that acts as scaffold for the assembly of HDAC-1 repressor complexes (Hassig *et al.*, 1997; Laherty *et al.*, 1997; Nagy *et al.*,

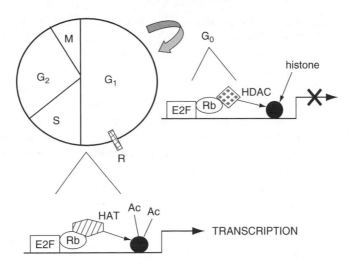

Figure 2.4 HDAC-1 is recruited to DNA by Rb, causing deacetylation and inhibition of transcription during G$_0$. At the G$_1$–S transition, this repression is relieved by the action of HAT, allowing transcription of genes necessary for DNA synthesis

1997). The occupancy of E2F-regulated promoters by HDAC-1 and mSin3B in pocket protein-deficient cells was recently assessed (Rayman *et al.*, 2002). These studies showed that recruitment of HDAC-1, but not mSin3B, was completely dependent upon p107 and p130 but not on pRb. This data suggests that specific E2F–Rb complexes are involved in recruitment of chromatin-modifying factors during G$_0$/G$_1$. There also is evidence that the tumour suppressor gene, transforming growth factor-β1 (TGF-β1), might function by recruitment of HDAC-1, since transgenic mice overexpressing TGF-β1 showed enhanced HDAC-1 binding to p130 compared to control animals (Bouzahzah, 2000). Thus, TGF-β1 may exert its growth inhibitory effects by recruitment of HDAC-1.

The model of Rb pocket proteins causing transcriptional repression by association with HDAC-1 is consistent with the model that pRb phosphorylation by cyclin D–CDK4(CDK6) complexes relieves E2F transcriptional inhibition. Phosphorylation of pRb by CDK4(CDK6) initiates intramolecular interactions between the carboxy-terminus of pRB and the pocket region, which displaces HDAC-1 from the pocket, thereby facilitating subsequent phosphorylation of pRb by CDK2-complexes, followed by disassociation from E2F. These results suggest a sequential phosphorylation of pRb by CDK4(CDK6) and CDK2 (Harbour *et al.*, 1999).

2.3.3 Role of the cyclin E–CDK2 complex in the G$_1$–S transition

As cells approach the G$_1$–S border, control of the cell cycle becomes dominated by cyclin E–CDK2 complexes. It has been demonstrated that overexpression of cyclin

E–CDK2 promotes S phase entry and blocking the kinase activity of this complex inhibits progression into S phase (Ohtsubo and Roberts, 1993; Tsai *et al.*, 1993; Resnitzky *et al.*, 1994). Consistent with its role in S phase, the cyclin E–CDK2 complex has been shown to be required for the initiation of DNA replication (Ohtsubo and Roberts, 1993; Tsai *et al.*, 1993; Resnitzky *et al.*, 1994). The importance of phosphorylation of pRb by cyclin E–CDK2 at the G_1–S border has already been discussed above (Section 2.3.1).

A recently discovered substrate for cyclin E–CDK2 has also been shown to be important for S phase entry. This substrate is nuclear protein that maps to the ATM locus (NPAT). NPAT was identified from a phage expression library, using cyclin E–CDK2 as a probe, and was shown by immunoprecipitation studies to associate with cyclin E–CDK2 *in vivo*. The NPAT protein was shown to be present at all stages of the cell cycle in synchronized cells; however, levels peaked at the G_1–S boundary and decreased as cells progressed through S. Overexpression of NPAT caused an increase in the number of S phase cells, suggesting that NPAT expression may be a rate limiting step for S phase entry (Zhao *et al.*, 1998).

Histone gene expression is a major event that occurs as cells pass into S phase. Histones form part of the nucleosomes that are a fundamental subunit of chromatin, and NPAT has been implicated in the regulation of histone gene expression. Both cyclin E and NPAT have been shown to localize to histone gene clusters at the G_1–S border, and phosphorylation of NPAT is required to activate histone gene expression (Ma *et al.*, 2000; Zhao *et al.*, 2000). Therefore, evidence exists to show that cyclin E–CDK2 regulates histone gene expression by phosphorylation of NPAT, a process required for entry into S phase (see Figure 2.5)

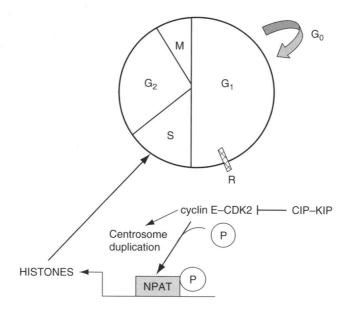

Figure 2.5   Activation of NPAT by cyclin E–CDK2 causes histone gene expression necessary for DNA synthesis and S phase progression

2.4 S phase

S phase is the point during the cell cycle at which a cell duplicates its chromosomes in readiness for mitosis and cell division (Alberts *et al.*, 2002). A number of checkpoints exist to ensure that DNA is replicated only once per cycle, that it is fully and correctly replicated and that replication occurs before cell division. Another important event during S phase, other than DNA replication, is centrosome duplication. The centrosomes are the primary microtubule organizing centre, and failure of cells to coordinate centrosome duplication with DNA replication leads to abnormal segregation of chromosomes, causing genomic instability, and this can promote cancer.

2.4.1 Role of the cyclin E–CDK2 complex in S phase progression

There is much evidence to suggest that DNA synthesis in higher eukaryotes is initiated by activation of CDK2 (Tsai *et al.*, 1993; reviewed in Nasmyth, 1996; Sherr, 1996). CDK2 associates with cyclin E just prior to the onset of S phase and the role of this complex in the activation of NPAT and histone gene expression has been discussed above (Section 2.3.4). A role for cyclin E–CDK2 in centrosome duplication has also been suggested (Lacey *et al.*, 1999). Tarapore *et al.* (2002) developed a cell-free centriole duplication system and demonstrated that centrosome duplication was dependent upon the presence of cyclin E–CDK2 complexes. Cyclin E–CDK2 was shown to phosphorylate nucleophosmin in this study, causing dissociation from centrosomes and subsequent initiation of centrosome duplication (Tarapore *et al.*, 2002).

2.4.2 Role of the cyclin A–CDK2 in S phase progression

The onset of S phase correlates with formation of cyclin A–CDK2 complexes. Microinjection of antibodies against CDK2 complexed with either cyclin A or cyclin E blocks the initiation of DNA synthesis in mammalian cells (Pagano *et al.*, 1992; Ohtsubo *et al.*, 1995). Cyclin A might be rate-limiting for DNA replication, since it can accelerate entry into S phase when overexpressed in cells (Resnitzky *et al.*, 1995). The fact that depletion of cyclin A by injection of anti-cyclin A antibody causes inhibition of DNA synthesis suggests that cyclin A plays a role in this process. It has been shown that CDC6 is an intracellular substrate for cyclin A–CDK2 (Peterson *et al.*, 1999). CDC6 is a protein required for formation of the initiation complex (see section on CDC6 and DNA replication, below) that is necessary for the onset of DNA replication, thereby providing one mechanism by which cyclin A–CDK2 may regulate DNA replication. CDC6 has been shown to be required for late firing of origins and this function may be achieved by phosphorylation by cyclin A–CDK2, suggesting that this complex may be required for continuation of DNA synthesis in addition to the initiation step (Herbig *et al.*, 2000). However, it was demonstrated that microinjection of cyclin A antibodies into cells already progressing through S phase caused accumulation of cells in G_2 (Pagano *et al.*, 1992), indicating that in this instance cyclin A was not required for cells to complete S phase, and CDC6 may therefore be regulated by other complexes.

2.4.3   Cell cycle control of DNA replication

Eukaryotic genomes are extremely large and can range from 10^7 to greater than 10^9 base pairs. Due to this large size, duplication of the eukaryotic genome occurs as a multiparallel process, with 10 000–100 000 parallel synthesis sites in human somatic cells (reviewed in Kelly and Brown, 2000; Nasheuer *et al*, 2002). The cells need to ensure that DNA replication occurs at the appropriate time in the cell cycle and also that re-replication does not occur before cells undergo mitosis and cytokinesis. Advancement in our understanding of the regulation of these sequential processes has come from numerous studies in yeast systems (reviewed in Kelly and Brown, 2000). These simple model systems have provided much information on the protein complexes involved in the activation and inhibition of DNA synthesis, and a number of homologues have since been identified in higher eukaryotes.

Early experiments carried out by Rao and Johnson (1970) showed that the initiation of DNA replication is believed to be a two-step process. These investigators showed that fusion of a G_1 cell with an S phase cell triggered DNA replication, but that G_2 cells were unable to undergo DNA initiation. This led to the notion that an S phase promoting factor was required to push cells from G_1 into S phase. The two-step process involves, first, the assembly of initiation factors at origins of replication, and second these complexes are triggered to activate DNA synthesis by the actions of protein kinases. The following sections will overview those molecules involved in driving DNA initiation and replication.

Origins of replication

DNA synthesis is known to occur at specific sites on the DNA, known as origins of replication. The best characterized origins of replication are those found in *Saccharomyces cerevisiae* and are known as autonomous replication sequences (ARSs) (Marahrens and Stillman, 1992). The ARS contains a highly conserved region of 100–200 base pairs, known as the ARS consensus sequence (ACS), and this is an essential component of the origin of replication, to which the origin recognition complex (ORC) binds. The ORC is conserved in all eukaryotes (Bell and Stillman, 1992).

Three ORC subunits have been identified in humans, HsORC 1, 2 and 4 (Gavin *et al.*, 1995), all of which are involved in the initiation of DNA replication by recruitment of specific factors to the DNA. Studies in *S. cerevisiae* have shown that ScORC interacts with a variety of other proteins, e.g. CDC6, minichromosome maintenance (MCM) proteins and CDC7 (Dutta *et al.*, 1991), and homologues for these proteins have also been identified in higher eukaryotes. The ORC may act to recruit these proteins to the origins of replication. Human ORC has also been shown to interact with a HAT and this may be involved in making the initiation site accessible, thereby facilitating replication (Iizuka and Stillman, 1999)

CDC6 and DNA replication

Another key regulator of DNA replication is CDC6 of *S. cerevisiae* and the *Schizosaccharomyces pombe* homologue, CDC18. CDC6/CDC18 are highly conserved throughout evolution, with related proteins being found in *Xenopus* and in human cells (Coleman *et al.*, 1996; Williams *et al.*, 1997; Hateboer *et al.*, 1998; Saha *et al.*, 1998a). Immunodepletion of CDC6 in human cells blocks S phase entry (Hateboer *et al.*, 1998; Yan *et al.*,1998) and has been shown to affect the interaction of ORC with MCM proteins, but not its interactions with DNA (Coleman *et al.*, 1996; Cook *et al.*, 2002). This data suggests that CDC6 may act as an adaptor protein for interactions of ORC with other proteins (e.g. MCM proteins). Levels of CDC6 in cycling human cells remain fairly stable during S phase, G_2 and mitosis (Saha *et al.*, 1998a; Jiang *et al.*, 1999b), but lower amounts are present in early G_1, when CDC6 is degraded by proteolysis (Mendez and Stillman, 2000; Peterson *et al.*, 2000). CDC6 does, however, change its subcellular localization during the cell cycle and it has been shown that nuclear CDC6 is phosphorylated during S phase and transported to the cytoplasm (Peterson *et al.*, 1999). Phosphorylation of CDC6 is carried out by cyclin A–CDK2 and also by Dbf–CDC7. Relocation of CDC6 may be one way in which cells ensure that re-replication does not occur. However, a substantial amount of CDC6 is found still associated with chromatin during S phase (Mendez and Stillman, 2000), suggesting that CDC6 might play roles other than assembly of proteins at the initiation site and may be required for continued synthesis. Due to the relocalization of CDC6 during S phase, CDC6 must be continually synthesized to account for the fraction associated with chromatin during S phase (Biermann *et al.*, 2002).

Minichromosome maintenance (MCM) proteins

The MCM proteins are a complex of six related proteins that form an essential component of the DNA initiation complex. Their requirement for DNA replication has been demonstrated by antibody injection and antisense oligonucleotide experiments (Kimura *et al.*, 1994; Todorov *et al.*, 1994; Fujita *et al.*, 1996). The six MCM proteins are not functionally redundant and deletion of any MCM protein in *S. cerevisiae* or *Sz. pombe* results in loss of cell viability. In most organisms, the MCM proteins are located in the nucleus throughout the cell cycle (Kearsey and Labib, 1998; Tye, 1999). In mammalian cells, MCM proteins associate with chromatin in G_1 but as cells progress through S phase they are phosphorylated and this reduces their affinity for chromatin (Kimura *et al.*, 1994; Todorov *et al.*, 1995). This may be one way in which cells ensure that replication occurs only once per cycle. In mammalian cells, some MCM proteins co-purify with DNA polymerase-α (Thommes *et al.*, 1992) and evidence exists to suggest that MCMs possess DNA helicase activity (Ishimi, 1997). Therefore, it is possible that association of the helicases with a primase forms a mobile primosome that drives discontinuous synthesis.

CDC45

CDC45 is essential for DNA replication in *S. cerevisiae* (Hopwood and Dalton, 1996; Hardy, 1997; Zou *et al.*, 1997) and this molecule has been shown to interact with MCM family members (Hardy, 1997; Zou *et al.*, 1997). A human homologue has been identified (Saha *et al.*, 1998b) and immunoprecipitation experiments indicate that it associates with chromatin periodically throughout the cell cycle. Association of CDC45 with chromatin may depend on cyclin–CDK complex activity at the G_1–S transition (Mimura and Takisawa, 1998; Zou and Stillman, 1998).

In summary, DNA replication begins with the assembly of the initiation complex at the origin of replication and a number of proteins are involved in the formation of this complex. Initially, the ORC (made up of six subunits) associates with chromatin at the initiation site. Association of CDC6 with ORC then acts as an adaptor protein for the association of ORC with MCM proteins. MCM proteins possess DNA helicase activity, which may form a primosome in association with DNA polymerase-α. Although regulated by phosphorylation, which changes the subcellular localization of CDC6, it may play a role later on in DNA replication, since its continued synthesis ensures its association with chromatin throughout S phase. Once the initiation complex has been assembled at the origin of replication, DNA synthesis is then triggered by the actions of various protein kinases, as described below.

Regulation of DNA initiation complexes

Two classes of protein kinases are essential for the initiation of replication, *viz.* the CDKs and Dbf4–CDC7 kinase.

CDKs   A role for CDK2 in the initiation of replication in higher eukaryotes has been demonstrated in a number of studies, e.g. microinjection of antibodies against certain cyclins and CDKs into mammalian cells inhibits S phase entry (Pagano *et al.*, 1993; Ohtsubo *et al.*, 1995).

As mentioned in Section 2.4.2, CDC6 is a substrate for cyclin A–CDK2, and phosphorylation of CDC6 by this complex possibly contributes to the prevention of re-initiation by causing export of CDC6 from the nucleus (Jiang *et al.*, 1999b; Peterson *et al.*, 1999).

MCM proteins also serve as substrates for certain CDKs and phosphorylation of MCM proteins causes dissociation from chromatin as cells progress through S phase. Thus, MCM proteins are substrates for the mitotic complex cyclin B–CDC2 (Hendrickson *et al.*, 1996) and this provides a link between mitotic cyclins and the inhibition of re-initiation, ensuring that DNA replication occurs only once before entry into mitosis. It has been shown that MCM 2 and 4 are phosphorylated in S phase and become hyperphosphorylated by G_2–M. MCM 2 and 4 are both good substrates for *in vitro* phosphorylation by cyclin B–CDC2 (Hendrickson *et al.*, 1996; Fujita *et al.*, 1998).

Dbf4/CDC7 kinase   CDC7 in *S. cerevisiae* and the *Sz. pombe* homologue, Hsk 1, have been shown to be essential for viability and are directly involved in DNA replication (Hartwell, 1973; Masai *et al.*, 1995). A human homologue of CDC7 also has been identified (Jiang and Hunter, 1997; Sato *et al.*, 1997; Hess *et al.*, 1998). The human homologue of Dbf4 is regulated transcriptionally (Kumagai *et al.*, 1999; Lepke *et al.*, 1999) with maximal expression during S phase, which also corresponds to the kinase activity of the Dbf4–CDC7 complex (Jiang *et al.*, 1999a; Kumagai *et al.*, 1999). Studies have shown that inactivation of CDC7 in early S phase prevents firing from replication origins, implicating CDC7 in the initiation of DNA replication (Bousset and Diffley, 1998; Donaldson *et al.*, 1998). Human MCM 2 and 3 are substrates for CDC7 *in vitro* (Sato *et al.*, 1997; Kumagai *et al.*, 1999).

2.4.4   DNA replication checkpoints

Various replication checkpoints serve to inhibit DNA replication in response to partially replicated DNA or DNA damage, to allow the cell sufficient time to repair the damage before undergoing mitosis (see Figure 2.1). Replication checkpoints have been extensively studied in yeast systems and homologues for the proteins involved have also been identified in higher eukaryotes.

The p53-dependent pathway

Several phosphatidylinositol (PI)-3-like kinase proteins are believed to be involved in the DNA replication checkpoint, including ATM, ATR and DNA-PK (Hartley *et al.*, 1995; Savitsky *et al.*, 1995; Bentley *et al.*, 1996; Keegan *et al.*, 1996), and these kinases have been shown to be activated by DNA *in vitro* (Gately *et al.*, 1998; Lakin *et al.*, 1999). The tumour suppressor protein p53 is a downstream target of ATM and immunoprecipitated ATM can phosphorylate p53 on Ser 15, a residue that is phosphorylated *in vivo* in response to DNA damage (Shieh *et al.*, 1997; Siliciano *et al.*, 1997; Banin *et al.*, 1998; Canman *et al.*, 1998). DNA damage, occurring for example in response to ionizing radiation, leads to stabilization and accumulation of p53, which is involved in a number of cellular responses such as cell cycle checkpoints, genomic stability, gene transactivation and apoptosis (Lakin *et al.*, 1999; Colman *et al.*, 2000; Ryan *et al.*, 2001; Taylor and Stark, 2001; Wahl and Carr, 2001). p53 is normally associated with the ubiquitin ligase, MDM2; phosphorylation of p53 on Ser 15 leads to its dissociation from MDM2, thereby stabilizing the p53 protein (Shieh *et al.*, 1997). Stabilization of p53 leads to trans-activation of the CDKI molecule, p21, which will lead to cell cycle arrest (Nayak and Das, 2002).

Other regulators of p53 include ATR and Pin 1, thus the ATR protein is capable of phosphorylating p53 on Ser 15 and may also play a part in activating the p53 checkpoint pathway in response to UV and ionizing radiation (Lakin *et al.*, 1999; Tibbetts *et al.*, 1999). Pin 1 has been shown to regulate the G_1–S, G_2–M and DNA replication checkpoints (Winkler *et al.*, 2000) and is overexpressed in many human cancers (Ryo *et al.*, 2001; Wulf *et al.*, 2001; Liou *et al.*, 2002). A recent report has shown that Pin 1 binds phosphorylated p53 and is involved in stabilization of the protein, probably by interfering with the MDM2 interaction, and is also involved with transactivation of p21 in response to DNA damage (Wulf *et al.*, 2002).

The p53-independent pathway

The p53-independent mechanism of cell cycle block in response to unreplicated DNA or DNA damage involves the Rad proteins (reviewed in Boddy and Russell, 1999; Kelly and Brown, 2000). These proteins were first identified in yeast and mammalian homologues also have been identified. The proteins involved in recognition and processing of the replication perturbation response are Rad 1, Rad 9, Rad 17 and Hus 1. The effects of these proteins are mediated by the protein kinases CDS 1 and CHK 1, which target proteins involved in cell cycle regulation, e.g. the CDC25 dual-specificity protein phosphatases (Section 2.5.2).

DNA-PK is the human homologue of the fission yeast (*Sz-pombe*) PI-3 like kinase, Rad 3, and is activated by proteins that detect sites of DNA strand breakage. Loss of function of these kinases results in inhibition of the checkpoint, suggesting that DNA-PK is important for sensing DNA damage and initiating the checkpoint mechanism (Chan *et al.*, 2002; Woo *et al.*, 2002).

Rad 1 has been shown to be similar to proliferating cell nuclear antigen (PCNA) and possesses exonuclease activity (Parker *et al.*, 1998; Thelen *et al.*, 1999). PCNA encircles the DNA during replication and retains the polymerase complex on the DNA. PCNA requires several factors in order to load onto the DNA, one of which is known as replication factor C (RFC). Rad 17 has been shown to share homology with RFC and also to interact with Rad 1 (Parker *et al.*, 1998). Rad 1, Rad 9, and Hus 1 have all been shown to physically interact in mammalian cells (Volkmer and Karintz, 1999; Hang and Lieberman, 2000) and it is believed that Rad 17 may serve as a recruitment complex for Rad 1, Rad 9 and Hus 1 to sites of DNA damage (Zou *et al.*, 2002). Indeed, a recent study has demonstrated that upon replication block, Rad 17 is recruited to the sites of DNA damage during late S phase and that it binds to the Rad1–Rad9–Hus1 complex, enabling its interaction with PCNA (Dahm and Hubscher, 2002).

The two downstream targets of the Rad proteins are the serine/threonine kinases CHK 1 and CDS 1. These kinases are activated differentially, such that CDS1 is involved in mediating responses to unreplicated DNA and CHK 1 is involved in the G$_2$ DNA damage response. CDS 1 has been shown to be phosphorylated by ATM (Brown *et al.*, 1999; Chaturvedi *et al.*, 1999) and following activation it phosphorylates and inhibits the mitotic activator, CDC25C (Matsuoka *et al.*, 1998; Brown *et al.*, 1999; Chaturvedi *et al.*, 1999), thereby mediating G$_2$ arrest. CHK 1 also phosphorylates CDC25C *in vitro* (Peng *et al.*, 1997). Phosphorylation of CDC25C by CDS 1 and CHK 1 creates a binding site for the 14-3-3 family of phosphoserine-binding proteins (Peng *et al.*, 1997; see section on the polo-like kinases, below). Binding of 14-3-3 has little effect on CDC25C activity and it is believed that 14-3-3 regulates CDC25C by sequestering it to the cytoplasm, thereby preventing interactions with cyclin B–CDC2, which is localized to the nucleus at the G$_2$–M transition (Peng *et al.*, 1997; Dalal *et al.*, 1999).

The mechanisms by which DNA replication and DNA damage checkpoints exert their effects on cell cycle progression are now becoming clearer. p53-Dependent and -independent mechanisms both exert their effects via complex pathways on key cell cycle regulatory molecules, such as p21 and the mitotic regulator CDC25C (see Figures 2.6(A) and 2.6(B)). These events occur at specific points in the cell cycle, ensuring that a cell does not proceed through mitosis without a full complement of replicated and intact DNA, thereby ensuring that the genome is passed equally to each of the daughter cells.

2.5 The G$_2$–M transition

The G$_2$ phase is another gap phase in the cell cycle, during which the cell assesses the state of chromosome replication and prepares to undergo mitosis and cytokinesis. Cyclin B–CDC2 is the key mitotic regulator of the G$_2$–M transition and was originally identified as the maturation promoting factor, a factor capable of inducing M phase in immature *Xenopus* oocytes (Masui and Market, 1971; Dunphy *et al.*, 1988; Gautier *et al.*, 1988). As is the case with other cyclin–CDK complexes, activation of the cyclin

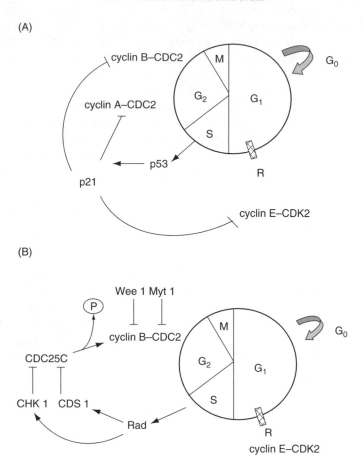

Figure 2.6   (A) Cell cycle arrest at G_1 and G_2 occurs following activation of p21, subsequent to p53 stabilization at the DNA checkpoint. (B) Activation of CHK1 and CDS1 by Rad causes cell cycle arrest in G_2 by inhibition of CDC25C

B–CDC2 complex is tightly regulated by phosphorylation and dephosphorylation events and also changes in subcellular localization (reviewed in Takizawa and Morgan, 2000; Smits and Medema, 2001). The molecules that regulate cyclin B–CDC2 activity receive signals from the checkpoint machinery (as described in the section on the p53-independent pathway, above). Cyclin A–CDK complexes also play a role in regulating the G_2–M transition.

2.5.1   Role of the cyclin B–CDC2 complex in the G_2–M transition

Cyclin B synthesis begins at the end of S phase (Pines and Hunter, 1989). Two cyclin B isoforms exist in mammalian cells, cyclin B1 and B2. Studies in cyclin B1- and

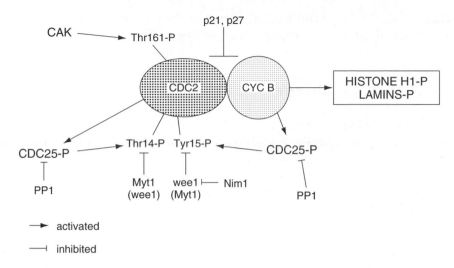

Figure 2.7 Regulation of the CDC2–cyclin B complex. The serine/tyrosine kinase Wee1 catalyses phosphorylation of Tyr15 on CDC2. Wee1 itself is phosphorylated and inactivated by Nim1 and other unidentified kinases to induce mitosis. Thr14 phosphorylation can be mediated by wee1, but only once Tyr15 has been phosphorylated. It appears that the thr/tyr kinase Myt1 is the critical kinase involved here. Inhibition of CDC2 by wee1 is counteracted by the CDC25 dual-specificity phosphatases. CDC25 is phosphorylated and activated by CDC2–cyclin B (amplification pathway). Protein phosphatase 1 inactivates CDC25 by dephosphorylation of the same residue that is phosphorylated by CDC2–cyclin B. Full activation of CDC2 requires Thr161 phosphorylation by CAK that then stabilizes CDC2 association with cyclin A (and B?)

cyclin B2-null mice have shown that cyclin B2 is non-essential for normal growth and development (Brandeis *et al.*, 1998). This isoform associates with the Golgi and may play a role in Golgi remodelling during mitosis (Jackman *et al.*, 1995; Brandeis *et al.*, 1998). In contrast to cyclin B2, cyclin B1 is thought to be responsible for most of the actions of CDC2 in the cytoplasm and nucleus, and it appears to compensate for the loss of cyclin B2 in B2-null mice, implying that cyclin B1 is capable of targeting CDC2 kinase to the essential substrates of cyclin B2 (Brandeis *et al.*, 1998).

Cyclin B–CDC2 complexes are regulated both positively and negatively by phosphorylation (see Figure 2.7). Phosphorylation of CDC2 on the conserved T-loop region (Thr 161) is required for activation, as is the case with all CDKs, and this phosphorylation event is mediated by CAK. During G$_2$, cyclin B–CDC2 complexes are held in an inactive state by phosphorylation on of CDC2 Thr 14 and Tyr 15. Phosphorylation on Thr 14 prevents ATP binding (Endicott *et al.*, 1994), whereas that on Tyr 15 interferes with phosphate transfer to the substrate, due to its positioning in the ATP-binding site on CDC2 (Atherton-Fessler *et al.*, 1993). These inhibitory phosphorylation events are carried out by the kinases, Wee 1 and Myt 1, where Wee 1 specifically phosphorylates Tyr 15 and Myt 1 phosphorylates both Tyr 15 and Thr 14, with a stronger affinity for Thr 14 (Lundgren *et al.*, 1991; Parker and Piwinca-Worms,

1992; Li *et al.*, 1995). Cyclin B–CDC2 becomes fully activated following dephosphorylation of these sites by the protein phosphatase CDC25C (Figure 2.7).

2.6   Mitosis (M-phase)

Mitosis (also called karyokinesis) and cytokinesis constitute the shortest phase of the eukaryotic cell cycle, typically taking around 1–2 h to complete in a mammalian cell. Mitosis itself comprises five distinct phases, as follows.

1. *Prophase*: this stage begins with condensation of the chromosomes in the nucleus and ends with breakdown of the nuclear envelope (this latter event occurs over a 1–2 min interval).

2. *Prometaphase*: at this stage the mitotic spindle forms. Three essential events must occur in prometaphase if cell division is to proceed normally: (a) the bipolar spindle axis must be established; (b) the daughter chromatids of each replicated chromosome must become committed to the opposing spindle poles; (c) the chromosomes must become aligned at, or near to, the spindle equator.

3. *Metaphase*: during this stage all chromosomes are bioriented and positioned near the spindle equator. All chromosomes align themselves along the metaphase plate.

4. *Anaphase*: the sister chromatids that comprise each chromosome separate to form two independent chromosomes. Anaphase is separated into anaphase A and anaphase B.

5. *Telophase*: this is the final stage of mitosis, in which the chromosomes decondense and a nuclear envelope forms around each set of chromatids. The contractile ring begins to form in readiness for the cell to split into two daughter cells, each with one nucleus.

A number of cell cycle-regulatory molecules play pivotal roles in promoting progression through mitosis, including the CDC25 protein phosphatases, the polo-like kinases (PLKs), the 14-3-3 proteins, mitotic cyclin–CDK complexes and the anaphase-promoting complex (APC). The roles that these individual groups of molecules play in mitosis are discussed in more detail below.

2.6.1   The CDC25 protein phosphatases

The mammalian CDC25 family of dual specificity phosphatases consists of three members, A, B and C (Sadhu *et al.*, 1990; Galaktinov and Beach, 1991). CDC25B and C are thought to be the main regulators of mitosis, whereas CDC25A plays a role in regulating the G_1–S transition. CDC25B may be involved in the initial dephosphorylation and activation of cyclin B–CDC2, which then initiates the positive feedback loop of CDC25C activation by CDC2 (Hoffmann *et al.*, 1993; Figure 2.7). CDC25B is also believed to play a role in centrosomal microtubule nucleation during mitosis, since overexpression of this molecule causes formation of minispindles in the cytoplasm (Gabrielli *et al.*, 1996). CDC25A expression is under transcriptional control of the E2F transcription factors in late G_1 (Vigo *et al.*, 1999), and is involved in activation of cyclin E–CDK2 and cyclin A–CDK2 complexes that regulate entry into S phase (Sections 2.4.1 and 2.4.2).

CDC25C is the protein phosphatase that is mainly responsible for dephosphory-lation and activation of the cyclin B–CDC2 complex (Russell and Nurse, 1986; Dunphy and Kumagai, 1991; Lee *et al.*, 1992; Sebastian *et al.*, 1993). Treatment of CDC25C with phosphatases *in vitro* led to reduced CDC25C phosphatase activity, indicating that hyperphosphorylation of CDC25C is required for phosphatase activity during mitosis (Izumi *et al.*, 1992; Kumagai and Dunphy, 1992; Hoffmann *et al.*, 1993). Cyclin B–CDC2 is able to phosphorylate CDC25C (Hoffmann *et al*, 1993; Izumi and Maller, 1993) (Figure 2.7) and this initiates a positive feedback loop, which induces rapid activation of cyclin B–CDC2 at the G_2–M transition. However, the initial trigger of CDC25C activation remains unclear, although CDC25C has been shown to be phosphorylated by cyclin E–CDK2 and cyclin A–CDK2 *in vitro* (Izumi and Maller, 1995). PLK-1 is another potential upstream regulatory kinase of CDC25C that might function *in vivo* (Roshak *et al.*, 2000).

The role of CDC25C as a key mediator of cyclin B–CDC2 activation has recently been questioned, due to studies performed in CDC25C knockout mice. These mice showed no phenotype with respect to regulation of mitosis and showed no differences in CDC2 phosphorylation (Chen *et al.*, 2001), suggesting that redundancy exists between CDC25 isoforms. Indeed, a role for CDC25A in CDC2 activation has recently been suggested. Destruction of CDC25A by the ubiquitin-mediated pathway serves to ensure cells do not undergo premature mitosis, and this is achieved by phosphorylation of CDC25A by CHK 1. However, Mailand *et al.* (2002) observed that once cells are committed to mitosis, the stability of CDC25A undergoes major changes at the G_2–M transition, due to phosphorylation on Ser 17 and Ser 115 that uncouple it from the ubiquitin-mediated degradation pathway. Phosphorylation of CDC25A on these specific residues is mediated by cyclin B–CDC2, and therefore forms part of a positive feedback activation loop, whereby CDC25A is stabilized by CDC2 and this is followed by dephosphorylation of CDC2 on Thr 14 and Tyr 15 (Mailand *et al.*, 2002).

The polo-like kinases (PLKs)

The polo-like kinases are a family of serine/threonine protein kinases, three of which have been described in mammalian cells, viz. PLK-1, PLK-2 and PLK-3 (Glover *et al.*, 1998). Each family member contains a homologous C-terminal domain called the 'polo box' that is required for directing subcellular localization of the kinase, since mutation of this region has been shown to disrupt localization of PLK-1 (Lee *et al.*, 1998). PLK protein levels and phosphorylation status are cell cycle-regulated. Thus, PLK is undetectable in cells at the G_1–S phase transition; however, levels rise during S phase and phosphorylation occurs during G_2 (Hamanaka *et al.*, 1995). Activation of PLK has been found to occur at a similar time to cyclin B–CDC2 activation and PLK is rapidly degraded after mitosis (Hamanaka *et al.*, 1995). A recent study by Roshak *et al.* (2000) demonstrated that human CDC25C is a substrate for PLK and that phosphorylation caused activation of the phosphatase and subsequent dephosphorylation of cyclin B–CDC2. The PLKs also have been implicated in the formation of the bipolar spindle (Lane and Nigg, 1996), thereby implicating them in the regulation of chromosome separation during prometaphase.

The 14-3-3 proteins

CDC25C phosphatase activity is regulated negatively by phosphorylation on a specific Ser 216 residue that creates a binding site for small phosphoserine-binding proteins, known as 14-3-3 (Peng *et al.*, 1998). A number of 14-3-3 proteins are known to exist, including 14-3-3 epsilon, gamma, beta, sigma, zeta and eta (Aitken *et al.*, 1992). CDC25C is localized to the cytoplasm during interphase and directed to the nucleus just prior to mitosis (Heald *et al.*, 1993; Dalal *et al.*, 1999). Binding of 14-3-3 may prevent nuclear localization of CDC25C by masking the NLS, which is in close proximity to the Ser 216 residue. In support of 14-3-3 sequestering CDC25C to the cytoplasm during interphase, Ogg *et al.* (1994) demonstrated that Ser 216 is the major phosphorylation site of CDC25C during interphase, but not during mitosis. Potential candidate kinases for phosphorylation of Ser 216 on CDC25C are CHK 1, CDS 1 and C-TAK 1 (Zeng *et al.*, 1998; Furnari *et al.*, 1999). CHK 1 and CDS 1 are both mediators of G_2 arrest in response to DNA damage or incomplete replication, as discussed in Section 2.4 above. Therefore, phosphorylation of CDC25C during interphase creates a binding site for 14-3-3, causing cytoplasmic retention. Dephosphorylation of Ser 216 (possibly by CDC25B) at the onset of mitosis results in nuclear localization and subsequent activation of cyclin B–CDC2, as described above.

2.6.2   Subcellular localization of cyclin B–CDC2 during G_2–M

During interphase, cyclin B–CDC2 complexes are found in the cytoplasm. However, by late prophase, the majority of cyclin B–CDC2 is found in the nucleus following breakdown of the nuclear envelope (Hagting *et al.*, 1999). Cyclin B has a cytoplasmic retention sequence (CRS) in the N-terminal region which, when deleted, causes localization of cyclin B to the nucleus (Pines and Hunter, 1994). A nuclear export signal (NES) also has been defined within the CRS and this binds to the export receptor,

CRM1 (Yang *et al.*, 1998). It has been that shown that a specific inhibitor of CRM1 causes accumulation of cyclin B in the nucleus (Yang *et al.*, 1998). During mitosis, cyclin B is hyperphosphorylated within the CRS region and this phosphorylation is thought to disrupt interactions with CRM1, holding cyclin B in the nucleus (Yang *et al.*, 1998; Hagting *et al.*, 1999).

The mechanism by which cyclin B enters the nucleus is less well understood. Both cyclin B and CDC2 lack an NLS. Cyclin B1 has, however, been shown to bind importin-β (Moore *et al.*, 1999), and this could be one mechanism that mediates cyclin B nuclear localization. The CRS of cyclin B1 is also known to interact with cyclin F, which is found predominantly within the nucleus and contains two NLSs (Kong *et al.*, 2000). Interestingly, overexpression of cyclin F causes relocation of cyclin B to the nucleus, suggesting that it may be involved in the import of cyclin B1.

Although cyclin B contains a CRS and an NES, both of which ensure that cyclin B is localized to the cytoplasm, phosphorylation of cyclin B can lead to association with other molecules that results in its relocation to the nucleus, thereby allowing access to nuclear substrates.

2.6.3   Function of cyclin B–CDC2 during mitosis

The cyclin B–CDC2 complex is involved in the initiation of a number of mitotic events in both the cytoplasm and the nucleus. During prophase, cyclin B–CDC2 is associated with duplicated centrosomes and promotes centrosome separation by phosphorylation of the centrosome-associated motor protein Eg5 (Blangy *et al.*, 1995). Cyclin B1–CDC2 and/or cyclin B2–CDC2 complexes are involved in the fragmentation of the Golgi network (Lowe *et al.*, 1998) and cyclin B–CDC2 is also involved in the breakdown of the nuclear lamina (Peter *et al.*, 1990) and cell rounding (Yamashiro *et al.*, 1990). Thus, the cyclin B–CDC2 complex is involved in completely reorganizing the cell architecture during mitosis.

2.6.4   Function of cyclin A–CDC2 during mitosis

Cyclin A plays important roles at two distinct phases of the cell cycle, viz. G_1–S (as discussed in Section 2.4.2) and also during G_2–M. These separate functions coincide with cyclin A binding to two different kinases, CDK2 at the G_1–S border and CDC2 during G_2. Cyclin A levels are undetectable during G_1 and levels begin to rise as the cells enter S phase. By the time a cell enters mitosis, cyclin A levels begin to decline; however, it still is present during prophase, where it is associated with the centrosomes, although by telophase cyclin A is undetectable (Pagano *et al.*, 1992). Cyclin A–CDC2 complexes are thought to play a role in activating cyclin B–CDC2 complexes, and recent reports suggest that cyclin A–CDK2 complexes may also act during the G_2 checkpoint (Furuno *et al.*, 1999; Goldstone *et al.*, 2001). Exit from mitosis requires degradation of both cyclin A and cyclin B, and this occurs via a ubiquitin-mediated pathway that is itself regulated by the anaphase-promoting complex (APC) pathway (see below).

2.6.5   The anaphase-promoting complex (APC)

Exit from mitosis requires ubiquitin-mediated degradation of mitotic cyclins via the 'cyclin destruction box' (Glotzer *et al.*, 1991), which is regulated by the APC ubiquitin ligase. APC is a multi-subunit ligase consisting of a number of protein subunits, such as APC1, APC2, CDC16 and CDC23 (reviewed in Harper *et al.*, 2002). APC is inactive in the S and G_2 phases of the cell cycle but becomes activated in mitosis as a result of phosphorylation, which is believed to be carried out by PLK-1 (Kotani *et al.*, 1998; Golan *et al.*, 2002) and/or cyclin B–CDC2 (Golan *et al.*, 2002). APC requires conversion to an active form by CDC20–Fizzy and this can only occur following phosphorylation of APC (Shteinberg *et al.*, 1999). APC is also required for sister chromatid separation during anaphase by causing destruction of securins, the proteins that hold the sister chromatids together.

2.7   Cytokinesis

At the end of mitosis the cell must ensure that division is taken to completion by a process called cytokinesis. This occurs following assembly of a cleavage furrow at the site of division that contains actin, myosin and other proteins that eventually form the contractile ring (Noguchi *et al.*, 2001). Following chromosome segregation, the microtubules bundle in the mid-region of the spindle, forming the spindle mid-zone. As the contactile ring contracts it creates a membrane barrier between each cell. The spindle mid-zone remains connected, forming a cytoplasmic bridge until this is finally cut during abcission. The mid-zone has been shown to contribute to actin ring assembly, since placement of an artificial barrier between the spindle mid-zone and the cell cortex during metaphase caused inhibition of the cleavage furrow, whereas if a barrier was created in early anaphase, cytokinesis proceeded without a problem (Cao and Wang, 1996).

Animal cells divide through the formation of an actomyosin contractile ring at the end of anaphase (Gratti *et al.*, 2000). As discussed above, the spindle mid-zone plays a role in contractile ring assembly. Two major classes of protein are believed to be important in signalling from the spindle mid-zone to the contractile ring. The chromosomal passenger proteins, e.g. inner centromere protein, are initially found localized to chromosomes and centromeres and then translocate to the mid-zone during anaphase (Adams *et al.*, 2001) and are involved in chromosome alignment, segregation and cytokinesis. The second class of proteins are the motor-associated proteins, e.g. Eg5, which are required to maintain the mid-zone. These proteins localize along the spindles during metaphase and concentrate in the spindle mid-zone during anaphase (Nislow *et al.*, 1990, 1992).

Specific CDKs also play a role in cytokinesis, and it has been shown that mammalian cells injected with a non-destructible form of cyclin B undergo anaphase and chromosome segregation but do not form a spindle midzone and fail to undergo cytokinesis (Wheatley *et al.*, 1997), suggesting a role for cyclin B in inhibiting cytokinesis. CDKs may also inhibit myosin, and thereby contractile ring, formation through inhibitory phosphorylation of the myosin regulatory light chain (RLC; Mishima and Mabuchi, 1996). Myosin RLC can be phosphorylated by CDC2, which inhibits its actin-activated ATPase activity *in vitro* (Sellers, 1991). This inhibitory phosphorylation increases in early mitosis and decreases in anaphase simultaneously with a decrease in CDC2 activation (Yamakita *et al.*, 1994).

2.8 Endoreduplication

In most eukaryotic cells, S phase and mitosis are coupled and occur only once during each cell cycle; however, occasionally the sequence of events is interrupted, such that the cell undergoes multiple rounds of DNA synthesis in the absence of mitosis. This process is called endoreduplication. Work in yeast has shown that endoreduplication can occur as a result of multiple initiations within S phase, recurring S phase or repeated S and G phases (Grafi, 1998). In addition, endoreduplication can result in either multiple DNA syntheses within a single nucleus, e.g. megakaryocytes, or in multi-nucleated cells, e.g. cardiac myocytes. Little is known about the molecular mechanisms responsible for endoreduplication in multinucleate cells, although a greater understanding of the processes involved might enable cell division to be initiated instead of endoreduplication, which would be useful for replacing damaged cells, and therefore avoid scarring in terminally differentiated tissues that contain cells such as cardiac myocytes and neurones.

2.9 Summary and conclusions

The eukaryotic cell cycle is a highly regulated, conserved and sequential process that is necessary for normal cell growth and development. Our understanding of the mechanisms involved in cell cycle regulation has increased significantly in recent years, as demonstrated by the award of the Nobel Prize for Physiology and Medicine to Leland Hartwell, Paul Nurse and Tim Hunt in 2001 for their seminal discoveries relating to the cell cycle machinery. Despite this increased understanding, there remains much to learn about the mechanisms involved in controlling growth and proliferation in specific cell types and organs. Extending our knowledge of cell cycle control in different cell types might help identify the causes of certain hyperproliferative diseases, including cancer and vascular disease. This could then lead to the development of new therapeutic agents that target specific cell cycle molecules that become altered in such disorders. Evidence exists to suggest that viruses can manipulate the host cell cycle to promote viral replication; the cell cycle may also, therefore, offer a suitable target for antiviral strategies (Paiardini *et al.*, 2001; Feuer *et al.*, 2002).

References

Abbott, D. W. and Holt, J. T. (1999). Mitogen-activated protein kinase kinase 2 activation is essential for progression through the G_2/M checkpoint arrest in cells exposed to ionizing radiation. *J Biol Chem* **274**, 2732–2742.

Adams, R. R., Maiato, H., Earnshaw, W. C. and Carmena, M. (2001). Essential roles of *Drosophila* inner centromere protein (INCENP) and aurora B in histone H3 phosphorylation, metaphase chromosome alignment, kinetochore disjunction, and chromosome segregation. *J Cell Biol* **153**, 865–880.

Aitken, A., Collinge, D. B., van Heusden, B. P., Isobe, T., Roseboom, P. H., Rosenfeld, G. and Soll, J. (1992). 14-3-3 proteins: a highly conserved, widespread family of eukaryotic proteins. *Trends Biochem Sci* **17**, 498–501.

Alberts, B., Johnson, A., Lewis, J., Raff, M., Roberts, K. and Walter, P. (2002). The cell cycle and programmed cell death. In *Molecular Biology of the Cell*, 4th edn, pp. 983–1026, B. Alberts, A. Johnson, J. Lewis, M. Raff, K. Roberts and P. Walter (eds). Garland Science:....

Alcorta, D. A., Xiong, Y., Phelps, D., Hannon, G., Beach, D. and Barrett, J. C. (1996). Involvement of the cyclin-dependent kinase inhibitor p16 (INK4a) in replicative senescence of normal human fibroblasts. *Proc Natl Acad Sci USA* **93**, 13742–13747.

Apostolova, M. D., Ivanova, I. A., Dagnino, C., D'Souza, S. J. and Dagnino, L. (2002). Active nuclear import and export pathways regulate E2F-5 subcellular localization. *J Biol Chem* **277**, 34471–34479.

Atherton-Fessler, S., Parker, L. L., Geahlen, R. L. and Piwnica-Worms, H. (1993). Mechanisms of p34cdc2 regulation. *Mol Cell Biol* **13**, 1675–1685.

Balmanno, K. and Cook, S. J. (1999). Sustained MAP kinase activation is required for the expression of cyclin D1, p21Cip1 and a subset of AP-1 proteins in CCL39 cells. *Oncogene* **18**, 3085–3097.

Banin, S., Moyal, L., Shieh, S., Taya, Y., Anderson, C. W., Chessa, L., Smorodinsky, N. I., Prives, C., Reiss, Y., Shiloh, Y. and Ziv, Y. (1998). Enhanced phosphorylation of p53 by ATM in response to DNA damage. *Science* **281**, 1674–1677.

Bell, S. P. and Stillman, B. (1992). ATP-dependent recognition of eukaryotic origins of DNA replication by a multiprotein complex. *Nature* **357**, 128–134.

Bentley, N. J., Holtzman, D. A., Flaggs, G., Keegan, K. S., DeMaggio, A., Ford, J. C., Hoekstra, M. and Carr, A. M. (1996). The *Schizosaccharomyces pombe rad3* checkpoint gene. *EMBO J* **15**, 6641–6651.

Bicknell, K. A., Surry, E. L. and Brooks, G. (2003). Targeting the cell cycle machinery for the treatment of cardiovascular disease. *J Pharm Pharmacol* **55**, 571–591.

Biermann, E., Baack, M., Kreitz, S. and Knippers, R. (2002). Synthesis and turnover of the replicative Cdc6 protein during the HeLa cell cycle. *Eur J Biochem* **269**, 1040–1046.

Blangy, A., Lane, H. A., d'Herin, P., Harper, M., Kress, M. and Nigg, E. A. (1995). Phosphorylation by p34cdc2 regulates spindle association of human Eg5, a kinesin-related motor essential for bipolar spindle formation *in vivo*. *Cell* **83**, 1159–1169.

Boddy, M. N. and Russell, P. (1999). DNA replication checkpoint control. *Front Biosci* **4**, D841–848.

Bottazzi, M. E., Zhu, X., Bohmer, R. M. and Assoisan, R. K. (1999). Regulation of p21 (Cip1) expression by growth factors and the extracellular matrix reveals a role for transient ERK activity in G1 phase. *J Cell Biol* **146**(6), 1255–1264.

Boussett, K. and Diffley, J. F. (1998). The Cdc7 protein kinase is required for origin firing during S phase. *Genes Dev* **12**(4), 480–490.

Bouzahzah, B., Fu, M., Iavarone, A., Factor, V. M., Thorgeirsson, S. S. and Pestell, R. G. (2000). Transforming growth factor-β1 recruits histone deacetylase 1 to a p130 repressor complex in transgenic mice. *Cancer Res* **60**, 4531–4537.

Brandeis, M., Roswell, I., Carrington, M., Crompton, T., Jacobs, M. A., Kirk, J., Gannon, J. and Hunt, T. (1998). Cyclin B2-null mice develop normally and are fertile whereas cyclin B1-null mice die in utero. *Proc Natl Acad Sci USA* **95**(8), 4344–4349.

Braun, K., Holzl, G., Soucek, T., Geisen, T., Moroy, T. and Hengstschlager, M. (1998). Investigation of the cell cycle regulation of cdk3-associated kinase activity and the role of cdk3 in proliferation and transformation. *Oncogene* **17**(17), 2259–2269.

Brehm, A., Miska, E. A., McCance, D. J., Reid, J. L., Bannister, A. J. and Kouzarides, T. (1998). Retinoblastoma protein recruits histone deacetylase to repress transcription. *Nature* **391**, 597–601.

Brooks, G., Poolman, R. A. and Li, J-M. (1998). Arresting developments in the cardiac myocyte cell cycle: Role of cyclin-dependent kinase inhibitors. *Cardiovascular Research* **39**, 301–311.

Brown, A. L., Lee, C. H., Scwarz, J. K., Mitiku, N., Piwinca-Worms, H. and Chung, J. H. (1999). A human Cds1-related kinse that functions downstream of ATM protein in the cellular response to DNA damage. *Proc Natl Acad Sci USA* **96**(7), 3745–3750.

Canman, C. E., Lim, D. S., Cimprach, K. A., Taya, K., Sakaguchi, K., Appella, E., Kastan, M. B. and Siliciano, J. D. (1998). Activation of the ATM kinase by ionising radiation and phosphorylation of p53. *Science* **281** (5383), 1677–1679.

Cartwright, P., Muller, H., Wagener, C., Holm, K. and Helin, K. (1998). E2F-6: a novel member of the E2F family is an inhibitor of E2F-dependent transcription. *Oncogene* **17**(5), 611–623.

Carmo-Fonseca, M., Mendes-Soares, L. and Campos, I. (2000). To be or not to be in the nucleolus. *Nat Cell Biol* **2**(6), E107–112.

Chan, D. W., Chen, B. P., Prithivirajsingh, S., Kurimasa, A., Story, M. D., Qin, J. and Chen, D. J. (2002). Autophosphorylation of the DNA-dependent protein kinase catalytic subunit is required for rejoining of DNA double-strand breaks. *Genes Dev* **16**, 2333–2338.

Chaturvedi, P., Eng, W. K., Zhu, Y., Mattern, M. R., Mishra, R., Hurle, M. R., Zhang, X., Annan, R. S., Lu, Q., Faucette, L. F., Scott, G. F., Li, X., Carr, S. A., Johnson, R. K., Winkler, J. D. and Zhou, B. B. (1999). Mammalian Chk2 is a downstream effector of the ATM-dependent DNA damage checkpoint pathway. *Oncogene* **18**, 4047–4054.

Chen, M. S., Hurov, J., White, L. S., Woodford-Thomas, T. and Piwnica-Worms, H. (2001). Absence of apparent phenotype in mice lacking Cdc25C protein phosphatase. *Mol Cell Biol* **21**, 3853–3861.

Cheng, M., Olivier, P., Diehl, J. A., Fero, M., Roussel, M. F., Roberts, J. M. and Sherr, C. J. (1999). The p21(Cip1) and p27(Kip1) CDK 'inhibitors' are essential activators of cyclin D-dependent kinases in murine fibroblasts. *EMBO J* **18**, 1571–1583.

Cheng, M., Sexl, V., Sherr, C. J. and Roussel, M. F. (1998). Assembly of cyclin D-dependent kinase and titration of p27Kip1 regulated by mitogen-activated protein kinase kinase (MEK1). *Proc Natl Acad Sci USA* **95**, 1091–1096.

Coleman, T. R., Carpenter, P. B. and Dunphy, W. G. (1996). The *Xenopus* Cdc6 protein is essential for the initiation of a single round of DNA replication in cell-free extracts. *Cell* **87**, 53–63.

Colman, M. S., Afshari, C. A. and Barrett, J. C. (2000). Regulation of *p53* stability and activity in response to genotoxic stress. *Mutat Res* **462**, 179–188.

Cook, J. G., Park, C. H., Burke, T. W., Leone, G., DeGregori, J., Engel, A. and Nevins, J. R. (2002). Analysis of Cdc6 function in the assembly of mammalian prereplication complexes. *Proc Natl Acad Sci USA* **99**, 1347–1352.

Cress, W. D. and Nevins, J. R. (1996). A role for a bent DNA structure in E2F-mediated transcription activation. *Mol Cell Biol* **16**, 2119–2127.

Dahm, K. and Hubscher, U. (2002). Colocalization of human Rad17 and PCNA in late S phase of the cell cycle upon replication block. *Oncogene* **21**, 7710–7719.

Dalal, S. N., Schweitzer, C. M., Gan, J. and DeCaprio, J. A. (1999). Cytoplasmic localization of human cdc25C during interphase requires an intact 14–3–3 binding site. *Mol Cell Biol* **19**, 4465–4479.

de Falco, G. and Giordano, A. (1998). CDK9 (PITALRE): a multifunctional cdc2-related kinase. *J Cell Physiol* **177**, 501–506.

Devault, A., Fesquet, D., Cavadore, J. C., Garrigues, A. M., Labbe, J. C., Lorca, T., Picard, A., Philippe, M. and Doree, M. (1992). Cyclin A potentiates maturation-promoting factor activation in the early *Xenopus* embryo via inhibition of the tyrosine kinase that phosphorylates cdc2. *J Cell Biol* **118**, 1109–1120.

Dhavan, R. and Tsai, L. H. (2001). A decade of CDK5. *Nat Rev Mol Cell Biol* **2**(10), 749–759.

Donaldson, A. D., Fangman, W. L. and Brewer, B. J. (1998). Cdc7 is required throughout the yeast S phase to activate replication origins. *Genes Dev* **12**, 491–501.

Dulic, V., Lees, E. and Reed, S. I. (1992). Association of human cyclin E with a periodic G_1–S phase protein kinase. *Science* **257**, 1958–1961.

Dunphy, W. G., Brizuela, L., Beach, D. and Newport, J. (1988). The *Xenopus* cdc2 protein is a component of MPF, a cytoplasmic regulator of mitosis. *Cell* **54**, 423–431.

Dunphy, W. G. and Kumagai, A. (1991). The cdc25 protein contains an intrinsic phosphatase activity. *Cell* **67**, 189–196.

Dutta, A., Din, S., Brill, S. J. and Stillman, B. (1991). Phosphorylation of replication protein A: a role for cdc2 kinase in G_1/S regulation. *Cold Spring Harb Symp Quant Biol* **56**, 315–324.

Dynlacht, B. D., Flores, O., Lees, J. A. and Harlow, E. (1994). Differential regulation of E2F *trans*-activation by cyclin/cdk2 complexes. *Genes Dev* **8**, 1772–1786.

Dynlacht, B. D., Moberg, K., Lees, J. A., Harlow, E. and Zhu, L. (1997). Specific regulation of E2F family members by cyclin-dependent kinases. *Mol Cell Biol* **17**, 3867–3875.

Dyson, N. (1998). The regulation of E2F by pRB-family proteins. *Genes Dev* **12**, 2245–2262.

el-Deiry, W. S., Tokino, T., Velculescu, V. E., Levy, D. B., Parsons, R., Trent, J. M., Lin, D., Mercer, W. E., Kinzler, K. W. and Vogelstein, B. (1993). WAF1, a potential mediator of *p53* tumor suppression. *Cell* **75**, 817–825.

Endicott, J. A., Nurse, P. and Johnson, L. N. (1994). Mutational analysis supports a structural model for the cell cycle protein kinase p34. *Protein Eng* **7**, 243–253.

Ewen, M. E. (1996). p53-dependent repression of cdk4 synthesis in transforming growth factor-beta-induced G_1 cell cycle arrest. *J Lab Clin Med* **128**, 355–360.

Fearon, E. R. (1997). Human cancer syndromes: clues to the origin and nature of cancer. *Science* **278**, 1043–1050.

Ferreira, R., Naguibneva, I., Mathieu, M., Ait-Si-Ali, S., Robin, P., Pritchard, L. L. and Harel-Bellan, A. (2001). Cell cycle-dependent recruitment of HDAC-1 correlates with deacetylation of histone H4 on an Rb-E2F target promoter. *EMBO Rep* **2**, 794–799.

Feuer, R., Mena, I., Pagarigan, R., Slifka, M. K. and Whitton, J. L. (2002). Cell cycle status affects Coxsackievirus replication, persistence, and reactivation *in vitro*. *J Virol* **76**, 4430–4440.

Fujita, M., Kiyono, T., Hayashi, Y. and Ishibashi, M. (1996). Inhibition of S-phase entry of human fibroblasts by an antisense oligomer against hCDC47. *Biochem Biophys Res Commun* **219**, 604–607.

Fujita, M., Yamada, C., Tsurumi, T., Hanaoka, F., Matsuzawa, K. and Inagaki, M. (1998). Cell cycle- and chromatin binding state-dependent phosphorylation of human MCM hetero-hexameric complexes. A role for cdc2 kinase. *J Biol Chem* **273**, 17095–17101.

Furnari, B., Blasina, A., Boddy, M. N., McGowan, C. H. and Russell, P. (1999). Cdc25 inhibited *in vivo* and *in vitro* by checkpoint kinases Cds1 and Chk1. *Mol Biol Cell* **10**, 833–845.

Furuno, N., den Elzen, N. and Pines, J. (1999). Human cyclin A is required for mitosis until mid-prophase. *J Cell Biol* **147**, 295–306.

Gabrielli, B. G., De Souza, C. P., Tonks, I. D., Clark, J. M., Hayward, N. K. and Ellem, K. A. (1996). Cytoplasmic accumulation of cdc25B phosphatase in mitosis triggers centrosomal microtubule nucleation in HeLa cells. *J Cell Sci* **109**, 1081–1093.

Galaktionov, K. and Beach, D. (1991). Specific activation of cdc25 tyrosine phosphatases by B-type cyclins: evidence for multiple roles of mitotic cyclins. *Cell* **67**, 1181–1194.

Gartel, A. L., Serfas, M. S. and Tyner, A. L. (1996). p21-Negative regulator of the cell cycle. *Proc Soc Exp Biol Med* **213**, 138–149.

Gately, D. P., Hittle, J. C., Chan, G. K. and Yen, T. J. (1998). Characterization of ATM expression, localization, and associated DNA-dependent protein kinase activity. *Mol Biol Cell* **9**, 2361–2374.

Gatti, M., Giansanti, M. G. and Bonaccorsi, S. (2000). Relationships between the central spindle and the contractile ring during cytokinesis in animal cells. *Microsc Res Tech* **49**, 202–208.

Gaubatz, S., Wood, J. G. and Livingston, D. M. (1998). Unusual proliferation arrest and transcriptional control properties of a newly discovered E2F family member, E2F-6. *Proc Natl Acad Sci USA* **95**, 9190–9195.

Gautier, J., Norbury, C., Lohka, M., Nurse, P. and Maller, J. (1988). Purified maturation-promoting factor contains the product of a *Xenopus* homolog of the fission yeast cell cycle control gene cdc2$^+$. *Cell* **54**, 433–439.

Gavin, K. A., Hidaka, M. and Stillman, B. (1995). Conserved initiator proteins in eukaryotes. *Science* **270**, 1667–1671.

Glotzer, M., Murray, A. W. and Kirschner, M. W. (1991). Cyclin is degraded by the ubiquitin pathway. *Nature* **349**, 132–138.

Glover, D. M., Hagan, I. M. and Tavares, A. A. (1998). Polo-like kinases: a team that plays throughout mitosis. *Genes Dev* **12**, 3777–3787.

Golan, A., Yudkovsky, Y. and Hershko, A. (2002). The cyclin–ubiquitin ligase activity of cyclosome/APC is jointly activated by protein kinases Cdk1-cyclin B and Plk. *J Biol Chem* **277**, 15552–15557.

Goldstone, S., Pavey, S., Forrest, A., Sinnamon, J. and Gabrielli, B. (2001). Cdc25-dependent activation of cyclin A/cdk2 is blocked in G_2 phase arrested cells independently of ATM/ATR. *Oncogene* **20**, 921–932.

Grafi, G. (1998). Cell cycle regulation of DNA replication: the endoreduplication perspective. *Exp Cell Res* **244**, 372–378.

Grana, X., Garriga, J. and Mayol, X. (1998). Role of the retinoblastoma protein family, pRB, p107 and p130 in the negative control of cell growth. *Oncogene* **17**, 3365–3383.

Hagting, A., Jackman, M., Simpson, K. and Pines, J. (1999). Translocation of cyclin B1 to the nucleus at prophase requires a phosphorylation-dependent nuclear import signal. *Curr Biol* **9**, 680–689.

Hagting, A., Karlsson, C., Clute, P., Jackman, M. and Pines, J. (1998). MPF localization is controlled by nuclear export. *EMBO J* **17**, 4127–4138.

Hamanaka, R., Smith, M. R., O'Connor, P. M., Maloid, S., Mihalic, K., Spivak, J. L., Longo, D. L. and Ferris, D. K. (1995). Polo-like kinase is a cell cycle-regulated kinase activated during mitosis. *J Biol Chem* **270**, 21086–21091.

Hang, H. and Lieberman, H. B. (2000). Physical interactions among human checkpoint control proteins HUS1p, RAD1p, and RAD9p, and implications for the regulation of cell cycle progression. *Genomics* **65**, 24–33.

Harbour, J. W., Luo, R. X., Dei Santi, A., Postigo, A. A. and Dean, D. C. (1999). Cdk phosphorylation triggers sequential intramolecular interactions that progressively block Rb functions as cells move through G_1. *Cell* **98**, 859–869.

Hardy, C. F. (1997). Identification of Cdc45p, an essential factor required for DNA replication. *Gene* **187**, 239–246.

Harlow, E. (1996). A research shortcut from a common cold virus to human cancer. *Cancer* **78**, 558–565.

Harper, J. W., Burton, J. L. and Solomon, M. J. (2002). The anaphase-promoting complex: it's not just for mitosis any more. *Genes Dev* **16**, 2179–2206.

Hartley, K. O., Gell, D., Smith, G. C., Zhang, H., Divecha, N., Connelly, M. A., Admon, A., Lees-Miller, S. P., Anderson, C. W. and Jackson, S. P. (1995). DNA-dependent protein kinase catalytic subunit: a relative of phosphatidylinositol 3-kinase and the ataxia telangiectasia gene product. *Cell* **82**, 849–856.

Hartwell, L. H. (1973). Three additional genes required for deoxyribonucleic acid synthesis in *Saccharomyces cerevisiae*. *J Bacteriol* **115**, 966–974.

Hassig, C. A., Fleischer, T. C., Billin, A. N., Schreiber, S. L. and Ayer, D. E. (1997). Histone deacetylase activity is required for full transcriptional repression by mSin3A. *Cell* **89**, 341–347.

Hateboer, G., Kerkhoven, R. M., Shvarts, A., Bernards, R. and Beijersbergen, R. L. (1996). Degradation of E2F by the ubiquitin–proteasome pathway: regulation by retinoblastoma family proteins and adenovirus transforming proteins. *Genes Dev* **10**, 2960–2970.

Hateboer, G., Wobst, A., Petersen, B. O., Le Cam, L., Vigo, E., Sardet, C. and Helin, K. (1998). Cell cycle-regulated expression of mammalian CDC6 is dependent on E2F. *Mol Cell Biol* **18**, 6679–6697.

Heald, R., McLoughlin, M. and McKeon, F. (1993). Human *wee1* maintains mitotic timing by protecting the nucleus from cytoplasmically activated Cdc2 kinase. *Cell* **74**, 463–474.

Helin, K., Wu, C. L., Fattaey, A. R., Lees, J. A., Dynlacht, B. D., Ngwu, C. and Harlow, E. (1993). Heterodimerization of the transcription factors E2F-1 and DP-1 leads to cooperative transactivation. *Genes Dev* **7**, 1850–1861.

Hendrickson, M., Madine, M., Dalton, S. and Gautier, J. (1996). Phosphorylation of MCM4 by *cdc2* protein kinase inhibits the activity of the minichromosome maintenance complex. *Proc Natl Acad Sci USA* **93**, 12223–12228.

Herbig, U., Griffith, J. W. and Fanning, E. (2000). Mutation of cyclin/cdk phosphorylation sites in HsCdc6 disrupts a late step in initiation of DNA replication in human cells. *Mol Biol Cell* **11**, 4117–4130.

Hess, G. F., Drong, R. F., Weiland, K. L., Slightom, J. L., Sclafani, R. A. and Hollingsworth, R. E. (1998). A human homolog of the yeast *CDC7* gene is overexpressed in some tumors and transformed cell lines. *Gene* **211**, 133–140.

Hiscox, J. A. (2002). The nucleolus – a gateway to viral infection? *Arch Virol* **147**, 1077–1089.

Hoffmann, I., Clarke, P. R., Marcote, M. J., Karsenti, E. and Draetta, G. (1993). Phosphorylation and activation of human cdc25-C by cdc2–cyclin B and its involvement in the self-amplification of MPF at mitosis. *EMBO J* **12**, 53–63.

Hofmann, F., Martelli, F., Livingston, D. M. and Wang, Z. (1996). The retinoblastoma gene product protects E2F-1 from degradation by the ubiquitin–proteasome pathway. *Genes Dev* **10**, 2949–2959.

Hopwood, B. and Dalton, S. (1996). Cdc45p assembles into a complex with Cdc46p/Mcm5p, is required for minichromosome maintenance, and is essential for chromosomal DNA replication. *Proc Natl Acad Sci USA* **93**, 12309–12314.

Hunter, T. and Pines, J. (1994). Cyclins and cancer. II: cyclin D and CDK inhibitors come of age. *Cell* **79**, 573–582.

Hurford, R. K. Jr, Cobrinik, D., Lee, M. H. and Dyson, N. (1997). pRB and p107/p130 are required for the regulated expression of different sets of E2F-responsive genes. *Genes Dev* **11**, 1447–1463.

Iizuka, M. and Stillman, B. (1999). Histone acetyltransferase HBO1 interacts with the ORC1 subunit of the human initiator protein. *J Biol Chem* **274**, 23027–25034.

Ishimi, Y. (1997). A DNA helicase activity is associated with an MCM4, −6, and −7 protein complex. *J Biol Chem* **272**, 24508–24513.

Izumi, T. and Maller, J. L. (1993). Elimination of cdc2 phosphorylation sites in the cdc25 phosphatase blocks initiation of M-phase. *Mol Biol Cell* **4**, 1337–1350.

Izumi, T. and Maller, J. L. (1995). Phosphorylation and activation of the *Xenopus* Cdc25 phosphatase in the absence of Cdc2 and Cdk2 kinase activity. *Mol Biol Cell* **6**, 215–226.

Izumi, T., Walker, D. H. and Maller, J. L. (1992). Periodic changes in phosphorylation of the *Xenopus* cdc25 phosphatase regulate its activity. *Mol Biol Cell* **3**, 927–939.

Jackman, M., Firth, M. and Pines, J. (1995). Human cyclins B1 and B2 are localized to strikingly different structures: B1 to microtubules, B2 primarily to the Golgi apparatus. *EMBO J* **14**, 1646–1654.

Jacks, T. and Weinberg, R. A. (1996). Cell-cycle control and its watchman. *Nature* **381**, 643–644.

Jiang, W. and Hunter, T. (1997). Identification and characterization of a human protein kinase related to budding yeast Cdc7p. *Proc Natl Acad Sci USA* **94**, 14320–14325.

Jiang, W., McDonald, D., Hope, T. J. and Hunter, T. (1999a). Mammalian Cdc7–Dbf4 protein kinase complex is essential for initiation of DNA replication. *EMBO J* **18**, 5703–5713.

Jiang, W., Wells, N. J. and Hunter, T. (1999b). Multistep regulation of DNA replication by Cdk phosphorylation of HsCdc6. *Proc Natl Acad Sci USA* **96**, 6193–6198.

Kamb, A., Gruis, N. A., Weaver-Feldhaus, J., Liu, Q., Harshman, K., Tavtigian, S. V., Stockert, E., Day, R. S. III, Johnson, B. E. and Skolnick, M. H. (1994). A cell cycle regulator potentially involved in genesis of many tumor types. *Science* **264**, 436–440.

Kasid, U., Suy, S., Dent, P., Ray, S., Whiteside, T. L. and Sturgill, T. W. (1996). Activation of Raf by ionizing radiation. *Nature* **382**, 813–816.

Kearsey, S. E. and Labib, K. (1998). MCM proteins: evolution, properties, and role in DNA replication. *Biochim Biophys Acta* **1398**, 113–136.

Keegan, K. S., Holtzman, D. A., Plug, A. W., Christenson, E. R., Brainerd, E. E., Flaggs, G., Bentley, N. J., Taylor, E. M., Meyn, M. S., Moss, S. B., Carr, A. M., Ashley, T. and

Hoekstra, M. F. (1996). The Atr and Atm protein kinases associate with different sites along meiotically pairing chromosomes. *Genes Dev* **10**, 2423–2437.

Kelly, T. J. and Brown, G. W. (2000). Regulation of chromosome replication. *Annu Rev Biochem* **69**, 829–880.

Keezer, S. M. and Gilbert, D. M. (2002). Evidence for a pre-restriction point Cdk3 activity. *J Cell Biochem* **85**(3), 545–552.

Kimura, H., Nozaki, N. and Sugimoto, K. (1994). DNA polymerase alpha associated protein P1, a murine homolog of yeast MCM3, changes its intranuclear distribution during the DNA synthetic period. *EMBO J* **13**, 4311–4320.

Knibiehler, M., Goubin, F., Escalas, N., Jonsson, Z. O., Mazarguil, H., Hubscher, U. and Ducommun, B. (1996). Interaction studies between the p21Cip1/Waf1 cyclin-dependent kinase inhibitor and proliferating cell nuclear antigen (PCNA) by surface plasmon resonance. *FEBS Lett* **391**, 66–70.

Koff, A., Giordano, A., Desai, D., Yamashita, K., Harper, J. W., Elledge, S., Nishimoto, T., Morgan, D. O., Franza, B. R. and Roberts, J. M. (1992). Formation and activation of a cyclin E–cdk2 complex during the G_1 phase of the human cell cycle. *Science* **257**, 1689–1694.

Kong, M., Barnes, E. A., Ollendorff, V. and Donoghue, D. J. (2000). Cyclin F regulates the nuclear localization of cyclin B1 through a cyclin–cyclin interaction. *EMBO J* **19**, 1378–1388.

Kotani, S., Tugendreich, S., Fujii, M., Jorgensen, P. M., Watanabe, N., Hoog, C., Hieter, P. and Todokoro, K. (1998). PKA and MPF-activated polo-like kinase regulate anaphase-promoting complex activity and mitosis progression. *Mol Cell* **1**, 371–380.

Krek, W., Ewen, M. E., Shirodkar, S., Arany, Z., Kaelin, W. G. Jr and Livingston, D. M. (1994). Negative regulation of the growth-promoting transcription factor E2F-1 by a stably bound cyclin A-dependent protein kinase. *Cell* **78**, 161–172.

Krek, W., Xu, G. and Livingston, D. M. (1995). Cyclin A-kinase regulation of E2F-1 DNA binding function underlies suppression of an S phase checkpoint. *Cell* **83**, 1149–1158.

Kumagai, A. and Dunphy, W. G. (1992). Regulation of the cdc25 protein during the cell cycle in *Xenopus* extracts. *Cell* **70**, 139–151.

Kumagai, H., Sato, N., Yamada, M., Mahony, D., Seghezzi, W., Lees, E., Arai, K. and Masai, H. (1999). A novel growth- and cell cycle-regulated protein, ASK, activates human Cdc7-related kinase and is essential for G_1/S transition in mammalian cells. *Mol Cell Biol* **19**, 5083–5095.

Lacey, K. R., Jackson, P. K. and Stearns, T. (1999). Cyclin-dependent kinase control of centrosome duplication. *Proc Natl Acad Sci USA* **96**, 2817–2822.

Laherty, C. D., Yang, W. M., Sun, J. M., Davie, J. R., Seto, E. and Eisenman, R. N. (1997). Histone deacetylases associated with the mSin3 corepressor mediate mad transcriptional repression. *Cell* **89**, 349–356.

Lakin, N. D., Hann, B. C. and Jackson, S. P. (1999). The ataxia-telangiectasia related protein ATR mediates DNA-dependent phosphorylation of p53. *Oncogene* **18**, 3989–3995.

Lane, H. A. and Nigg, E. A. (1996). Antibody microinjection reveals an essential role for human polo-like kinase 1 (Plk1) in the functional maturation of mitotic centrosomes. *J Cell Biol* **135**, 1701–1713.

Lavoie, J. N., L'Allemain, G., Brunet, A., Muller, R. and Pouyssegur, J. (1996). Cyclin D1 expression is regulated positively by the p42/p44MAPK and negatively by the p38/HOGMAPK pathway. *J Biol Chem* **271**, 20608–20616.

Leclerc, V. and Leopold, P. (1996). The cyclin C/Cdk8 kinase. *Prog Cell Cycle Res* **2**, 197–204.

Lee, K. S., Grenfell, T. Z., Yarm, F. R. and Erikson, R. L. (1998). Mutation of the polo-box disrupts localization and mitotic functions of the mammalian polo kinase Plk. *Proc Natl Acad Sci USA* **95**, 9301–9306.

Lee, M. S., Ogg, S., Xu, M., Parker, L. L., Donoghue, D. J., Maller, J. L. and Piwnica-Worms, H. (1992). cdc25⁺ encodes a protein phosphatase that dephosphorylates p34cdc2. *Mol Biol Cell* **3**, 73–84.

Lepke, M., Putter, V., Staib, C., Kneissl, M., Berger, C., Hoehn, K., Nanda, I., Schmid, M. and Grummt, F. (1999). Identification, characterization and chromosomal localization of the cognate human and murine DBF4 genes. *Mol Gen Genet* **262**, 220–229.

Levine, A. J. (1997). p53, the cellular gatekeeper for growth and division. *Cell* **88**, 323–331.

Li, J., Meyer, A. N. and Donoghue, D. J. (1995). Requirement for phosphorylation of cyclin B1 for *Xenopus* oocyte maturation. *Mol Biol Cell* **6**, 1111–1124.

Lin, A. W., Barradas, M., Stone, J. C., van Aelst, L., Serrano, M. and Lowe, S. W. (1998). Premature senescence involving p53 and p16 is activated in response to constitutive MEK/MAPK mitogenic signaling. *Genes Dev* **12**, 3008–3019.

Lindeman, G. J., Gaubatz, S., Livingston, D. M. and Ginsberg, D. (1997). The subcellular localization of E2F-4 is cell cycle-dependent. *Proc Natl Acad Sci USA* **94**, 5095–5100.

Liou, Y. C., Ryo, A., Huang, H. K., Lu, P. J., Bronson, R., Fujimori, F., Uchida, T., Hunter, T. and Lu, K. P. (2002). Loss of Pin1 function in the mouse causes phenotypes resembling cyclin D1-null phenotypes. *Proc Natl Acad Sci USA* **99**, 1335–1340.

Liu, J. J., Chao, J. R., Jiang, M. C., Ng, S. Y., Yen, J. J. and Yang-Yen, H. F. (1995). Ras transformation results in an elevated level of cyclin D1 and acceleration of G_1 progression in NIH 3T3 cells. *Mol Cell Biol* **15**, 3654–3663.

Lowe, M., Rabouille, C., Nakamura, N., Watson, R., Jackman, M., Jamsa, E., Rahman, D., Pappin, D. J. and Warren, G. (1998). Cdc2 kinase directly phosphorylates the *cis*-Golgi matrix protein GM130 and is required for Golgi fragmentation in mitosis. *Cell* **94**, 783–793.

Loyola, A., LeRoy, G., Wang, Y.-H. and Reinberg, D. (2001). Reconstitution of recombinant chromatin establishes a requirement for histone-tail modifications during chromatin assembly and transcription. *Genes Dev* **15**, 2837–2851.

Lundgren, K., Walworth, N., Booher, R., Dembski, M., Kirschner, M. and Beach, D. (1991). mik1 and wee1 cooperate in the inhibitory tyrosine phosphorylation of cdc2. *Cell* **64**, 1111–1122.

Luo, R. X., Postigo, A. A. and Dean, D. C. (1998). Rb interacts with histone deacetylase to repress transcription. *Cell* **92**, 463–473.

Luo, Y., Hurwitz, J. and Massague, J. (1995). Cell-cycle inhibition by independent CDK and PCNA binding domains in p21Cip1. *Nature* **375**, 159–161.

Ma, T., Van Tine, B. A., Wei, Y., Garrett, M. D., Nelson, D., Adams, P. D., Wang, J., Qin, J., Chow, L. T. and Harper, J. W. (2000). Cell cycle-regulated phosphorylation of p220(NPAT) by cyclin E/Cdk2 in Cajal bodies promotes histone gene transcription. *Genes Dev* **14**, 2298–2313.

Magae, J., Wu, C. L., Illenye, S., Harlow, E. and Heintz, N. H. (1996). Nuclear localization of DP and E2F transcription factors by heterodimeric partners and retinoblastoma protein family members. *J Cell Sci* **109**, 1717–1726.

Magnaghi-Jaulin, L., Groisman, R., Naguibneva, I., Robin, P., Lorain, S., Le Villain, J. P., Troalen, F., Trouche, D. and Harel-Bellan, A. (1998). Retinoblastoma protein represses transcription by recruiting a histone deacetylase. *Nature* **391**, 601–605.

Mailand, N., Podtelejnikov, A. V., Groth, A., Mann, M., Bartek, J. and Lukas, J. (2002). Regulation of G_2/M events by Cdc25A through phosphorylation-dependent modulation of its stability. *EMBO J* **21**, 5911–5920.

Marahrens, Y. and Stillman, B. (1992). A yeast chromosomal origin of DNA replication defined by multiple functional elements. *Science* **255**, 817–823.

Marshall, M. S. (1995). Ras target proteins in eukaryotic cells. *FASEB J* **9**, 1311–1318.

Martinez, A. M., Afshar, M., Martin, F., Cavadore, J. C., Labbe, J. C. and Doree, M. (1997). Dual phosphorylation of the T-loop in cdk7: its role in controlling cyclin H binding and CAK activity. *EMBO J* **16**, 343–354.

Masai, H., Miyake, T. and Arai, K. (1995). hsk1$^+$, a *Schizosaccharomyces pombe* gene related to *Saccharomyces cerevisiae* CDC7, is required for chromosomal replication. *EMBO J* **14**, 3094–3104.

Masui, Y. and Markert, C. L. (1971). Cytoplasmic control of nuclear behavior during meiotic maturation of frog oocytes. *J Exp Zool* **177**, 129–145.

Matsuoka, S., Huang, M. and Elledge, S. J. (1998). Linkage of ATM to cell cycle regulation by the Chk2 protein kinase. *Science* **282**, 1893–1897.

McGill, C. J. and Brooks, G. (1995). Cell cycle control mechanisms and their role in cardiac growth. *Cardiovasc Res* **30**, 557–569.

Mendez, J. and Stillman, B. (2000). Chromatin association of human origin recognition complex, cdc6, and minichromosome maintenance proteins during the cell cycle: assembly of prerepliaction complexes in late mitosis. *Mol Cell Biol* **20**, 8602–8612.

Meyer, C. A., Jacobs, H. W. and Lehner, C. F. (2002). Cyclin D-cdk4 is not a master regulator of cell multiplication in *Drosophila* embryos. *Curr Biol* **12**, 661–666.

Miller, M. E. and Cross, F. R. (2001). Cyclin specificity: how many wheels do you need on a unicycle? *J Cell Sci* **114**, 1811–1820.

Mimura, S. and Takisawa, H. (1998). *Xenopus* Cdc45-dependent loading of DNA polymerase alpha onto chromatin under the control of S-phase Cdk. *EMBO J* **17**, 5699–5707.

Mishima, M. and Mabuchi, I. (1996). Cell cycle-dependent phosphorylation of smooth muscle myosin light chain in sea urchin egg extracts. *J Biochem (Tokyo)* **119**, 906–913.

Moore, J. D., Yang, J., Truant, R. and Kornbluth, S. (1999). Nuclear import of Cdk/cyclin complexes: identification of distinct mechanisms for import of Cdk2/cyclin E and Cdc2/cyclin B1. *J Cell Biol* **144**, 213–224.

Morkel, M., Wenkel, J., Bannister, A. J., Kouzarides, T. and Hagemeier, C. (1997). An E2F-like repressor of transcription. *Nature* **390**, 567–568.

Nagy, L., Kao, H. Y., Chakravarti, D., Lin, R. J., Hassig, C. A., Ayer, D. E., Schreiber, S. L. and Evans, R. M. (1997). Nuclear receptor repression mediated by a complex containing SMRT, mSin3A, and histone deacetylase. *Cell* **89**, 373–380.

Napolitano, G., Majello, B. and Lania, L. (2002). Role of cyclinT/Cdk9 complex in basal and regulated transcription [review]. *Int J Oncol* **21**, 171–177.

Nasheuer, H. P., Smith, R., Bauerschmidt, C., Grosse, F. and Weisshart, K. (2002). Initiation of eukaryotic DNA replication: regulation and mechanisms. *Prog Nucleic Acid Res Mol Biol* **72**, 41–94.

Nasmyth, K. (1996). Viewpoint: putting the cell cycle in order. *Science* **274**, 1643–5.

Nayak, B. K. and Das, G. M. (2002). Stabilization of p53 and *trans*-activation of its target genes in response to replication blockade. *Oncogene* **21**, 7226–7229.

Nevins, J. R. (1992). E2F: a link between the Rb tumor suppressor protein and viral oncoproteins. *Science* **258**, 424–429.

Nigg, E.A. (1996). Cyclin-dependent kinase 7: at the cross-roads of transcription, DNA repair and cell cycle control? *Curr Opin Cell Biol* **8**(3), 312–317.

Nigg, E. A. (1998). Polo-like kinases: positive regulators of cell division from start to finish. *Curr Opin Cell Biol* **10**, 776–783.

Nislow, C., Lombillo, V. A., Kuriyama, R. and McIntosh, J. R. (1992). A plus-end-directed motor enzyme that moves antiparallel microtubules *in vitro* localizes to the interzone of mitotic spindles. *Nature* **359**, 543–547.

Nislow, C., Sellitto, C., Kuriyama, R. and McIntosh, J. R. (1990). A monoclonal antibody to a mitotic microtubule-associated protein blocks mitotic progression. *J Cell Biol* **111**, 511–522.

Nobori, T., Miura, K., Wu, D. J., Lois, A., Takabayashi, K. and Carson, D. A. (1994). Deletions of the cyclin-dependent kinase-4 inhibitor gene in multiple human cancers. *Nature* **368**, 753–756.

Noguchi, T., Arai, R., Motegi, F., Nakano, K. and Mabuchi, I. (2001). Contractile ring formation in *Xenopus* egg and fission yeast. *Cell Struct Funct* **26**, 545–554.

Ogg, S., Gabrielli, B. and Piwnica-Worms, H. (1994). Purification of a serine kinase that associates with and phosphorylates human Cdc25C on serine 216. *J Biol Chem* **269**, 30461–30469.

Ohtsubo, M. and Roberts, J. M. (1993). Cyclin-dependent regulation of G_1 in mammalian fibroblasts. *Science* **259**, 1908–1912.

Ohtsubo, M., Theodoras, A. M., Schumacher, J., Roberts, J. M. and Pagano, M. (1995). Human cyclin E, a nuclear protein essential for the G_1-to-S phase transition. *Mol Cell Biol* **15**, 2612–2624.

Pagano, M., Pepperkok, R., Lukas, J., Baldin, V., Ansorge, W., Bartek, J. and Draetta, G. (1993). Regulation of the cell cycle by the cdk2 protein kinase in cultured human fibroblasts. *J Cell Biol* **121**, 101–111.

Pagano, M., Pepperkok, R., Verde, F., Ansorge, W. and Draetta, G. (1992). Cyclin A is required at two points in the human cell cycle. *EMBO J* **11**, 961–971.

Paiardini, M., Galati, D., Cervasi, B., Cannavo, G., Galluzzi, L., Montroni, M., Guetard, D., Magnani, M., Piedimonte, G. and Silvestri, G. (2001). Exogenous interleukin-2 administration corrects the cell cycle perturbation of lymphocytes from human immunodeficiency virus-infected individuals. *J Virol* **75**, 10843–10855.

Parker, A. E., Van de Weyer, I., Laus, M. C., Oostveen, I., Yon, J., Verhasselt, P. and Luyten, W. H. (1998). A human homologue of the *Schizosaccharomyces pombe* rad1$^+$ checkpoint gene encodes an exonuclease. *J Biol Chem* **273**, 18332–18339.

Parker, L. L. and Piwnica-Worms, H. (1992). Inactivation of the p34cdc2-cyclin B complex by the human WEE1 tyrosine kinase. *Science* **257**, 1955–1957.

Pavletich, N. P. (1999). Mechanisms of cyclin dependent kinase regulation: structures of Cdks, their cyclin activators, and Cip and INK4 inhibitors. *J Mol Biol* **287**, 821–828.

Peeper, D. S., Parker, L. L., Ewen, M. E., Toebes, M., Hall, F. L., Xu, M., Zantema, A., van der Eb, A. J. and Piwnica-Worms, H. (1993). A- and B-type cyclins differentially modulate substrate specificity of cyclin–cdk complexes. *EMBO J* **12**, 1947–1954.

Peng, C. Y., Graves, P. R., Ogg, S., Thoma, R. S., Byrnes, M. J III, Wu, Z., Stephenson, M. T. and Piwnica-Worms, H. (1998). C-TAK1 protein kinase phosphorylates human Cdc25C on serine 216 and promotes 14-3-3 protein binding. *Cell Growth Diff* **9**, 197–208.

Peng, C. Y., Graves, P. R., Thoma, R. S., Wu, Z., Shaw, A. S. and Piwnica-Worms, H. (1997). Mitotic and G_2 checkpoint control: regulation of 14–3–3 protein binding by phosphorylation of Cdc25C on serine-216. *Science* **277**, 1501–1505.

Peter, M., Nakagawa, J., Doree, M., Labbe, J. C. and Nigg, E. A. (1990). *In vitro* disassembly of the nuclear lamina and M phase-specific phosphorylation of lamins by cdc2 kinase. *Cell* **61**, 591–602.

Petersen, B. O., Lukas, J., Sorensen, C. S., Bartek, J. and Helin, K. (1999). Phosphorylation of mammalian CDC6 by cyclin A/CDK2 regulates its subcellular localization. *EMBO J* **18**, 396–410.

Petersen, B. O., Wagener, C., Marinoni, F., Kramer, E. R., Melixetian, M., Denchi, E. L., Gieffers, C., Matteucci, C., Peters, J. M. and Helin, K. (2000). Cell cycle- and cell growth-regulated proteolysis of mammalian CDC6 is dependent on APC-CDH1. *Genes Dev* **14**, 2330–2343.

Pines, J. (1997). Cyclin-dependent kinase inhibitors: the age of crystals. *Biochim Biophys Acta* **1332**, M39–42.

Pines, J. and Hunter, T. (1989). Isolation of a human cyclin cDNA: evidence for cyclin mRNA and protein regulation in the cell cycle and for interaction with p34cdc2. *Cell* **58**, 833–846.

Pines, J. and Hunter, T. (1994). The differential localization of human cyclins A and B is due to a cytoplasmic retention signal in cyclin B. *EMBO J* **13**, 3772–3781.

Porter, P. L., Malone, K. E., Heagerty, P. J., Alexander, G. M., Gatti, L. A., Firpo, E. J., Daling, J. R. and Roberts, J. M. (1997). Expression of cell-cycle regulators p27Kip1 and cyclin E, alone and in combination, correlate with survival in young breast cancer patients. *Nat Med* **3**, 222–225.

Prives, C. (1993). Doing the right thing: feedback control and p53. *Curr Opin Cell Biol* **5**, 214–218.

Rao, P. N. and Johnson, R. T. (1970). Mammalian cell fusion: studies on the regulation of DNA synthesis and mitosis. *Nature* **225**, 159–164.

Rayman, J. B., Takahashi, Y., Indjeian, V. B., Dannenberg, J. H., Catchpole, S., Watson, R. J., te Riele, H. and Dynlacht, B. D. (2002). E2F mediates cell cycle-dependent transcriptional repression *in vivo* by recruitment of an HDAC1/mSin3B corepressor complex. *Genes Dev* **16**, 933–947.

Reed, S. I., Bailly, E., Dulic, V., Hengst, L., Resnitzky, D. and Slingerland, J. (1994). G_1 control in mammalian cells. *J Cell Sci Suppl* **18**, 69–73.

Resnitzky, D., Gossen, M., Bujard, H. and Reed, S. I. (1994). Acceleration of the G_1/S phase transition by expression of cyclins D1 and E with an inducible system. *Mol Cell Biol* **14**, 1669–1679.

Resnitzky, D., Hengst, L. and Reed, S. I. (1995). Cyclin A-associated kinase activity is rate limiting for entrance into S phase and is negatively regulated. *Mol Cell Biol* **15**, 4347–4352.

Reynisdottir, I., Polyak, K., Iavarone, A. and Massague, J. (1995). Kip/Cip and Ink4 Cdk inhibitors cooperate to induce cell cycle arrest in response to TGF-β. *Genes Dev* **9**, 1831–1845.

Rivard, N., Boucher, M. J., Asselin, C. and L'Allemain, G. (1999). MAP kinase cascade is required for p27 downregulation and S phase entry in fibroblasts and epithelial cells. *Am J Physiol* **277**, C652–664.

Roshak, A. K., Capper, E. A., Imburgia, C., Fornwald, J., Scott, G. and Marshall, L. A. (2000). The human polo-like kinase, PLK, regulates cdc2/cyclin B through phosphorylation and activation of the cdc25C phosphatase. *Cell Signal* **12**, 405–411.

Russell, P. and Nurse, P. (1986). Cdc25 + functions as an inducer in the mitotic control of fission yeast. *Cell* **4**, 145–153.

Russo, A. A., Jeffrey, P. D. and Pavletich, N. P. (1996). Structural basis of cyclin-dependent kinase activation by phosphorylation. *Nat Struct Biol* **3**, 696–700.

Ryan, K. M., Phillips, A. C. and Vousden, K. H. (2001). Regulation and function of the p53 tumor suppressor protein. *Curr Opin Cell Biol* **13**, 332–337.

Ryo, A., Nakamura, M., Wulf, G., Liou, Y. C. and Lu, K. P. (2001). Pin1 regulates turnover and subcellular localization of β-catenin by inhibiting its interaction with APC. *Nat Cell Biol* **3**, 793–801.

Sadhu, K., Reed, S. I., Richardson, H. and Russell, P. (1990). Human homolog of fission yeast cdc25 mitotic inducer is predominantly expressed in G_2. *Proc Natl Acad Sci USA* **87**, 5139–5143.

Saha, P., Chen, J., Thome, K. C., Lawlis, S. J., Hou, Z. H., Hendricks, M., Parvin, J. D. and Dutta, A. (1998a). Human CDC6/Cdc18 associates with Orc1 and cyclin–cdk and is selectively eliminated from the nucleus at the onset of S phase. *Mol Cell Biol* **18**, 2758–2767.

Saha, P., Thome, K. C., Yamaguchi, R., Hou, Z., Weremowicz, S. and Dutta, A. (1998b). The human homolog of *Saccharomyces cerevisiae* CDC45. *J Biol Chem* **273**, 18205–18209.

Sano, M. and Schneider M. D. (2003). Cyclins that don't cycle: cyclin T/cyclin-dependent kinase 9 determines cardiac muscle size. *Cell Cycle* **2**(2), 99–104.

Sato, N., Arai, K. and Masai, H. (1997). Human and *Xenopus* cDNAs encoding budding yeast Cdc7-related kinases: *in vitro* phosphorylation of MCM subunits by a putative human homologue of Cdc7. *EMBO J* **16**, 4340–4351.

Savitsky, K., Bar-Shira, A., Gilad, S., Rotman, G., Ziv, Y., Vanagaite, L., Tagle, D. A., Smith, S., Uziel, T., Sfez, S. *et al.* (1995). A single ataxia telangiectasia gene with a product similar to PI-3 kinase. *Science* **268**, 1749–1753.

Sebastian, B., Kakizuka, A. and Hunter, T. (1993). Cdc25M2 activation of cyclin-dependent kinases by dephosphorylation of threonine-14 and tyrosine-15. *Proc Natl Acad Sci USA* **90**, 3521–3524.

Sellers, J. R. (1991). Regulation of cytoplasmic and smooth muscle myosin. *Curr Opin Cell Biol* **3**, 98–104.

Serrano, M. (1997). The tumor suppressor protein p16INK4a. *Exp Cell Res* **237**, 7–13.

Serrano, M., Gomez-Lahoz, E., DePinho, R. A., Beach, D. and Bar-Sagi, D. (1995). Inhibition of ras-induced proliferation and cellular transformation by p16INK4. *Science* **267**, 249–252.

Serrano, M., Lin, A. W., McCurrach, M. E., Beach, D. and Lowe, S. W. (1997). Oncogenic ras provokes premature cell senescence associated with accumulation of p53 and p16INK4a. *Cell* **88**, 593–602.

Sheaff, R. J., Groudine, M., Gordon, M., Roberts, J. M. and Clurman, B. E. (1997). Cyclin E–CDK2 is a regulator of p27Kip1. *Genes Dev* **11**, 1464–1478.

Sherr, C. J. (1996). Cancer cell cycles. *Science* **274**, 1672–1677.

Sherr, C. J. and Roberts, J. M. (1995). Inhibitors of mammalian G_1 cyclin-dependent kinases. *Genes Dev* **9**, 1149–1163.

Sherr, C. J. and Roberts, J. M. (1999). CDK inhibitors: positive and negative regulators of G_1-phase progression. *Genes Dev* **13**, 1501–1512.

Shieh, S. Y., Ikeda, M., Taya, Y. and Prives, C. (1997). DNA damage-induced phosphorylation of p53 alleviates inhibition by MDM2. *Cell* **91**, 325–334.

Shteinberg, M., Protopopov, Y., Listovsky, T., Brandeis, M. and Hershko, A. (1999). Phosphorylation of the cyclosome is required for its stimulation by Fizzy/cdc20. *Biochem Biophys Res Commun* **260**, 193–198.

Siliciano, J. D., Canman, C. E., Taya, Y., Sakaguchi, K., Appella, E. and Kastan, M. B. (1997). DNA damage induces phosphorylation of the amino-terminus of p53. *Genes Dev* **11**, 3471–3481.

Simone, C. and Giordano, A. (2001). New insight in cdk9 function: from Tat to MyoD. *Front Biosci* **6**, D1073–1082.

Sklar, M. D. (1988). The ras oncogenes increase the intrinsic resistance of NIH 3T3 cells to ionizing radiation. *Science* **239**, 645–647.

Smits, V. A. and Medema, R. H. (2001). Checking out the G_2/M transition. *Biochim Biophys Acta* **1519**, 1–12.

Takahashi, Y., Rayman, J. B. and Dynlacht, B. D. (2000). Analysis of promoter binding by the E2F and pRB families in vivo: distinct E2F proteins mediate activation and repression. *Genes Dev* **14**, 804–816.

Takizawa, C. G. and Morgan, D. O. (2000). Control of mitosis by changes in the subcellular location of cyclin-B1–Cdk1 and Cdc25C. *Curr Opin Cell Biol* **12**, 658–665.

Tarapore, P., Okuda, M. and Fukasawa, K. (2002). A mammalian in vitro centriole duplication system: evidence for involvement of CDK2/cyclin E and nucleophosmin/B23 in centrosome duplication. *Cell Cycle* **1**, 75–81.

Taylor, W. R. and Stark, G. R. (2001). Regulation of the G_2/M transition by p53. *Oncogene* **20**, 1803–1815.

Thelen, M. P., Venclovas, C. and Fidelis, K. (1999). A sliding clamp model for the Rad1 family of cell cycle checkpoint proteins. *Cell* **96**, 769–770.

Thommes, P., Fett, R., Schray, B., Burkhart, R., Barnes, M., Kennedy, C., Brown, N. C. and Knippers, R. (1992). Properties of the nuclear P1 protein, a mammalian homologue of the yeast Mcm3 replication protein. *Nucleic Acids Res* **20**, 1069–1074.

Tibbetts, R. S., Brumbaugh, K. M., Williams, J. M., Sarkaria, J. N., Cliby, W. A., Shieh, S. Y., Taya, Y., Prives, C. and Abraham, R. T. (1999). A role for ATR in the DNA damage-induced phosphorylation of p53. *Genes Dev* **13**, 152–157.

Todorov, I. T., Attaran, A. and Kearsey, S. E. (1995). BM28, a human member of the MCM2–3–5 family, is displaced from chromatin during DNA replication. *J Cell Biol* **129**, 1433–1445.

Todorov, I. T., Pepperkok, R., Philipova, R. N., Kearsey, S. E., Ansorge, W. and Werner, D. (1994). A human nuclear protein with sequence homology to a family of early S phase proteins is required for entry into S phase and for cell division. *J Cell Sci* **107**, 253–265.

Trimarchi, J. M., Fairchild, B., Verona, R., Moberg, K., Andon, N. and Lees, J. A. (1998). E2F-6, a member of the E2F family that can behave as a transcriptional repressor. *Proc Natl Acad Sci USA* **95**, 2850–2855.

Trouche, D., Cook, A. and Kouzarides, T. (1996). The CBP co-activator stimulates E2F1/DP1 activity. *Nucleic Acids Res* **24**, 4139–4145.

Tsai, L. H., Lees, E., Faha, B., Harlow, E. and Riabowol, K. (1993). The cdk2 kinase is required for the G_1-to-S transition in mammalian cells. *Oncogene* **8**, 1593–1602.

Tye, B. K. (1999). MCM proteins in DNA replication. *Annu Rev Biochem* **68**, 649–86.

Tyson, J. J., Novak, B., Odell, G. M., Chen, K. and Thron, C. D. (1996). Chemical kinetic theory: understanding cell-cycle regulation. *Trends Biochem Sci* **21**(3), 89–96.

Verona, R., Moberg, K., Estes, S., Starz, M., Vernon, J. P. and Lees, J. A. (1997). E2F activity is regulated by cell cycle-dependent changes in subcellular localization. *Mol Cell Biol* **17**, 7268–7282.

Vidal, A. and Koff, A. (2000). Cell-cycle inhibitors: three families united by a common cause. *Gene* **247**, 1–15.

Vigo, E., Muller, H., Prosperini, E., Hateboer, G., Cartwright, P., Moroni, M. C. and Helin, K. (1999). CDC25A phosphatase is a target of E2F and is required for efficient E2F-induced S phase. *Mol Cell Biol* **19**, 6379–6395.

Volkmer, E. and Karnitz, L. M. (1999). Human homologs of *Schizosaccharomyces pombe rad1, hus1,* and *rad9* form a DNA damage-responsive protein complex. *J Biol Chem* **274**, 567–570.

Vousden, K. H. (1995). Regulation of the cell cycle by viral oncoproteins. *Semin Cancer Biol* **6**, 109–116.

Waga, S., Hannon, G. J., Beach, D. and Stillman, B. (1994). The p21 inhibitor of cyclin-dependent kinases controls DNA replication by interaction with PCNA. *Nature* **369**, 574–578.

Wahl, G. M. and Carr, A. M. (2001). The evolution of diverse biological responses to DNA damage: insights from yeast and p53. *Nat Cell Biol* **3**, E277–286.

Wang, L., Liu, L. and Berger, S. L. (1998). Critical residues for histone acetylation by Gcn5, functioning in Ada and SAGA complexes, are also required for transcriptional function *in vivo*. *Genes Dev* **12**, 640–653.

Weinberg, R. A. (1995). The retinoblastoma protein and cell cycle control. *Cell* **81**, 323–330.

Wheatley, S. P., Hinchcliffe, E. H., Glotzer, M., Hyman, A. A., Sluder, G. and Wang, Y. (1997). CDK1 inactivation regulates anaphase spindle dynamics and cytokinesis *in vivo*. *J Cell Biol* **138**, 385–393.

Williams, R. S., Shohet, R. V. and Stillman, B. (1997). A human protein related to yeast Cdc6p. *Proc Natl Acad Sci USA* **94**, 142–147.

Winkler, K. E., Swenson, K. I., Kornbluth, S. and Means, A. R. (2000). Requirement of the prolyl isomerase Pin1 for the replication checkpoint. *Science* **287**, 1644–1647.

Wolfel, T., Hauer, M., Schneider, J., Serrano, M., Wolfel, C., Klehmann-Hieb, E., De Plaen, E., Hankeln, T., Meyer zum Buschenfelde, K. H. and Beach, D. (1995). A p16INK4a-insensitive CDK4 mutant targeted by cytolytic T lymphocytes in a human melanoma. *Science* **269**, 1281–1284.

Woo, M. S., Sanchez, I. and Dynlacht, B. D. (1997). p130 and p107 use a conserved domain to inhibit cellular cyclin-dependent kinase activity. *Mol Cell Biol* **17**, 3566–3579.

Woo, R. A., Jack, M. T., Xu, Y., Burma, S., Chen, D. J. and Lee, P. W. (2002). DNA damage-induced apoptosis requires the DNA-dependent protein kinase, and is mediated by the latent population of p53. *EMBO J* **21**, 3000–3008.

Wright, J. H., Munar, E., Jameson, D. R., Andreassen, P. R., Margolis, R. L., Seger, R. and Krebs, E. G. (1999). Mitogen-activated protein kinase kinase activity is required for the G(2)/M transition of the cell cycle in mammalian fibroblasts. *Proc Natl Acad Sci USA* **96**, 11335–11340.

Wulf, G. M., Liou, Y. C., Ryo, A., Lee, S. W. and Lu, K. P. (2002). Role of Pin1 in the regulation of p53 stability and p21 *trans*-activation, and cell cycle checkpoints in response to DNA damage. *J Biol Chem* **17**, 17.

Wulf, G. M., Ryo, A., Wulf, G. G., Lee, S. W., Niu, T., Petkova, V. and Lu, K. P. (2001). Pin1 is overexpressed in breast cancer and cooperates with Ras signaling in increasing the transcriptional activity of c-Jun towards cyclin D1. *EMBO J* **20**, 3459–3472.

Yamakita, Y., Yamashiro, S. and Matsumura, F. (1994). *In vivo* phosphorylation of regulatory light chain of myosin II during mitosis of cultured cells. *J Cell Biol* **124**, 129–137.

Yamashiro, S., Yamakita, Y., Ishikawa, R. and Matsumura, F. (1990). Mitosis-specific phosphorylation causes 83K non-muscle caldesmon to dissociate from microfilaments. *Nature* **344**, 675–678.

Yan, Z., DeGregori, J., Shohet, R., Leone, G., Stillman, B., Nevins, J. R. and Williams, R. S. (1998). Cdc6 is regulated by E2F and is essential for DNA replication in mammalian cells. *Proc Natl Acad Sci USA* **95**, 3603–3608.

Yang, J., Bardes, E. S., Moore, J. D., Brennan, J., Powers, M. A. and Kornbluth, S. (1998). Control of cyclin B1 localization through regulated binding of the nuclear export factor CRM1. *Genes Dev* **12**, 2131–2143.

Zeng, Y., Forbes, K. C., Wu, Z., Moreno, S., Piwnica-Worms, H. and Enoch, T. (1998). Replication checkpoint requires phosphorylation of the phosphatase Cdc25 by Cds1 or Chk1. *Nature* **395**, 507–510.

Zhang, H., Hannon, G. J. and Beach, D. (1994). p21-containing cyclin kinases exist in both active and inactive states. *Genes Dev* **8**, 1750–1758.

Zhao, J., Dynlacht, B., Imai, T., Hori, T. and Harlow, E. (1998). Expression of NPAT, a novel substrate of cyclin E–CDK2, promotes S-phase entry. *Genes Dev* **12**, 456–461.

Zhao, J., Kennedy, B. K., Lawrence, B. D., Barbie, D. A., Matera, A. G., Fletcher, J. A. and Harlow, E. (2000). NPAT links cyclin E–Cdk2 to the regulation of replication-dependent histone gene transcription. *Genes Dev* **14**, 2283–2297.

Zou, L., Cortez, D. and Elledge, S. J. (2002). Regulation of ATR substrate selection by Rad17-dependent loading of Rad9 complexes onto chromatin. *Genes Dev* **16**, 198–208.

Zou, L., Mitchell, J. and Stillman, B. (1997). CDC45, a novel yeast gene that functions with the origin recogniton complex Mcm proteins in initiation of DNA replication. *Mol Cell Biol* **17**, 553–563.

Zou, L. and Stillman, B. (1998). Formation of preinitiation complex by S-phase cyclin CDK-dependent loading of Cdc45p onto chromatin. *Science* **280**, 593–596.

Zuo, L., Weger, J., Yang, Q., Goldstein, A. M., Tucker, M. A., Walker, G. J., Hayward, N. and Dracopoli, N. C. (1996). Germline mutations in the p16INK4a binding domain of CDK4 in familial melanoma. *Nat Genet* **12**, 97–99.

3 DNA Viruses and the Nucleus

G. Eric Blair and Nicola James

Molecular Cell Biology Research Group, School of Biochemistry and Microbiology, Faculty of Biological Sciences, University of Leeds, Leeds, UK

3.1 Introduction

DNA viruses synthesize regulatory proteins that subvert the transcription and RNA processing systems of the host cell to effect viral gene expression. These viruses exhibit early and late phases of infection, where the early phase comprises the synthesis of regulatory proteins, e.g. those required for transcriptional regulation (such as the human papillomavirus E2, adenovirus E1A and SV40 T-antigen proteins) and DNA replication (such as the adenovirus DNA polymerase, E2A DNA-binding protein and terminal protein). In addition, certain DNA viruses encode proteins that have the ability to immortalize and transform cells, e.g. adenoviruses, papovaviruses and some herpesviruses. Many of these viral regulatory proteins are targeted to the nucleus, where they affect the organization of nuclear substructures (Hiscox, 2002). The late phase of infection follows the onset of viral DNA replication and is the phase of capsid protein gene expression and viral assembly, both processes taking place in the nucleus. The reorganization of nuclear structures may facilitate late gene expression and viral assembly (Everett, 2001). This chapter will briefly review the structure and composition of the major nuclear compartments and substructures (described more extensively in Chapter 1) and then discuss nuclear targeting of proteins encoded by the principal groups of DNA viruses that replicate in the nucleus, giving consideration to the biological role that viral protein targeting to particular nuclear substructures may play in the life cycle of the virus.

Viruses and the Nucleus   Edited by Julian A. Hiscox
© 2006 John Wiley & Sons, Ltd

3.2 The structure of the nucleus

Mammalian nuclei are cellular organelles that contain the machinery required for gene expression. Characterization of the nuclear architecture remains incomplete; however, the existence of specialized domains and organelles that correspond to assemblies of proteins and nucleic acids involved in mediating nuclear processes has been recognized (Dundr and Misteli, 2001). While the nucleolus is the best characterized subnuclear structure, other structures with less well-defined functions include Cajal (or coiled) bodies, promyelocytic leukaemia (PML) bodies (or PODs, PML oncogenic domains), speckles, autonomous parvovirus-associated replication (APAR) bodies, and gems (Fox *et al.*, 2002; Grande *et al.*, 1996; Lamond and Earnshaw, 1998; Matera, 1999). There is good evidence for the exchange of constituents between these different compartments.

3.2.1   The nuclear envelope and lamina

The nucleus is bounded by a double-membraned envelope that is in a continuum with the rough endoplasmic reticulum. In places, the outer and inner membranes are fused, forming portals for the transit of components between the cytoplasm and the nucleus (Stoffler *et al.*, 1999). Regulation of nuclear envelope structure and the anchoring of interphase chromatin to the nuclear periphery is thought to be mediated by the peripheral nuclear lamina, which comprises lamins A/C and B. In interphase, chromosomes occupy distinct areas within the nucleoplasm, known as chromosome territories (Schardin *et al.*, 1985), with active genes tending to associate with the boundaries (Kurz *et al.*, 1996; Wansink *et al.*, 1996). Heterochromatin (inactive chromatin) is located in close proximity to the nuclear lamina as well as more internally. PcG bodies, containing polycomb group proteins, are associated with pericentromeric heterochromatin (Saurin *et al.*, 1998). These domains vary in both number (from two to several hundred) and size (0.2–1.5 μm) and may either be storage compartments or function in gene silencing. Chromosome territories are separated by interchromosomal domain channels into which RNA transcripts are released. These channels are linked to the nuclear pores which serve to enable export of nascent RNAs (Kramer *et al.*, 1994).

3.2.2   The nucleolus

The nucleolus is the most extensively characterized dynamic subnuclear organelle involved in ribosome subunit biogenesis, RNA processing and possibly also in cell cycle control. Nucleoli range in diameter from 0.5 to 5 μm and most mammalian cells contain around one to five. A proteomic analysis of the nucleolus revealed that it is composed of some 271 proteins (Andersen *et al.*, 2002), although recent data suggests that this figure is nearer 400 (http://www.lamondlab.com/f5nucleolus.htm). The major nucleolar proteins are nucleolin, B23 and fibrillarin. Localized around chromosomal loci containing ribosomal DNA repeats, the nucleolus is responsible for the synthesis and processing of rRNA (including methylation and pseudouridylation) and the assembly of ribosome subunits. Nucleoli contain three distinct regions; fibrillar centres, dense fibrillar components (which surround and invade the fibrillar centres) and granular regions. Transcription and processing of rRNA is thought to occur in the dense fibrillar component, while the granular region comprises large and small ribosomal subunits and pre-ribosomal particles of varying degrees of maturation.

Further work has revealed additional functions of nucleoli, e.g. studies of UV-induced inactivation of nucleolar function have proposed a role in RNA export or degradation (Harris *et al.*, 1969). Furthermore, both perinucleolar compartments (PNC) and coiled bodies (discussed below) also interact with nucleoli, suggesting that more functions may yet be discovered.

Many viruses encode proteins that localize to the nucleolus and examples can be found from DNA, RNA and retroviruses (Hiscox, 2002). This is discussed below for the DNA viruses. Although these viruses have quite different replication strategies, they may need to recruit nucleolar factors to facilitate virus replication and/or to subvert nucleolar function and thus inhibit host cell macromolecular synthesis.

3.2.3 The perinucleolar compartment and SAM68 nuclear bodies

The perinucleolar compartment (PNC) and SAM68 nuclear bodies are distinct structures that are 0.25–1 μm in diameter and number one to ten per nucleus. These structures associate with nucleoli and are rarely observed in non-transformed cells. Their functions are unknown; however, RNAs transcribed by RNA polymerase III and RNA-binding proteins are present in PNCs (Ghetti *et al.*, 1992). SAM68 bodies have RNA-binding proteins containing a GSG (GRP33, Sam68, GLD-1) or STAR (signal transduction and activation of RNA) domain (Huang, 2000).

3.2.4 Speckles

Indirect immunofluorescence of nucleoplasmic RNA–protein complexes called snRNPs (small nuclear ribonucleoproteins), which are involved in splicing introns from mRNAs, produces a punctate pattern of 25–50 nuclear speckles against a diffuse background (Lamond and Carmo-Fonseca, 1993; Spector, 2001). The punctate pattern arises from the association of snRNPs with a variety of structures, including interchromatin granules (ICG) that contain other splicing factors, the interchromatin granule-associated zones that flank them; perichromatin fibrils containing nascent mRNA precursors and processed mRNA and coiled bodies. However, the term 'speckles' is used specifically to refer to the larger speckles (0.8–1.8 μm diameter) that correspond to ICG composed of apparently interconnected particles of 20–25 nm.

Speckles are dynamic structures that contain snRNPs and other splicing factors. Although speckles can be labelled with poly d(T) or polyd(U) oligonucleotides, they do not incorporate labelled uridine analogues and are devoid of DNA. Several thousand transcription sites are distributed throughout the nucleoplasm, including those associated with ICG clusters. Speckles may thus be involved in the assembly and/or modification of pre-mRNA splicing factors. In addition, perichromatin fibril transcription sites recruit factors from nuclear speckles; thus, speckles may effect storage and delivery of splicing factors to active gene loci. In addition, an involvement in mRNA export has been proposed.

3.2.5 OPT domains

Oct1/PTF/transcription domains contain high levels of transcription factors and nascent transcripts but lack factors involved in RNA processing (Grande *et al.*, 1997). Interestingly, OPT domains appear often in proximity to nucleoli during G_1 but are absent during S phase.

3.2.6 Cajal bodies

The coiled body (CB, also termed the Cajal body) was discovered in 1903 by Santiago Ramon y Cajal, who noted their frequent association with nucleoli. When viewed by electron microscopy, Cajal bodies consist of tangles of coiled threads with diameters of 0.15–1.5 μm. CBs contain several characteristic proteins and factors required for splicing, ribosome biogenesis and transcription (Gall, 2000; Lamond and Carmo-Fonseca, 1993; Ogg and Lamond 2002). In addition to snRNPs, they are highly enriched in basal transcription and cell cycle factors. They are, however, unlikely to be sites of transcription or pre-mRNA splicing, since they contain neither DNA nor non-snRNP protein-splicing factors. Instead, they are more likely to be involved in coordinating the assembly of macromolecular complexes, particularly of nuclear ribonucleoproteins (Ogg and Lamond, 2002). Consistent with this is the observation that CBs are dynamic structures, with recent live-cell imaging studies showing migration of CBs and CB-associated proteins within the nucleus (Sleeman *et al.*, 2003).

The predominant component of the Cajal body is the protein p80 coilin, which was originally identified as a human autoantigen and provides the best means of identifying these structures (Andrade *et al.*, 1991). Other proteins associated with CBs include fibrillarin and SMN (survival of motor neuron) (Tucker *et al.*, 2001), while CB proteins are also found in other nuclear structures (Leung and Lamond, 2002). A simple procedure for the large-scale isolation of CBs from HeLa nuclei has recently been described (Lam *et al.*, 2002), opening up the prospect of biochemical and proteomic analysis of Cajal bodies.

In addition to being involved in RNP maturation and snRNA modification, it is also possible that Cajal bodies have other roles in the nucleus. They may also coordinate a broad range of macromolecular assembly reactions, including the preassembly of transcription factors (Gall, 2000). They may also play a role in feedback regulation of gene expression (Matera, 1998). A regulatory function could explain the observation that CBs can interact *in vivo* with specific gene loci, including snRNA and histone genes. Cajal bodies are associated with chromatin in a manner that requires active transcription and ATP (Platani *et al.*, 2002), and a link has been established between Cajal bodies and transcription of histone genes. Cajal bodies contain the cyclin E–Cdk2 substrate p220 (NPAT) in a cell cycle-regulated manner. Over-expression of p220 in U2OS cells promotes G_1–S transition and transcription from histone H2B and H4 luciferase reporter constructs (Wei *et al.*, 2003).

3.2.7 Gems

Gemini of Cajal bodies (gems) are identified by the presence of the survival of motor neuron protein (SMN) and associated Gemin2 (Matera, 1998). Mutation of SMN results in spinal muscular atrophy. Distinct cytoplasmic and nuclear pools of these proteins have been identified and are implicated in snRNP assembly and maturation, respectively. Gems either co-localize with or adjoin Cajal bodies. The biochemical relationship between Cajal bodies and gems was demonstrated in a study which showed that localization of SMN to either Cajal bodies or gems is determined by the methylation of p80 coilin (Hebert *et al.*, 2002).

3.2.8 PML bodies

PML bodies, also termed PODs (PML oncogenic domains) and ND10 (nuclear domains of around 10 per nucleus), are so named due to the PML protein component. A full review of the structure and function of PML bodies is given in Chapter 8. These bodies are tightly associated with the nuclear matrix (see below), vary in diameter (0.3–1 μm) and often display a doughnut-like morphology. A typical nucleus contains 10–30 bodies. The functions of PML bodies remain to be elucidated but they are known to contain regulators of proliferation, apoptosis and genome stability (Maul *et al.*, 2000; Negorev and Maul, 2001). The constitutive proteins PML and Sp100 are required for the structural integrity of PML (Sternsdorf *et al.*, 1995). PML belongs to a family of proteins characterized by a Cys/His rich cluster called a RING finger. PML protein exists as a nucleoplasmic free form and as a complex bound to SUMO1 (small ubiquitin-like modifier). The association of PML with SUMO1 promotes the assembly of PML bodies.

   PML bodies have been proposed to function as sites of transcriptional activity, since they associate with sites of active transcription in a cell cycle-dependent fashion

(Kiesslich *et al.*, 2002). However, they do not contain snRNPs or splicing factors. Chromatin-modifying enzymes including acetyl transferases and deacetylases accumulate in PML bodies. The association of PML bodies with active gene loci may serve to effect gene expression regulated by histone acetylation. The presence of these factors suggests that PML bodies may function to silence/activate specific genes.

The PML protein functions as a tumour suppressor. Acute promyelocytic leukaemia (APL) is linked to disruption of the PML gene by chromosomal translocation and fusion with the retinoic acid receptor, which is responsible for the malignant phenotype of these cells. In relation to tumour suppression, it is noteworthy that p53 is transiently recruited to PML bodies. Furthermore PML bodies are targeted by DNA tumour viruses, some of whose oncoproteins bind and inactivate p53, giving further weight to the notion that the PML protein may have a role in tumour suppression (Carvalho *et al.*, 1995). This may be linked to the activation of p53-dependent transcription via HIPK2 (homeodomain-interacting protein kinase) domains (Moller *et al.*, 2003). The serine/threonine kinase HIPK2 (see HIPK domains) (D'Orazi *et al.*, 2002; Hofmann *et al.*, 2002; Salomoni and Pandolfi, 2002) also interacts transiently with PML bodies. HIPK2 can effect apoptosis via its ability to phosphorylate p53 (Moller *et al.*, 2003). Thus, PML bodies may be required for the execution of p53-mediated cell cycle arrest.

3.2.9   HIPK domains

PML proteins occur in at least seven different isoforms that are all capable of targeting p53 to PML bodies; however, only PML-IV can regulate p53 activity (Moller *et al.*, 2003). The N-terminus of p53 is phosphorylated by HIPK2, leading to the induction of p53-dependent gene expression and cell cycle arrest and apoptosis (D'Orazi et al., 2002; Hofmann *et al.*, 2002). The majority of HIPK2 localizes to nuclear speckles (Moller *et al.*, 2003); however, it appears that HIPK2 can exist as a distinct nuclear compartment. Nevertheless, PML bodies are required for the antiproliferative action of HIPK2. Only a subset of PML bodies display transcriptional activity; it is not known whether HIPK2 is associated with this subset of PMLs or, if not, whether HIPK2/p53-mediated transcription occurs outside of PML bodies.

3.2.10   Nuclear matrix

The nuclear matrix (NM) has been defined as the amorphous fibrogranular structure that remains subsequent to DNAse digestion and salt extraction of nuclei (Belgrader *et al.*, 1991; described in detail in Chapter 1). The existence of the NM as an independent structure is widely accepted, despite the reservation that it may result from the preparation method itself. Two distinct structures are apparent; a mesh-like internal nuclear matrix (INM) and a nuclear shell that connects the INM to the nuclear membrane. In addition, the INM is coated with nuclear matrix-associated proteins. A recent study has described the identification of 398 nuclear matrix proteins (Mika and Rost, 2005). Recruitment of active genes to the nuclear scaffold is exemplified by scaffold attachment factor A (SAF-A) (Romig *et al.*, 1992). SAF-A is an RNA-binding protein which also binds the transcriptional co-activator p300 (Martens *et al.*, 2002). In addition, assembly of transcriptional apparatus is thought to be mediated via SAF-B protein (Nayler *et al.*, 1998). Conversely, silencing of genes by NM proteins is exemplified by attachment region binding protein (ARBP). ARBP binds methylated DNA and recruits a histone deacetylase (Stratling and Yu, 1999). SATB1 provides an example of a multifunctional nuclear matrix protein that recruits histone deacetylase and RNA polymerase regulating nucleosome positioning (Durrin and Krontiris, 2002; Yasui *et al.*, 2002).

3.3 DNA viruses and the nucleus

3.3.1 Parvoviruses

Infection of cells with the parvovirus, minute virus of mice (MVM), results in reorganization of intranuclear structure (Bashir *et al.*, 2000). A key to this reorganization seems to be the expression of the large non-structural protein NS1 of MVM, a protein essential for viral replication and a potent transcriptional activator. During the course of MVM infection, the initially distinct Cajal bodies, PML bodies, speckles and autonomous parvovirus-associated replication (APAR) bodies aggregate to form new large nuclear bodies (termed SMN-associated APAR bodies) that are active sites of viral replication and viral capsid assembly. Following transient NS1 expression, NS1 co-localizes with the SMN protein within Cajal bodies (Young *et al.*, 2002).

3.3.2 Papovaviruses

The papovaviruses consist of the papillomaviruses, polyomaviruses and vacuolating viruses such as simian virus 40 (SV40). This group of viruses replicates and assembles virus particles in the nucleus of permissive cells. However, they can also transform cells, forming immortalized clones of cells that may form tumours *in vivo*.

Papillomaviruses have received intense interest due to the association of 'high risk' papillomaviruses, e.g. human papillomavirus type 16, HPV16 with cervical and other ano-genital cancers (zur Hausen, 2000). Approximately 100 human papillomaviruses (HPV) have been described (de Villiers *et al.*, 2004). HPVs infect squamous epithelial cells of the skin, laryngeal and genital mucosa. They can be classified according to both DNA sequence relationships and site of infection (cutaneous or genital HPVs) and whether they are associated with a low (e.g. HPV6) or high risk (e.g. HPV16, 18) of progression to malignancy. Following viral entry and migration to the nucleus,

HPV DNA viral genomes are episomal; however, in the case of high-grade malignant lesions, viral integration into host chromosomes takes place, resulting in constitutive expression of only two HPV proteins, E6 and E7, which are necessary and sufficient for tumour formation in the human host and also for immortalization of human and rodent cells *in vitro*. In productive HPV infection, other HPV proteins play a role in DNA replication (E1 and E2) and in regulation of transcription (E2). The E2 protein binds to the viral origin of replication in the circular double-stranded viral DNA and also binds host cell chromosomes, ensuring partition of viral episomes on cell division (Van Tine *et al.*, 2004). The E2 protein of HPV11 is also associated with the nuclear matrix (Zou *et al.*, 2000). The E7 protein binds pRb1, the retinoblastoma susceptibility protein, which negatively regulates the transcription factor E2F; release of E2F activates transcription of genes required for the S phase of the cell cycle. The E7 protein is localized to the nucleus, although differing subnuclear locations have been described for E7. High-risk (HPV16) E7 has been localized to the nucleolus of the HPV16-transformed cell line CaSki, using immuno-gold labelling and electron microscopy (Zatsepina *et al.*, 1997). Interestingly, variation in nucleolar localization of E7 was noted in this study, with lower levels of E7 being associated with the nucleolus in the G_2 phase of the cell cycle. pRb1 was also found to co-localize with E7 in the nucleolus, as well being present throughout the nucleus. Guccione *et al.* (2002) utilized E7 fusions with green fluorescent protein (E7–GFP) to localize E7 in transient expression assays. In this approach, the E7–GFP fusion protein of the low-risk HPV6 was found to have a punctate nuclear location (which was not further identified), whereas the high-risk HPV16 E7–GFP had a diffuse nuclear location. Finally, in transduction of human fibroblasts with an E7-expressing retrovirus, HPV16 E7 has been found to localize to PML bodies, where it suppresses cellular senescence induced by over-expressed PML IV isoform (Bischof *et al.*, 2005). These diverse subnuclear locations of E7 may be due to the methods of delivery of E7 to human cells and the human cell type studied. In addition, in the case of immunochemical identification of E7, certain immunodominant epitopes on E7 may be masked due to protein–protein complex formation or self-oligomerization (Kanda *et al.*, 1991). A more intriguing possibility is that these locations could be indicative of a dynamic interchange of E7 between different subnuclear structures. The high-risk E6 protein recruits p53 for ubiquitination and destruction by the proteasome (Mantovani and Banks, 1999). Both high-risk (HPV16) and low-risk (HPV11) E6 co-localize to PML bodies in transfected CasKi and osteosarcoma U2-OS cells, and both oncoproteins can overcome PML IV-mediated senescence in primary baby rat kidney cells (Guccione *et al.*, 2004) in a manner reminiscent of HPV16 E7, discussed above.

The integration of HPV DNA into host chromosomes appears to take place close to nuclear matrix attachment regions (MARs) in both HPV-positive cell lines established from cervical tumours and in primary tumours *in situ* (Shera *et al.*, 2001). Interestingly, other small DNA tumour viruses (SV40, hepatitis B virus) and retroviruses (Human T cell leukaemia virus type 1, HTLV-1) also integrate into MARs in neoplastic cells, while integration in non-neoplastic human cells appears to take place outside of MARs. This implies an association of integrated tumour viral sequences with MARs in oncogenic cell lines, which may permit access of transcription factors to viral promoters and ensure constitutive expression of mRNAs encoding viral oncoproteins.

The large T-antigens of SV40 and polyomaviruses are required for viral DNA replication and transcription, as well as cell transformation. SV40 T-antigen binds pRb1 and p53 (Saenz-Robles *et al.*, 2001) and is a nuclear protein that targets chromatin and the nuclear matrix in viral transformation (Deppert *et al.*, 2000). In SV40 replication, viral DNA replication and transcription both take place adjacent to PML (ND10) bodies, although T-antigen expression alone did not target to ND10 bodies (Tang *et al.*, 2000), indicating that association of T-antigen and viral transcripts with ND10 may be a secondary event, linked to their presence at sites of replication. Polyomavirus large T-antigen has been detected in nuclei of transformed cells in association with nuclear pore-like structures (but was excluded from nucleoli) by immunocytochemical techniques and in discrete, unidentified areas in polyoma virus-infected cells (Dilworth *et al.*, 1986).

3.3.3 Adenoviruses

Adenoviruses (Ads) have been detected in most vertebrate species and infect epithelial cells that line the respiratory and gastroenteric tracts. Some adenoviruses are oncogenic and can induce tumours in heterologous hosts, e.g. human Ad12 induces tumours in rodents such as rats, hamsters and mice. There are currently 51 distinct human adeno-virus serotypes, which have been classified into six subgroups (A–F) according to various properties, including the oncogenicity of viruses in newborn rodents (Shenk, 2001). All human adenoviruses studied so far can transform primary rodent cells in culture; however, only cells transformed by viruses of subgroup A and B are oncogenic in new-born rodents, paralleling the oncogenic properties of the parental viruses (Blair and Blair Zajdel, 2004). The viral genes required for transformation are the E1A and E1B genes, which are linked together at the left 11% of the approximately 36 kb linear genome. The E1A gene encodes two major mRNAs, which are generated by alternative RNA splicing and direct synthesis of two proteins (289 and 243 amino acid residues, R, in Ad2 and Ad5, 266R and 235R in Ad12) that are identical except for an internal peptide segment (Boulanger and Blair, 1991; Gallimore and Turnell, 2001). Comparison of predicted E1A protein sequences from several viral serotypes led to the identification of three conserved regions, CR1, CR2 and CR3. The CR3 region is unique to the 289R and 266R proteins and comprises a zinc finger domain that is responsible for transcriptional activation of other viral early promoters (E1B, E2, E3 and E4) as well as the major late promoter (MLP, located at around 16.4% on the viral genome). The MLP directs transcription of late viral gene products, which mainly comprise structural proteins of the viral capsid. The larger E1A proteins do not bind DNA directly, but appear to form heterodimers with basal and ubiquitous transcription factors, thus favouring formation of transcriptional pre-initiation complexes. The CR1 and CR2 regions are involved in cell transformation, by binding the retinoblastoma tumour suppressor gene product, pRb1 (in common with HPV E7 and SV40 T-antigen, as described above) or an Rb-related protein, p107 (to CR2) and a transcriptional co-activator protein, p300, also termed CBP or CREB-binding protein (to the amino-terminal region and CR1). The p300/CBP protein possesses histone acetyltransferase (HAT) activity and is involved in chromatin remodelling; E1A binding to p300/CBP both activates and represses transcription of target genes (Gallimore and Turnell, 2001; Sang *et al.*, 2002). Further cellular proteins involved in transcriptional regulation of the cell cycle, e.g. p400 and TRRAP, are targets for binding and modulation by E1A (Frisch and Mymryk, 2002). In addition, E1A downregulates transcription of cyclin-dependent kinase inhibitor proteins such as $p21^{Cip1/Waf1}$ and $p27^{Kip1}$, leading to increased cell cycle progression. Thus, E1A proteins have the capacity to act as either transcriptional activators or repressors of target cellular genes.

Consistent with their role in cell cycle modulation and transcriptional regulation, the E1A proteins are targeted to the nucleus via a nuclear localization sequence (NLS) located at their carboxy-termini. An initial report using antipeptide antibodies against Ad2 E1A indicated that E1A proteins had a diffuse nuclear distribution in infected cells, although they seemed to be excluded from nucleoli (Schmitt *et al.*, 1987). The 289R protein appeared to more highly enriched in a nuclear matrix fraction than the 243R. A subsequent analysis of Ad2-infected cells with anti-E1A monoclonal antibodies showed five types of intranuclear staining: diffuse, reticular, nucleolar, punctate and peripheral (White *et al.*, 1988); in addition, the 70 kDa heat shock protein (hsp70), which is induced by E1A, co-localized with E1A in all these intranuclear compartments. E1A has also been localized to PML bodies (Carvalho *et al.*, 1995), using carefully controlled methods to fix and detergent-extract infected cells. Further studies using transfection of mutant E1A plasmid constructs showed that deletions in the CR2 domain reduced E1A targeting to PML structures, although point mutations in the binding sites for pRb or Rb-related proteins did not reduce interactions between E1A and PML (Carvalho *et al.*, 1995). Therefore, PML targeting is probably not linked to cell transformation or cell cycle regulation mediated by E1A and the biological function of this targeting event remains to be established.

The E1B gene encodes two major proteins of 58 kDa (also termed the 55 kDa protein) and 19 kDa that play important roles in viral infection and transformation. The 19 kDa protein possesses similarities to the bcl-2 family of anti-apoptosis proteins and binds to pro-apoptotic proteins such as Bax, Bak and Bik, leading to suppression of apoptosis (Chinnadurai, 1998; Cuconati and White, 2002). The Ad5 58 kDa protein forms a stable complex with the p53 tumour suppressor gene product in transformed cells (Sarnow *et al.*, 1982) while, in contrast, the corresponding Ad12 protein (of 54 kDa) forms a very weak or unstable complex with p53, although p53 is stabilized in Ad12-, as well as Ad5-transformed cells (Grand *et al.*, 1999). The presence of either the Ad5 E1B-58 kDa protein or its Ad12 counterpart is sufficient to abrogate the transcriptional activation properties of p53 (Wienzek *et al.*, 2000). In infected cells, the 58 kDa protein forms complexes with the E4 ORF3 and ORF6 gene products and functions in mRNA transport (Dobner and Kzhyshkowska, 2001). Furthermore, the 58 kDa protein also forms a complex with the E4 ORF6 gene product and p53, an interaction that recruits a Cullin-containing complex that stimulates ubiquinitation of p53 and targets it for degradation by the proteasome (Querido *et al.*, 2001). The half-life of p53 in untransformed cells is short (around 30 minutes) and p53 is stabilized when cellular DNA is damaged, leading to transcriptional activation of cell cycle-regulatory genes such as p21$^{Cip1/Waf1}$ and either cell cycle arrest at the G_1–S phase transition or apoptosis (Woods and Vousden, 2001).

The E1B-19 kDa protein localizes to the nuclear lamina in infected cells as well as intracellular cytoplasmic membranes and intermediate filaments (White and Cipriani, 1989). Further studies showed that E1B-19 kDa interacted with lamin A/C and the use of lamin and E1B-19 kDa mutants led Rao *et al.* (1997) to conclude that interaction between the 19 kDa protein and lamin A/C is necessary to direct E1B-19 kDa to the nuclear envelope and suppress apoptosis. Over-expression of E1B-19 kDa disaggregates the nuclear lamina (White and Cipriani, 1989) and could play a role in viral egress from infected nuclei in the late phase of infection, by analogy with corresponding processes in herpesvirus-infected cells (see below).

The E1B-58 kDa protein interacts with two proteins encoded in the E4 region, E4-ORF3 and E4-ORF6, and targets PML structures in infected cells, which then become reorganized into replication and transcription centres (reviewed extensively in Chapter 8). Interaction between E1B-58 kDa and E4-ORF6 appears to be required for selective transport of viral mRNA through the nucleus (Dobner and Kzhyshkowska, 2001), and mutations in the 58 kDa protein that impair its location in viral replication centres also lead to increased levels of nuclear transcripts encoding late viral mRNAs (Gonzalez and Flint, 2002), providing a link between targeted localization of 58 kDa and viral mRNA export. Both E1B-58 kDa and ORF6 are shuttle proteins, containing leucine-rich nuclear export signals (NES) and mutations in either NES reduce export of viral mRNA from the nucleus (Dobner and Kzhyshkowska, 2001). This suggests that adenovirus late transcripts may be exported from the nucleus by a pathway utilizing the transport receptor CRM1; however, recent data suggest that treatment of infected cells with leptomycin B (a CRM1 inhibitor) delays but does not prevent late mRNA accumulation (Carter *et al.*, 2003), leaving open the precise mechanism of mRNA export. E1B-58 kDa also interacts with a heterogeneous nuclear protein family member, E1B-AP5 (Gabler *et al.*, 1998), which may link with the cellular TAP protein, another pathway of mRNA export (Dobner and Kzhyshkowska, 2001).

Infection of cells with adenoviruses causes dissociation of Cajal bodies into microfoci in the late phase of infection (Rebelo *et al.*, 1996; Rodrigues *et al.*, 1996), although it is not clear whether this is a direct effect of adenovirus infection or whether it is an indirect effect due to viral shut-off of host protein synthesis, since this effect could be mimicked by inhibition of protein synthesis in uninfected cells (Rebelo *et al.*, 1996).

Several adenovirus proteins are targeted to the nucleolus. The E4 ORF4 protein has recently been shown to locate to the nucleolus via an arginine-rich nucleolar localization signal (NoLS), inducing cell death in human cancer cells (Miron *et al.*, 2004). Structural proteins of the viral capsid locate to the nucleolus, namely protein V (Matthews and Russell, 1998; Matthews, 2001) and the precursor to the basic core protein Mu (Lee *et al.*, 2004) via arginine-rich NoLS. Interestingly, expression of preMu suppressed expression of late genes, probably by blocking expression of the precursor to the terminal protein (pTP), which is required for DNA replication and transition to the late phase of infection (Lee *et al.*, 2004). Protein V probably links the outer shell of the viral capsid with the nucleoprotein core; expression of V–EGFP fusion proteins leads to accumulation in the nucleolus, accompanied by exclusion of the host cell nucleolar proteins B23 and nucleolin and their relocation to the cytoplasm (Matthews, 2001). B23 has been identified as a factor that stimulates replication of adenovirus DNA complexed to with the viral basic core proteins *in vitro* (Okuwaki *et al.*, 2001). It might be speculated that protein V may displace B23 from the nucleolus in order to facilitate replication of the viral core in replication centres. The other basic protein of the core, pVII, contains sequences that target the protein to the nucleus and nucleolus; in addition, the protein also contains two distinct domains that mediate binding of VII to host chromosomes, one domain sharing homology with histone H3 (Lee *et al.*, 2003). Since adenovirus DNA is not maintained as an episome, VII interaction with chromosomes is unlikely to function analogously with the HPV E2 protein; rather, it may reflect an interaction with viral chromatin, which is maintained throughout the early phase of infection (Xue *et al.*, 2005). Protein VII–viral DNA complexes are the substrate for transcriptional activation by E1A and a direct interaction between VII and E1A has been demonstrated *in vitro* (Johnson *et al.*, 2004).

3.3.4   Herpesviruses

The herpesviruses target the nucleus for replication and a number of herpesvirus gene products interact with PML bodies during the transcription and replication phases of their infectious cycles. This area has been extensively reviewed (Maul, 1998; Everett, 2001) and is considered in detail in Chapter 8. Certain herpesviruses establish latent and persistent infections and some can transform cells, forming immortalized and oncogenic cell clones. In addition, during productive infection, herpesviruses also require to egress from the nucleus of the infected cell. These aspects of the herpesvirus life cycles involve specific nuclear processes and will be considered here.

The gammaherpesviruses establish latent infections and maintain their genomes as episomes. The prototype gamma-2 herpesvirus, the primate Herpesvirus Saimiri (HVS), encodes a protein termed ORF73, which tethers episomes to host cell chromosomes (Calderwood *et al.*, 2004) in a manner analogous to the HPV E2 protein (discussed above). A protein of the related human virus HHV8 or Kaposi's sarcoma herpesvirus (KSHV) encodes a protein termed latency-associated nuclear antigen 1 (LANA-1), which fulfils a similar role (Viejo-Borbolla *et al.*, 2003). In addition to its tethering role, LANA-1 also acts as a transcriptional regulator and binds p53 and pRb1. The ORF 57 protein of HVS locates to the nucleus and nucleolus (Goodwin and Whitehouse, 2001) and is required for viral mRNA export via the TAP pathway (Williams *et al.*, 2004).

Several herpesviruses encode proteins that interact with nuclear lamins and are required for viral egress from the nuclei of infected cells. The gamma-herpesvirus Epstein–Barr virus nuclear antigen EBNA-1 and the UL53 protein of human cytomegalovirus (CMV, a beta-herpesvirus) co-localize with lamin B (Ito *et al.*, 2003; Dal Monte *et al.*, 2002). The M50 product of murine CMV inserts into the nuclear envelope and recruits protein kinase C to phosphorylate lamins (Muranyi *et al.*, 2002). The M50 protein of MCMV is essential for virus production and has homologues in the gamma-herpesviruses (ORF67) and alpha-herpesviruses (UL34). Studies on the alpha-herpesvirus HSV-1 revealed redistribution of lamins and the lamin B receptor of the inner nuclear membrane, as well as degradation of nuclear lamins in infected cells (Scott and O'Hare, 2001). This perturbation of the nuclear lamina is mediated by a complex of UL31 and UL34 (Reynolds *et al.*, 2004; Simpson-Holley *et al.*, 2004). This probably degrades the nuclear lamina, allowing the virus to access the inner nuclear membrane, from where the virus can bud into the cytoplasm.

3.4 Concluding comments and future directions

It is clear that DNA viruses can access most nuclear substructures to facilitate their replication and also to interfere with or subvert the cell cycle in the case of viral proteins with nuclear components. Therefore, the interaction of viral proteins with the nucleus is a critical but as yct poorly explored area of molecular virology. In the near future, proteomic analysis will be a powerful tool to be applied in the study of interactions between viral and host nuclear proteins. Furthermore, the recent development of techniques for biochemical purification and proteomic characterization of nuclear substructures (such as the Cajal body and the nucleolus) provides the basis for studies on these structures in virus-infected cells.

Acknowledgements

We would like to thank our colleagues Julian Hiscox, Adrian Whitehouse and John Walker for comments. Research in the authors' laboratory is supported by BBSRC and Yorkshire Cancer Research.

References

Andersen, J. S., Lyon, C. E., Fox, A. H., Leung, A. K., Lam, Y. W., Steen, H., Mann, M. and Lamond, A. I. (2002). Directed proteomic analysis of the human nucleolus. *Curr Biol* **12**, 1–11.

Andrade, L. E., Chan, E. K., Raska, I., Peebles, C. L., Roos, G. and Tan, E. M. (1991). Human autoantibody to a novel protein of the nuclear coiled body: immunological characterization and cDNA cloning of p80-coilin. *J Exp Med* **173**, 1407–1419.

Bashir, T., Horlein, R., Rommelaere, J. and Willwand, K. (2000). Cyclin A activates the DNA polymerase delta-dependent elongation machinery *in vitro*: a parvovirus DNA replication model. *Proc Natl Acad Sci USA* **97**, 5522–5527.

Belgrader, P., Siegel, A. J. and Berezney, R. (1991). A comprehensive study on the isolation and characterization of the HeLa S3 nuclear matrix. *J Cell Sci* **98**(3), 281–291.

Bischof, O., Nacerddine, K. and Dejean, A. (2005). Human papillomavirus oncoprotein E7 targets the promyelocytic leukemia protein and circumvents cellular senescence via the Rb and p53 tumor suppressor pathways. *Mol Cell Biol* **25**, 1013–1024.

Blair, G. E. and Blair-Zajdel M. E. (2004). Evasion of the immune system by adenoviruses. *Curr Top Microbiol Immunol* **273**, 3–28.

Boulanger, P. A. and Blair, G. E. (1991). Expression and interactions of human adenovirus oncoproteins. *Biochem J* **275**, 281–299.

Calderwood, M. A., White, R. E. and Whitehouse, A. (2004). Development of herpesvirus-based episomally maintained gene delivery vectors. *Expert Opin Biol Ther* **4**, 493–505.

Carter, C. C., Izadpanah, R. and Bridge, E. (2003). Evaluating the role of CRM1-mediated export for adenovirus gene expression. *Virology* **315**, 224–233.

Carvalho, T., Seeler, J. S., Ohman, K., Jordan, P., Pettersson, U., Akusjarvi, G., Carmo-Fonseca, M. and Dejean, A. (1995). Targeting of adenovirus E1A and E4-ORF3 proteins to nuclear matrix-associated PML bodies. *J Cell Biol* **131**, 45–56.

Chinnadurai, G. (1998). Control of apoptosis by human adenovirus genes. *Semin Virol* **8**, 399–408.

Cuconati, A. and White, E. (2002). Viral homologs of BCL-2: role of apoptosis in the regulation of virus infection. *Genes Dev* **16**, 2465–2478.

Dal Monte, P., Pignatelli, S., Zini, N., Maraldi, N. M., Perret, E., Prevost, M. C. and Landini, M. P. (2002). Analysis of intracellular and intraviral localization of the human cytomegalovirus UL53 protein. *J Gen Virol* **83**, 1005–1012.

de Villiers, E. M., Fauquet, C., Broker, T. R., Bernard, H. U. and zur Hausen, H. (2004). Classification of papillomaviruses. *Virology* **324**, 17–27.

Deppert, W., Gohler, T., Koga, H. and Kim, E. (2000). Mutant p53: 'gain of function' through perturbation of nuclear structure and function? *J Cell Biochem Suppl* **35**, 115–122.

Dilworth, S. M., Hansson, H. A., Darnfors, C., Bjursell, G., Streuli, C. H. and Griffin, B. E. (1986). Subcellular localisation of the middle and large T-antigens of polyomavirus. *EMBO J.* **5**, 491–499.

Dobner, T. and Kzhyshkowska, J. (2001). Nuclear export of adenovirus RNA. *Curr Top Microbiol Immunol* **259**, 25–54.

D'Orazi, G., Cecchinelli, B., Bruno, T., Manni, I., Higashimoto, Y., Saito, S., Gostissa, M., Coen, S., Marchetti, A., Del Sal, G., Piaggio, G., Fanciulli, M., Appella, E. and Soddu, S. (2002). Homeodomain-interacting protein kinase-2 phosphorylates p53 at Ser 46 and mediates apoptosis. *Nat Cell Biol* **4**, 11–19.

Dundr, M. and Misteli, T. (2001). Functional architecture in the cell nucleus. *Biochem J* **356**, 297–310.

Durrin, L. K. and Krontiris, T. G. (2002). The thymocyte-specific MAR binding protein, SATB1, interacts *in vitro* with a novel variant of DNA-directed RNA polymerase II, subunit 11. *Genomics* **79**, 809–817.

Everett, R. D. (2001). DNA viruses and viral proteins that interact with PML nuclear bodies. *Oncogene* **20**, 7266–7273.

Fox, A. H., Lam, Y. W., Leung, A. K., Lyon, C. E., Andersen, J., Mann, M. and Lamond, A. I. (2002). Paraspeckles: a novel nuclear domain. *Curr Biol* **12**, 13–25.

Frisch, S. M. and Mymryk, J. S. (2002). Adenovirus-5 E1A: paradox and paradigm. *Nat Rev Mol Cell Biol* **3**, 441–452.

Gabler, S., Schutt, H., Groitl, P., Wolf, H., Shenk, T. and Dobner, T. (1998). E1B 55-kiloDalton-associated protein: a cellular protein with RNA-binding activity implicated in nucleocytoplasmic transport of adenovirus and cellular mRNAs. *J Virol* **72**, 7960–7971.

Gall, J. G. (2000). Cajal bodies: the first 100 years. *Annu Rev Cell Dev Biol* **16**, 273–300.

Gallimore, P. H. and Turnell, A. S. (2001). Adenovirus E1A: remodelling the host cell, a life or death experience. *Oncogene* **20**, 7824–7835.

Ghetti, A., Pinol-Roma, S., Michael, W. M., Morandi, C. and Dreyfuss, G. (1992). hnRNP I, the polypyrimidine tract-binding protein: distinct nuclear localization and association with hnRNAs. *Nucleic Acids Res* **20**, 3671–3678.

Gonzalez, R. A. and Flint, S. J. (2002). Effects of mutations in the adenoviral E1B 55-kiloDalton protein coding sequence on viral late mRNA metabolism. *J Virol* **76**, 4507–4519.

Goodwin, D. J. and Whitehouse, A. (2001). A gamma-2 herpesvirus nucleocytoplasmic shuttle protein interacts with importin alpha 1 and alpha 5. *J Biol Chem* **276**, 19905–19912.

Grand, R. J., Parkhill, J., Szestak, T., Rookes, S. M., Roberts, S. and Gallimore, P. H. (1999). Definition of a major p53 binding site on Ad2E1B58K protein and a possible nuclear localization signal on the Ad12E1B54K protein. *Oncogene* **18**, 955–965.

Grande, M. A., van der Kraan, I., van Steensel, B., Schul, W., de The, H., van der Voort, H. T., de Jong, L. and van Driel, R. (1996). PML-containing nuclear bodies: their spatial distribution in relation to other nuclear components. *J Cell Biochem* **63**, 280–291.

Grande, M. A., van der Kraan, I., de Jong, L. and van Driel, R. (1997). Nuclear distribution of transcription factors in relation to sites of transcription and RNA polymerase II. *J Cell Sci* **110** (15), 1781–1791.

Guccione, E., Massimi, P., Bernat, A. and Banks, L. (2002). Comparative analysis of the intracellular location of the high- and low-risk human papillomavirus oncoproteins. *Virology* **293**, 20–25.

Guccione, E., Lethbridge, K. J., Killick, N., Leppard, K. N. and Banks, L. (2004). HPV E6 proteins interact with specific PML isoforms and allow distinctions to be made between different POD structures. *Oncogene* **23**, 4662–4672.

Harris, H., Sidebottom, E., Grace, D. M. and Bramwell, M. E. (1969). The expression of genetic information: a study with hybrid animal cells. *J Cell Sci* **4**, 499–525.

Hebert, M. D., Shpargel, K. B., Ospina, J. K., Tucker, K. E. and Matera, A. G. (2002). Coilin methylation regulates nuclear body formation. *Dev Cell* **3**, 329–337.

Hiscox, J. A. (2002). The nucleolus—a gateway to viral infection? *Arch Virol* **147**; 1077–1089.

Hofmann, T. G., Moller, A., Sirma, H., Zentgraf, H., Taya, Y., Droge, W., Will, H. and Schmitz, M. L. (2002). Regulation of p53 activity by its interaction with homeodomain-interacting protein kinase-2. *Nat Cell Biol* **4**, 1–10.

Huang, S. (2000). Review: perinucleolar structures. *J Struct Biol* **129**, 233–240.

Ito, S., Eda, H., Ban, F. and Yanagi, K. (2003). Epstein–Barr virus nuclear antigen-1 colocalizes with lamin B1 in the nucleoplasm and along the nuclear rim. *Arch Virol* **148**, 1633–1642.

Johnson, J. S., Osheim, Y. N., Xue, Y., Emanuel, M. R., Lewis, P. W., Bankovich, A., Beyer, A. L. and Engel, D. A. (2004). Adenovirus protein VII condenses DNA, represses transcription, and associates with transcriptional activator E1A. *J Virol* **78**, 6459–6468.

Kanda, T., Zanma, S., Watanabe, S., Furuno, A. and Yoshiike, K. (1991). Two immunodominant regions of the human papillomavirus type 16 E7 protein are masked in the nuclei of monkey COS-1 cells. *Virology* **182**, 723–731.

Kiesslich, A., von Mikecz, A. and Hemmerich, P. (2002). Cell cycle-dependent association of PML bodies with sites of active transcription in nuclei of mammalian cells. *J Struct Biol* **140**, 167–179.

Kramer, J., Zachar, Z. and Bingham, P. M. (1994). Nuclear pre-mRNA metabolism: channels and tracks. *Trends Cell Biol* **4**, 35–37.

Kurz, A., Lampel, S., Nickolenko, J. E., Bradl, J., Benner, A., Zirbel, R. M., Cremer, T. and Lichter, P. (1996). Active and inactive genes localize preferentially in the periphery of chromosome territories. *J.Cell Biol* **135**, 1195–1205.

Lam, Y. W., Lyon, C. E. and Lamond, A. I. (2002). Large-scale isolation of Cajal bodies from HeLa cells. *Mol Biol Cell* **13**, 2461–2473.

Lamond, A. I. and Carmo-Fonseca, M. (1993). The coiled body. *Trends Cell Biol* **3**, 198–204.

Lamond, A. I. and Earnshaw, W. C. (1998). Structure and function in the nucleus. *Science* **280**, 547–553.

Lee, T. W., Blair, G. E. and Matthews, D. A. (2003). Adenovirus core protein VII contains distinct sequences that mediate targeting to the nucleus and nucleolus, and colocalization with human chromosomes. *J Gen Virol* **84**, 3423–3428.

Lee, T. W., Lawrence, F. J., Dauksaite, V., Akusjarvi, G., Blair, G. E. and Matthews, D. A. (2004). Precursor of human adenovirus core polypeptide Mu targets the nucleolus and modulates the expression of E2 proteins. *J Gen Virol* **85**, 185–196.

Leung, A. K. and Lamond, A. I. (2002). *In vivo* analysis of NHPX reveals a novel nucleolar localization pathway involving a transient accumulation in splicing speckles. *J.Cell Biol* **157**, 615–629.

Mantovani, F. and Banks, L. (1999). The interaction between p53 and papillomaviruses. *Semin Cancer Biol* **9**, 387–395.

Martens, J. H., Verlaan, M., Kalkhoven, E., Dorsman, J. C. and Zantema, A. (2002). Scaffold/matrix attachment region elements interact with a p300-scaffold attachment factor A complex and are bound by acetylated nucleosomes. *Mol Cell Biol* **22**, 2598–2606.

Matera, A. G. (1998). Of coiled bodies, gems, and salmon. *J Cell Biochem* **70**, 181–192.

Matera, A. G. (1999). Nuclear bodies: multifaceted subdomains of the interchromatin space. *Trends Cell Biol* **9**, 302–309.

Matthews, D. A. and Russell, W. C. (1998). Adenovirus core protein V is delivered by the invading virus to the nucleus of the infected cell and later in infection is associated with nucleoli. *J Gen Virol* **79**(7), 1671–1675.

Matthews, D. A. (2001). Adenovirus protein V induces redistribution of nucleolin and B23 from nucleolus to cytoplasm. *J Virol* **75**, 1031–1038.

Maul, G. G. (1998). Nuclear domain 10, the site of DNA virus transcription and replication. *Bioessays* **20**, 660–667.

Maul, G. G., Guldner, H. H. and Spivack, J. G. (1993). Modification of discrete nuclear domains induced by herpes simplex virus type 1 immediate early gene 1 product (ICP0). *J Gen Virol* **74** (12), 2679–2690.

Maul, G. G., Negorev, D., Bell, P. and Ishov, A. M. (2000). Review: properties and assembly mechanisms of ND10, PML bodies, or PODs. *J Struct Biol* **129**, 278–287.

Mika, S. and Rost, B. (2005). NMPdb: Database of Nuclear Matrix Proteins. *Nucleic Acids Res* **33** (Database Issue), D160–D163.

Miron, M. J., Gallouzi, I. E., Lavoie, J. N. and Branton, P. E. (2004). Nuclear localization of the adenovirus E4orf4 protein is mediated through an arginine-rich motif and correlates with cell death. *Oncogene* **23**, 7458–7468.

Moller, A., Sirma, H., Hofmann, T. G., Rueffer, S., Klimczak, E., Droge, W., Will, H. and Schmitz, M. L. (2003). PML is required for homeodomain-interacting protein kinase 2 (HIPK2)-mediated p53 phosphorylation and cell cycle arrest but is dispensable for the formation of HIPK domains. *Cancer Res* **63**, 4310–4314.

Muranyi, W., Haas, J., Wagner, M., Krohne, G. and Koszinowski, U. H. (2002). Cytomegalovirus recruitment of cellular kinases to dissolve the nuclear lamina. *Science* **297**, 854–857.

Nayler, O., Stratling, W., Bourquin, J. P., Stagljar, I., Lindemann, L., Jasper, H., Hartmann, A. M., Fackelmayer, F. O., Ullrich, A. and Stamm, S. (1998). SAF-B protein couples transcription and pre-mRNA splicing to SAR/MAR elements. *Nucleic Acids Res* **26**, 3542–3549.

Negorev, D. and Maul, G. G. (2001). Cellular proteins localized at and interacting within ND10/PML nuclear bodies/PODs suggest functions of a nuclear depot. *Oncogene* **20**, 7234–7242.

Ogg, S. C. and Lamond, A. I. (2002). Cajal bodies and coilin—moving towards function. *J Cell Biol* **159**, 17–21.

Okuwaki, M., Iwamatsu, A., Tsujimoto, M. and Nagata, K. (2001). Identification of nucleo-phosmin/B23, an acidic nucleolar protein, as a stimulatory factor for *in vitro* replication of adenovirus DNA complexed with viral basic core proteins. *J Mol Biol* **311**, 41–55.

Platani, M., Goldberg, I., Lamond, A. I. and Swedlow, J. R. (2002). Cajal body dynamics and association with chromatin are ATP-dependent. *Nat Cell Biol* **4**, 502–508.

Querido, E., Blanchette, P., Yan, Q., Kamura, T., Morrison, M., Boivin, D., Kaelin, W. G., Conaway, R. C., Conaway, J. W. and Branton, P. E. (2001). Degradation of p53 by adenovirus E4orf6 and E1B55K proteins occurs via a novel mechanism involving a Cullin-containing complex. *Genes Dev* **15**, 3104–3117.

Rao, L., Modha, D. and White, E. (1997). The E1B 19K protein associates with lamins *in vivo* and its proper localization is required for inhibition of apoptosis. *Oncogene* **15**, 1587–1597.

Rebelo, L., Almeida, F., Ramos, C., Bohmann, K., Lamond, A. I. and Carmo-Fonseca, M. (1996). The dynamics of coiled bodies in the nucleus of adenovirus-infected cells. *Mol Biol Cell* **7**, 1137–1151.

Reynolds, A. E., Liang, L. and Baines, J. D. (2004). Conformational changes in the nuclear lamina induced by herpes simplex virus type 1 require genes U(L)31 and U(L)34. *J Virol* **78**, 5564–5575.

Rodrigues, S. H., Silva, N. P., Delicio, L. R., Granato, C. and Andrade, L. E. (1996). The behavior of the coiled body in cells infected with adenovirus *in vitro*. *Mol Biol Rep* **23**, 183–189.

Romig, H., Fackelmayer, F. O., Renz, A., Ramsperger, U. and Richter, A. (1992). Characteri zation of SAF-A, a novel nuclear DNA binding protein from HeLa cells with high affinity for nuclear matrix/scaffold attachment DNA elements. *EMBO J* **11**, 3431–3440.

Saenz-Robles, M. T., Sullivan, C. S. and Pipas, J. M. (2001). Transforming functions of Simian virus 40. *Oncogene* **20**, 7899–7907.

Salomoni, P. and Pandolfi, P. P. (2002). The role of PML in tumor suppression. *Cell* **108**, 165–170.

Sang, N., Caro, J. and Giordano, A. (2002). Adenoviral E1A: everlasting tool, versatile applications, continuous contributions and new hypotheses. *Front Biosci* **7**, d407–d413.

Samow, P., Ho, Y. S., Williams, J. and Levine, A. J. (1982). Adenovirus E1b-58 kDa tumor antigen and SV40 large tumor antigen are physically associated with the same 54 kDa cellular protein in transformed cells. *Cell* **28**, 387–394.

Saurin, A. J., Shiels, C., Williamson, J., Satijn, D. P., Otte, A. P., Sheer, D. and Freemont, P. S. (1998). The human polycomb group complex associates with pericentromeric heterochromatin to form a novel nuclear domain. *J Cell Biol* **142**, 887–898.

Schardin, M., Cremer, T., Hager, H. D. and Lang, M. (1985). Specific staining of human chromosomes in Chinese hamster × man hybrid cell lines demonstrates interphase chromosome territories. *Hum Genet* **71**, 281–287.

Schmitt, R. C., Fahnestock, M. L. and Lewis, J. B. (1987). Differential nuclear localization of the major adenovirus type 2 E1a proteins. *J Virol* **61**, 247–255.

Scott, E. S. and O'Hare, P. (2001). Fate of the inner nuclear membrane protein lamin B receptor and nuclear lamins in herpes simplex virus type 1 infection. *J Virol* **75**, 8818–8830.

Shenk T. (2001). Adenoviridae: the viruses and their replication. In *Fields' Virology*, D. M. Knipe, P. M. Howley, D. E. Griffin, R. A. Lamb, M. A. Martin, B. Roizman, and S. E. Straus (eds). vol 2, pp. 2111–2148, Lippincott Williams & Wilkins.

Shera, K. A., Shera, C. A. and McDougall, J. K. (2001). Small tumor virus genomes are integrated near nuclear matrix attachment regions in transformed cells. *J Virol* **75**, 12339–12346.

Shiels, C., Islam, S. A., Vatcheva, R., Sasieni, P., Sternberg, M. J., Freemont, P. S. and Sheer, D. (2001). PML bodies associate specifically with the MHC gene cluster in interphase nuclei. *J Cell Sci* **114**, 3705–3716.

Simpson-Holley, M., Baines, J., Roller, R. and Knipe, D. M. (2004). Herpes simplex virus 1 U(L)31 and U(L)34 gene products promote the late maturation of viral replication compartments to the nuclear periphery. *J Virol* **78**, 5591–5600.

Sleeman, J. E., Trinkle-Mulcahy, L., Prescott, A. R., Ogg, S. C. and Lamond, A. I. (2003). Cajal body proteins SMN and coilin show differential dynamic behaviour *in vivo. J Cell Sci* **116**, 2039–2050.

Spector, D. L. (2001). Nuclear domains. *J Cell Sci* **114**, 2891–2893.

Sternsdorf, T., Guldner, H. H., Szostecki, C., Grotzinger, T. and Will, H. (1995). Two nuclear dot-associated proteins, PML and Sp100, are often co-autoimmunogenic in patients with primary biliary cirrhosis. *Scand J Immunol* **42**, 257–268.

Stoffler, D., Fahrenkrog, B. and Aebi, U. (1999). The nuclear pore complex: from molecular architecture to functional dynamics. *Curr Opin Cell Biol* **11**, 391–401.

Stratling, W. H. and Yu, F. (1999). Origin and roles of nuclear matrix proteins. Specific functions of the MAR-binding protein MeCP2/ARBP. *Crit Rev Eukaryot Gene Expr* **9**, 311–318.

Tang, Q., Bell, P., Tegtmeyer, P. and Maul, G. G. (2000). Replication but not transcription of simian virus 40 DNA is dependent in nuclear domain 10. *J Virol* **74**, 9694–9700.

Tucker, K. E., Berciano, M. T., Jacobs, E. Y., LePage, D. F., Shpargel, K. B., Rossire, J. J., Chan, E. K., Lafarga, M., Conlon, R. A. and Matera, A. G. (2001). Residual Cajal bodies in coilin knockout mice fail to recruit Sm snRNPs and SMN, the spinal muscular atrophy gene product. *J Cell Biol* **154**, 293–307.

Van Tine, B. A., Dao, L. D., Wu, S. Y., Sonbuchner, T. M., Lin, B. Y., Zou, N., Chiang, C. M., Broker, T. R. and Chow, L. T. (2004). Human papillomavirus (HPV) origin-binding protein associates with mitotic spindles to enable viral DNA partitioning. *Proc Natl Acad Sci USA* **101**, 4030–4035.

Viejo-Borbolla, A., Kati, E., Sheldon, J. A., Nathan, K., Mattsson, K., Szekely, L. and Schulz, T. F. (2003). A domain in the C-terminal region of latency-associated nuclear antigen 1 of Kaposi's sarcoma-associated *Herpesvirus* affects transcriptional activation and binding to nuclear heterochromatin. *J Virol* **77**, 7093–7100.

Wansink, D. G., Sibon, O. C., Cremers, F. F., van Driel, R. and de Jong, L. (1996). Ultrastructural localization of active genes in nuclei of A431 cells. *J Cell Biochem* **62**, 10–18.

Wei, Y., Jin, J. and Harper, J. W. (2003). The cyclin E/Cdk2 substrate and Cajal body component p220(NPAT) activates histone transcription through a novel LisH-like domain. *Mol Cell Biol* **23**, 3669–3680.

White, E., Spector, D. and Welch, W. (1988). Differential distribution of the adenovirus E1A proteins and colocalization of E1A with the 70-kilodalton cellular heat shock protein in infected cells. *J Virol* **62**, 4153–4166.

White, E. and Cipriani, R. (1989). Specific disruption of intermediate filaments and the nuclear lamina by the 19-kDa product of the adenovirus E1B oncogene. *Proc Natl Acad Sci USA* **86**, 9886–9890.

Wienzek, S., Roth, J. and Dobbelstein, M. (2000). E1B 55-kilodalton oncoproteins of adenovirus types 5 and 12 inactivate and relocalize p53, but not p51 or p73, and cooperate with E4orf6 proteins to destabilize p53. *J Virol* **74**, 193–202.

Williams, B. J., Boyne, J. R., Goodwin, D. J., Roaden, L., Hautbergue, G. M., Wilson, S. A. and Whitehouse, A. (2004). The prototype gamma-2 herpesvirus nucleocytoplasmic shuttling protein, ORF 57, transports viral RNA via the cellular mRNA export pathway. *Biochem J* **387**, 295–308.

Woods, D. B. and Vousden, K. H. (2001). Regulation of p53 function. *Exp Cell Res* **264**, 56–66.

Xue, Y., Johnson, J. S., Ornelles, D. A., Lieberman, J. and Engel, D. A. (2005). Adenovirus protein VII functions throughout early phase and interacts with cellular proteins SET and pp32. *J Virol* **79**, 2474–2483.

Yasui, D., Miyano, M., Cai, S., Varga-Weisz, P. and Kohwi-Shigematsu, T. (2002). SATB1 targets chromatin remodelling to regulate genes over long distances. *Nature* **419**, 641–645.

Young, P. J., Jensen, K. T., Burger, L. R., Pintel, D. J. and Lorson, C. L. (2002). Minute virus of mice NS1 interacts with the SMN protein, and they colocalize in novel nuclear bodies induced by parvovirus infection. *J Virol* **76**, 3892–3904.

Zatsepina, O., Braspenning, J., Robberson, D., Hajibagheri, M. A., Blight, K. J., Ely, S., Hibma, M., Spitkovsky, D., Trendelenburg, M., Crawford, L. and Tommasino, M. (1997). The human papillomavirus type 16 E7 protein is associated with the nucleolus in mammalian and yeast cells. *Oncogene* **14**, 1137–1145.

Zou, N., Lin, B. Y., Duan, F., Lee, K. Y., Jin, G., Guan, R., Yao, G., Lefkowitz, E., Broker, T. R. and Chow, L. T. (2000). The hinge of the human papillomavirus type 11 E2 protein contains major determinants for nuclear localization and nuclear matrix association. *J Virol* **74**, 3761–3770.

zur Hausen, H. (2000). Papillomaviruses causing cancer: evasion from host-cell control in early events in carcinogenesis. *J Natl Cancer Inst* **92**, 690–698.

4 Retroviruses and the Nucleus

Carlos de Noronha[1],* and Warner C. Greene[1,2,3]

[1]Gladstone Institute of Virology and Immunology and [2]Departments of Medicine and of [3]Microbiology and Immunology, University of California, San Francisco, USA

Retroviral replication is characterized by the transcription of the positive-strand RNA genome into a double-stranded DNA and subsequent insertion of this DNA into host cell chromatin, thereby forming the provirus. This process relies upon numerous components and functions within the cell nucleus. First, the uncoated viral preintegration complex (PIC) must access the host genome, either entering the nucleus after nuclear envelope breakdown in dividing cells or, in the case of the primate lentivirus, by active nuclear import across the nuclear pore complex of non-dividing cells. The DNA copy of the viral genome must then be spliced into the cellular chromatin, with the assistance of nuclear factors, before efficient transcription can begin.

Overlapping genes are a hallmark of retroviruses and allow their relatively small genomes to directly produce many different proteins (Figure 4.1). The quantity and timing of protein production are determined not only by the action of cellular transcription factors, but also by the interaction of cellular and viral components that regulate the splicing, export, and translation of viral mRNA.

In this chapter, we examine the nuclear components involved in the retroviral replication cycle, focusing on those of the nuclear envelope. We also discuss the import of the PIC and other viral components, the export of viral RNAs and proteins and other nuclear processes, including transcription, RNA processing and the integration of viral DNA.

* Present address Center for Immunology and Microbial Disease, Albany Medical College, Albany, NY 12208-3479, USA.

Viruses and the Nucleus   Edited by Julian A. Hiscox
© 2006 John Wiley & Sons, Ltd

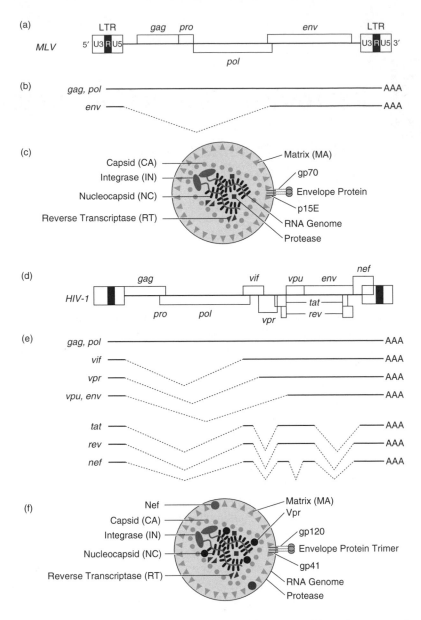

Figure 4.1   A comparison of genome structures and virion composition of representative simple and complex retroviruses. The representative simple retrovirus, murine leukaemia virus (MLV), has three overlapping reading frames, one that encodes the structural Gag proteins and a protease, a second, *pol*, that encodes the reverse transcriptase function and integrase, and a third, *env*, that encodes the envelope glycoprotein (a). Gag/protease and Gag/protease/Pol polyproteins are both translated from the same unspliced, genome-length RNA that is, in fact, also used as the viral genome. The Gag/protease/Pol polyprotein is produced only when either in some viruses the ribosome slides backwards to resume translation of an otherwise shifted reading frame, or a stop

4.1 Retroviral structure

Retroviral positive-strand RNA genomes are transcribed into double-stranded DNA before integration into the host chromatin. All retroviral genomes encode at least three proteins, Gag, Pol and Env, which are proteolytically processed to produce structural, enzymatic and viral surface components, respectively. The simple retroviruses, such as gamma-retroviruses and oncoretroviruses, are limited to these gene products, while others, such as the primate lentiviruses, possess additional genes, including *tat*, *rev*, *vif*, *vpu*, *vpr* and *nef*. Interestingly, HIV-2 and SIVmac express a tenth gene, *vpx*, but lack *vpu* (Figure 4.1).

4.2 The retroviral life cycle

Retroviral infection begins with the binding of viral envelope proteins to cognate cell-surface receptors (Figure 4.2). These receptors represent a diverse range of proteins, including: a cationic amino acid transporter for ecotropic murine leukemia virus; phosphate symporters for amphotropic murine leukaemia virus, gibbon ape leukaemia virus, and subgroup B feline leukaemia virus; a low-density lipoprotein receptor-related protein for subgroup A avian sarcoma/leucosis virus; and CD4 in conjunction with chemokine receptors for HIV and SIV.

The HIV entry process is among the best characterized. Viral envelope trimers decorate the virions. Each of the trimer components consists of two glycoproteins, a gp120 SU protein and a gp41 transmembrance protein that anchors the complex in the cytoplasmic membrane-derived lipid bilayer that envelops the virion. Contact between gp120 and the receptor CD4 promotes a conformational change in gp120 that unmasks a co-receptor binding sequence. The co-receptors include chemokine receptors CXCR4 for syncytium-inducing strains and CCR5 for non-syncytium-inducing strains. Engagement of the receptors induces conformational alterations in gp41 that allow insertion of the gp41 trimer into the target cell membrane. This process brings the viral and cellular membranes into sufficient proximity for fusion to occur.

Figure 4.1 (*continued*)
codon between the reading frames is suppressed by a transfer RNA. The envelope polyprotein is translated from a singly spliced, subgenomic RNA (b). All of the polyproteins are later proteolytically processed. MLV virion composition is illustrated in (c). The HIV-1 genome is much more complex (d). First, in addition to the *gag*, *pro*, *pol* and *env* reading frames, HIV-1 and other complex retroviruses encode a number of other proteins, including Tat and Rev and so-called 'accessory' proteins, including Vif, Vpr, Vpu and Nef. In order to encode these proteins, a complex RNA splicing pattern and overlapping reading frames are employed (e). The composition of these virions is also more complex (f)

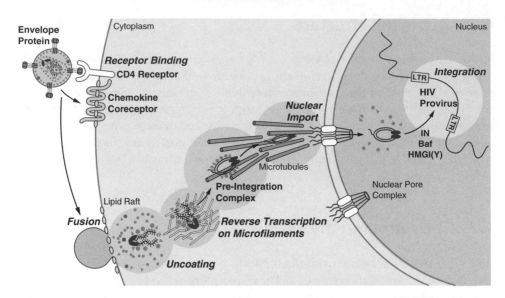

Figure 4.2 Retroviral infection from virus entry to establishment of the provirus. The infection is initiated when the virus, HIV-1 in this illustration, engages its cognate receptor. This contact promotes conformational alterations in the envelope glycoproteins that allow fusion of the lipid bilayers surrounding the virion with those enclosing the cell. The contents of the virus are then released into the cytoplasm in an as-yet poorly defined process known as uncoating. The viral genome, which has likely already begun the reverse transcription process, remains associated with viral proteins as it attaches to microtubules in the cytoplasm that help to guide it towards the cell nucleus. Once at the nucleus, the complex is, in the case of complex retroviruses, imported into the nucleus. Simple retroviruses rely on cell division for infection and probably gain access to cellular chromatin at the time of nuclear envelope breakdown. Once inside the nucleus, the viral protein IN mediates insertion of the fully reverse-transcribed genome into the host chromatin. Details of the integration process are provided in the text and in Figure 4.5

After virus–cell fusion, the virus particle undergoes a poorly characterized process known as uncoating, during which the viral envelope and associated lipid bilayer are shed at the cytoplasmic membrane. Other viral and cellular proteins are abandoned as well. The identity of components remaining in the uncoated viral complex has been the subject of debate and is thought to differ between viruses. The murine leukemia virus PIC, for example, retains the *gag*-encoded capsid protein (Bowerman *et al*., 1989; Bushman, 1994; Lee and Craigie, 1994), whereas HIV-1 does not (Farnet and Haseltine, 1999; Karageorgos *et al*., 1993; Bukrinsky *et al*., 1993). Thus, the uncoated HIV-1 particle is composed of viral nucleic acids, tRNALys primer, viral proteins including integrase (IN), matrix (MA), viral protein R (Vpr) and reverse transcriptase (Karageorgos *et al*., 1993). The cellular nuclear envelope component barrier-to-autointegration factor and HMGI(Y) (Farnet and Bushman, 1997) are thought to join the PIC *en route* to the nucleus (Lee and Craigie, 1994).

Reverse transcription, the process through which viral RNA is copied into double-stranded DNA, may begin within the virion (Lori *et al.*, 1992; Trono, 1992; Zack *et al.*, 1992; Zhu and Cunningham, 1993) but is largely completed within the target cell (Figure 4.3). Most retroviruses complete reverse transcription in the cytoplasm; avian leukaemia virus, however, requires nuclear entry (Varmus *et al.*, 1978; Lee and Coffin, 1991). The usual criterion for completion of reverse transcription is whether the product can integrate into DNA in an *in vitro* integration assay.

Recent work has shown that many cells targeted by HIV and other complex retroviruses produce an enzyme, APOBEC3G, which interferes with reverse transcription. This cytidine deaminase (Harris *et al.*, 2002) can be packaged into virions (Mariani *et al.*, 2003; Stopak *et al.*, 2003) and acts, during reverse transcription, on the first (negative) DNA strand transcribed from the viral RNA (Harris *et al.*, 2003; Mangeat *et al.*, 2003; Zhang *et al.*, 2003; Lecossier *et al.*, 2003). It causes C → U mutations that can either be excised by uracil-N-glycosylase or become fixed as G → A mutations of the coding strand. Uracil excision may destabilize the viral genome and thereby interfere with reverse transcription. Further, hypermutation of genomes that complete reverse transcription likely interferes with the production of infectious virus. To combat this natural antiretroviral function, all lentiviruses, with the exception of equine infectious anaemia virus (EIAV), produce virion infectivity factor or Vif. Vif prevents APOBEC3G from being incorporated into virions in virus-producing cells (Mariani *et al.*, 2003) by interfering with its translation, enhancing its turnover (Stopak *et al.*, 2003), and sequestering it from the site of viral assembly. Interestingly, despite the rather ubiquitous expression of APOBEC3G, some retroviruses, such as murine leukaemia virus, which do not express Vif, appear to be relatively resistant to its effects. Finally, Vif's inhibition of APOBEC3G is quite species-specific (Mariani *et al.*, 2003).

After entering the host cell, HIV-1 associates with microfilaments in a process that enhances reverse transcription (Bukrinskaya *et al.*, 1998). Like herpes simplex virus (Smith *et al.*, 2001) and adenovirus (Suomalainen *et al.*, 2001), HIV utilizes microtubules to direct the reverse transcription complex to the nucleus (McDonald *et al.*, 2002).

4.3 Entering the nucleus

Entry into the nucleus is accomplished differently by the PICs of many simple retroviruses and the more complex lentiviruses. The former require cell replication and probably nuclear envelope breakdown to enter the nucleus and gain access to host chromatin (Roe *et al.*, 1993; Lewis and Emerman, 1994). On the other hand, the cellular host range of lentiviruses extends beyond actively dividing cells (Lewis and Emerman, 1994). HIV-1 readily infects non-dividing macrophages (Weinberg *et al.*, 1991) and resting and naive CD4 T cells, all of which form important viral reservoirs (Zhang *et al.*, 1999; Ostrowski *et al.*, 1999; Blaak *et al.*, 2000). Lentiviral transduction

(a) Import

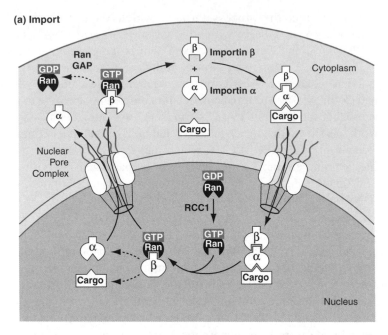

(b) Export

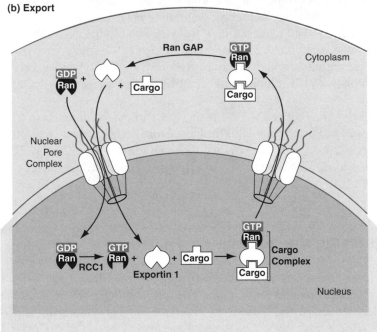

Figure 4.3 Reverse transcription of simple and complex retroviruses. These diagrams outline the step-by-step process of reverse transcription. A and I represent integrated proviruses, the arrow (⤳) indicates the point of transcription initiation, and B and J represent the viral genomes. Black lines represent DNA, grey represents RNA and broken grey lines represent RNase-degraded RNA

vectors have been used to introduce genetic material into a variety of non-dividing cells, including terminally differentiated neurons, myocytes, retinal cells and hepatocytes (Amado and Chen, 1999).

4.3.1 How does the retroviral PIC gain entry into the cell nucleus?

The nucleus and the cytoplasm are separated by a double lipid bilayer, the outer of which is contiguous with the endoplasmic reticulum. Molecular trafficking between the nucleus and the cytoplasm occurs through nuclear pore complexes, which are 125 MDa macromolecular complexes of 50–100 polypeptides that are distributed over the nuclear surface (reviewed Mattaj and Englmeier, 1998). The inner diameter of the pores provides a 9 nm aqueous channel for passive diffusion of small molecules or proteins, probably in specific conformations, of up to approximately 60 kDa. During active transport, the effective inner diameter of the channel increases to an estimated 25 nm. Thus, the HIV-1 PIC, with an estimated Stokes diameter of 56 nm, is quite large relative to the inner dimensions of the nuclear pore.

Transport across nuclear pores is an active process mediated by proteins designated karyopherins and classified as importins and exportins (reviewed in Mattaj and Englmeier, 1998; Weis, 1998). The best-characterized import pathway is that which targets proteins containing the lysine-rich SV40 large T antigen nuclear import signal (PKKKRKV) (Kalderon *et al.*, 1984) or the bipartite nucleoplasmin nuclear import signal (150-KRPAAIKKAGQAKKKK-170) (Robbins *et al.*, 1991) for import into the nucleus. This basic sequence binds to importin-α which, in turn, engages importin-β through an interaction mediated by its amino-terminal 66 amino acid importin-β-binding domain (Gorlich *et al.*, 1996; Weis *et al.*, 1996). Importin-β then directly engages nucleoporins, which are constituents of the nuclear pore (Moroianu *et al.*, 1995; Gorlich *et al.*, 1995; Imamoto *et al.*, 1995; Radu *et al.*, 1995). The trimeric complex is then passed through the pore involving a series of binding and release reactions that require energy (Figure 4.3).

The directionality of transport is provided by a release factor gradient. The release factor, Ran, is a member of the Ras family of small GTPases (Mattaj and Englmeier, 1998; Ohno *et al.*, 1998; Gorlich and Kutay, 1999; Yoneda, 2000) that associates primarily with GTP within the nucleus and with GDP in the cytoplasm. This gradient is established by the nuclear, chromatin-bound factor RCC1 (Chi *et al.*, 1995), a GDP–GTP exchange factor (RanGEF) that promotes the release of GDP from RanGDP and thereby permits GTP binding. In the cytoplasm, two related RanGTP-binding proteins, RanBP1 and RanBP2, cooperate with the Ran GTPase-activating protein (RanGAP) to promote rapid Ran-mediated GTP hydrolysis (Mattaj and Englmeier, 1998; Ohno *et al.*, 1998; Gorlich and Kutay, 1999; Yoneda, 2000). RanGTP, but not RanGDP, binds to importin-β. This physical interaction both inhibits the function of RanGTPase (Bischoff and Gorlich, 1997) and promotes the release of cargo (Izaurralde *et al.*, 1997). Nuclear export, which is mediated by the importin-β family member Crm1 (exportin-1) (Ossareh-Nazari *et al.*, 1997), is also dependent on release factor gradients. It binds and exports proteins with leucine-rich nuclear export signals in the presence of nuclear RanGTP (Richards *et al.*, 1997; Fornerod *et al.*, 1997) and releases cargo in the presence of RanGTP-binding proteins, RanBP1 and RanBP2 (Kehlenbach *et al.*, 1999).

Partially purified PICs contain IN, MA (p17^{Gag}), nucleocapsid (p7^{Gag}), Vpr and reverse transcriptase (reviewed in Fouchier and Malim, 1999), as well as the cellular proteins barrier-to-autointegration factor (Chen and Engelman, 1998) and HMGI(Y) (Farnet and Bushman, 1997) that assemble with the PIC within the host cell. Of the viral proteins, IN, Vpr and MA each display karyophilic properties (Bukrinsky *et al.*, 1993; Heinzinger *et al.*, 1994; Gallay *et al.*, 1996, 1997; Fouchier *et al.*, 1998; Pluymers *et al.*, 1999; Vodicka *et al.*, 1998), and an intermediate of the reverse transcription complex, known as the central flap, has also been implicated in nuclear import (Zennou *et al.*, 2000).

4.3.2 Matrix

The 19 kDa, Gag-encoded protein MA appears to have several functions in the viral life cycle. It targets Gag proteins and unspliced viral RNA to the site of virus assembly (Krausslich and Welker, 1996), facilitates the incorporation of envelope glycoproteins into nascent virions (Yu *et al.*, 1993; Dorfman *et al.*, 1994), and helps target the viral PIC to the nucleus. Both an amino-terminal myristylation (Bryant and Ratner, 1990; Gottlinger *et al.*, 1989) and a basic region (Zhou *et al.*, 1994) direct this protein to the cytoplasmic membrane, where it serves as a structural member of the virion. MA is found in the cell nucleus, importantly, with other constituents of the viral PIC (Bukrinsky *et al.*, 1993a). MA contains a classical, SV40 large T antigen-like, nuclear localization signal (NLS), in the same basic region that promotes targeting of MA to the cytoplasmic membrane (Bukrinsky *et al.*, 1993b). The switch that determines whether MA will be a component of the budding virion or a nucleus-targeting component of the PIC is an amino-terminal tyrosine phosphorylation. This modification, which affects only a small fraction of the molecules, retargets the protein from the cytoplasmic membrane to the nucleus (Gallay *et al.*, 1995b) and induces MA to bind to the core domain of IN (Gallay *et al.*, 1995a). These results, however, are somewhat controversial (Freed *et al.*, 1997). More recently, MA was found to have a nuclear export signal (Dupont *et al.*, 1999) that was reported to be important for viral replication. Apparently, the export of MA helps to assure that the unspliced viral RNA, which is to be packaged into virions, is correctly targeted to the site of viral assembly.

Two recent studies dispute the importance of HIV MA for nuclear import and for the infection of non-dividing cells. Fouchier *et al.* (1997) reported that while the amino-terminus of MA is important for the processing of Gag, this region is dispensable for mediating post-entry nuclear import of the HIV-1 PIC. Reil *et al.* (1998) demonstrated that large parts of Gag are unnecessary for the production of infectious viral particles. In fact, the entire globular domain, containing the afore-mentioned basic NLS and membrane targeting signal, is dispensable for the infection of terminally differentiated, non-dividing cells and probably for nuclear import as well.

While MA contains an NLS, it is not the only NLS found in the virus. The requirement for this and others may depend upon the cellular context and on the availability of other nuclear import signals.

4.3.3 Central flap

The central flap is a triple-stranded intermediate of reverse transcription resulting from the overlap of two plus strand segments. This structure has been reported to participate in the nuclear import of the HIV PIC (Zennou *et al.*, 2000).

The viral genome, a full-length transcript from the provirus (Figure 4.4(a), (i)), presents a template with identical, repeated sequences at the 5′ and 3′ termini (R) (Figure 4.4(b), (j)). The reverse transcription process will, upon completion, duplicate a sequence designated U5, which is directly downstream of the 5′ R, and attach it directly downstream of the 3′ R. Similarly, a sequence directly upstream of the 3′ R, designated U3, will be duplicated at the 5′ terminus of the reverse transcription product, resulting in a DNA transcript with extended duplications known as long terminal repeats (LTRs) (Figure 4.4(a), (i)).

HIV-1 carries out reverse transcription in a slightly different manner than has been described for the simple oncoretroviruses (Figure 4.4). In both schemes, reverse transcription begins with the interaction of a tRNA to the primerbinding site. Elongation proceeds to the 5′ end of the viral genome, and the viral Pol protein RNaseH degrades the RNA template. The DNA product, which includes sequence complementary to the R and U5, known as minus-strand strong-stop DNA, is transferred to the 3′ R, where it is elongated to include sequences complementary to the viral genome through the primary binding site. Both HIV-1 and the oncoretrovirus RNAs have polypurine tracts (PPT) located directly upstream of U3 that are relatively resistant to RNAseH and therefore remain to act as primers for the synthesis of positive-strand DNA. HIV-1, unlike oncoretroviruses, has an additional PPT (cPPT) within the IN coding sequences. The cPPT promotes the generation of a second positive-strand primer, making bipartite positive-strand synthesis possible. This synthesis proceeds to the 5′ end of the minus strand, where the primer binding site is transcribed from the remaining tRNA template, which is then degraded by RNAseH. During both HIV-1 and oncoretrovirus reverse transcription, the primer binding site sequence of the positive and negative strands anneal, forming a circular structure in which the strands provide templates for their mutual completion. In the case of HIV-1, however, the positive strand initiated at the 3′ PPT extends to the positive strand initiated at the cPPT, where it displaces the second positive strand until the reverse transcriptase is released upon encountering the central termination sequence after an additional

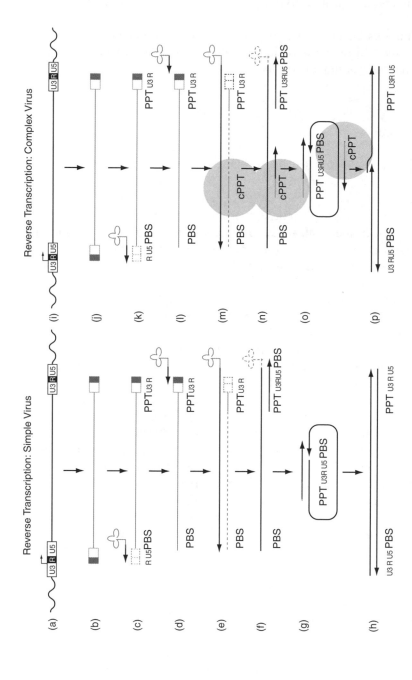

Figure 4.4 Nuclear transport. Similar strategies are central to both nuclear import and export of large molecules. The importin-α/β nuclear import pathway is depicted here (a). The basic nuclear import signal binds to importin-α, which in turn binds importin-β, which engages the nuclear pore. Energy-dependent transport throught the pore delivers the cargo into the nucleus, where RanGTP binding to importin-β promotes release of the cargo and importin-α. Directionality of this process depends upon higher concentrations of RanGTP in the nucleus than in the cytoplasm. Nuclear RCC1 is responsible for converting RanGTP to RanGTP, and cytoplasmic RanGAP converts RanGTP to RanGDP. The directionality of nuclear export also depends on the RanGDP–GTP gradient. In the presence of RanGTP, the canonical leucine-rich export signal assembles with exportin 1 (Crm1), which engages the nuclear pore to mediate transport to the cytoplasm. There, the complex is dissociated when RanGAP hydrolyses the RanGTP to RanGDP. The components of the transport systems are recycled for subsequent use

99 nucleotides (Charneau *et al.*, 1994). This process results in the triple-stranded central DNA flap.

Preservation of the central flap was first shown to be important for viral replication (Hungnes *et al.*, 1992; Charneau *et al.*, 1992) and then more specifically for nuclear import of the PIC (Zennou *et al.*, 2000). How this DNA structure mediates nuclear import is unclear. In a nuclear import model, where the diameter of the nuclear pore remains unaltered, the PIC or a portion thereof would be elongated to fit through the pore. In the case of mRNA export, as Zennou *et al.* (2000) indicate, the molecule is guided from the nucleus by the 5′ cap structure (Hamm and Mattaj, 1990). Nonetheless, it is likely that the flap mediates a direct interaction with either nuclear pore proteins or a cellular protein intermediate that facilitates import.

Dvorin *et al.* (2002) published contradictory findings. They reported that preservation of the cPPT, and therefore bipartite positive-strand synthesis and central DNA flap formation, enhances HIV-1 replication slightly, but is not important for viral replication in non-dividing cells. This work does not explain the findings by Zennou *et al.* and others, who have shown that incorporation of the cPPT into retroviral vectors enhances the infection of non-dividing cells (Demaison *et al.*, 2002; Lewis *et al.*, 2001; Mautino *et al.*, 2000; Mautino and Morgan, 2002; Follenzi *et al.*, 2000).

4.3.4 Vpr

HIV-1 Vpr is a nucleophilic, 96 amino acid, 11 kDa protein that is expressed late in the viral life cycle in a Rev-dependent manner. This protein is incorporated into virions through a physical association with the amino-terminal p6 component of the Gag polyprotein (Lu *et al.*, 1995) and appears to remain associated with the PIC as it enters the nucleus. Early studies indicated that equimolar quantities of Gag and Vpr (1000–2000 copies) are packaged into each virion; however, recent experiments suggest that only tens of Vpr molecules are incorporated (Singh *et al.*, 2001).

HIV-1 Vpr has two known functions: it enhances viral replication in terminally differentiated macrophages (Bukrinsky *et al.*, 1993; Gallay *et al.*, 1996, 1997; Zennou *et al.*, 2000; von Schwedler *et al.*, 1994; Bouyac-Bertoia *et al.*, 2001); and it promotes cell cycle arrest at or near the G_2 checkpoint (Jowett *et al.*, 1995). The former function, often attributed to the nucleophilic character of Vpr, is relevant to nuclear entry. The latter, which may improve viral replication, will be discussed below. Interestingly, Vpr has only been identified in primate lentiviruses. In the case of HIV-2 and SIV_{agm}, two functions of HIV-1 Vpr are segregated between two adjacently encoded proteins, Vpx and Vpr. Vpx is incorporated into virions and facilitates the infection of non-dividing cells, while Vpr is responsible for cell-cycle arrest (Fletcher *et al.*, 1996).

The nuclear import of Vpr is not blocked by NLS peptides, suggesting that this protein contains a non-conventional nuclear import signal (Gallay *et al.*, 1996). Mapping of Vpr's NLSs indicates that nuclear import is inhibited by different mutations in various regions of the protein (Di Marzio *et al.*, 1995; Lu *et al.*, 1993; Mahalingam *et al.*, 1995, 1997; Zhou *et al.*, 1998; Yao *et al.*, 1995). Jenkins *et al.* (1998) showed that Vpr has at least two distinct import signals that do not rely on the classical importin-α/β-mediated nuclear import pathway. One import signal was subsequently mapped to the leucine-rich sequences. Of note is the finding that the leucine-rich sequence present in the second helix also participates as a signal for Crm-1-dependent nuclear export (Sherman *et al.*, 2001). The second nuclear import signal is a bipartite arginine-rich sequence located within the relatively unstructured carboxy-terminal region of Vpr (Sherman *et al.*, 2001). The import mechanisms used by the two NLSs have not yet been completely characterized. Jenkins *et al.* (1998) extracted soluble nuclear import factors from cells with digitonin and then reconstituted the nuclear import system with defined components. These experiments revealed that neither signal required GTP hydrolysis and that little or no energy was required for nuclear import by either signal. These studies further suggested that Vpr engages the import machinery downstream of importin-α or the M9 import signal. These observations are consistent with a direct interaction between Vpr and the nucleoporins Nsp 1p (Vodicka *et al.*, 1998) and POM 121 (Fouchier *et al.*, 1998). As such, Vpr may act as an importin-β analogue that directly engages the nuclear pore. Vpr binds importin-α (Vodicka *et al.*, 1998; Popor *et al.*, 1998a, 1998b).

Vpr exhibits a number of characteristics that may facilitate nuclear import of the HIV PIC. Vpr remains associated with the PIC, contains two NLSs, and is required for replication in many non-dividing cells where nuclear integrity is likely to be maintained. The majority of evidence that Vpr is important for the infection of non-dividing cells derives from the analysis of monocyte-derived macrophages (Heinzinger *et al.*, 1994; Popor *et al.*, 1998a, 1998b; Bukrinsky *et al.*, 1992; Connor *et al.*, 1995), where the effects of Vpr are most pronounced after low-multiplicity infections (Subbramanian *et al.*, 1998). These findings suggest that Vpr, rather than being essential for infection, improves the overall efficiency. This could play an important role in HIV-1 transmission, where the viral titre may be low. Vpr is dispensable for the infection of quiescent, naive T cells in lymphoid histocultures but plays an important role in tissue macrophages (Ecksvein *et al.*, 2001). Similar results have been obtained in various cell types arrested in G_2 by γ-irradiation (Bouyac-Bertoia *et al.*, 2001).

4.3.5   Integrase

To insert the reverse-transcribed retroviral genome into the host chromosome, the viral enzyme IN must accompany the viral PIC into the cell nucleus. Lentiviruses do not rely on mitosis to gain access to the host cell chromatin and must import IN into the nucleus with the other components of the PIC. Nuclear import signals have been described in HIV-1 IN (Gallay *et al.*, 1997; Tsurutani *et al.*, 2000) and in avian sarcoma virus IN (Kukolj *et al.*, 1997, 1998). Avian sarcoma virus appears to be unique among the oncoretroviruses, in that it can infect non-dividing calls, albeit less efficiently than HIV-1 (Hatzüoannou and Goff, 2001).

In summary, while oncoretroviruses, with the exception of avian sarcoma virus, rely on mitosis to gain entry into the cell nucleus, viral access to nuclear contents may not be sufficient for successful infection (Bouyac-Bertoia *et al.*, 2001; Seamon *et al.*, 2002). The lentiviruses appear to have redundant PIC-associated NLSs in the form of MA, IN and the central DNA flap, in addition to Vpr in the primate lentiviruses. These import signals could either act in concert or differ in their importance and role, depending on the type of target cell. These scenarios of interdependent nuclear import signals are beginning to be addressed. For example, Fassati *et al.* (2003) examined the import of the entire HIV-1 PIC in a defined nuclear import system, rather than determining the import factor dependence of the individual components. This work revealed that importin 7, an import receptor for ribosomal proteins and histone H1, plays a major role in import of the PIC. This original finding can now be further dissected in other defined systems in the presence and absence of various viral components.

4.4   Inside the nucleus

4.4.1   Integration

The next step in the retroviral life cycle is integration, the insertion of the double-stranded viral genome into the host chromosome. This process, as described above, occurs in the nucleus and depends on nuclear import of the PIC or its entry into the nucleus during mitosis.

Several lines of investigation have shown that the cell cycle and/or nuclear entry can affect integration by interfering with reverse transcription. HIV-1 can enter quiescent T cells, but reverse transcription is not completed until cells are allowed to transit to the G_{1b} phase of the cell cycle, just before host DNA replication in the S phase (Korin and Zack, 1998). However, both HIV-1 and murine sarcoma virus can enter the nucleus before reverse transcription is completed (Lee and Coffin, 1991; Miller *et al.*, 1995). Bouyac-Bertoia *et al.* (2001) showed that HIV-1 can enter the nucleus during mitosis but appears to rely on import to establish a productive infection. This finding suggests that engagement of the nuclear pores provides a signal to the PIC that is required for integration. Murine leukaemia virus relies on mitosis for nuclear entry but cannot integrate until some point after this phase has been completed (Roe *et al.*, 1993); however, it completes reverse transcription and processing of the genome in preparation for integration while still in the cytoplasm (Roe *et al.*, 1993, 1997; Brown *et al.*, 1987, 1989; Fujiwara and Mizunchi, 1988; Roth *et al.*, 1989) Finally, Rous sarcoma virus (RSV) reverse transcription largely occurs in the cytoplasm (Varmus *et al.*, 1974) but also appears to be dependent on nuclear entry for completion (Lee and Coffin, 1991).

Although they are dependent on mitosis for entry into the nucleus, oncoretroviruses, like lentiviruses, may be influenced by the cell cycle. The availability of DNA synthesis precursors, cellular co-factors, and accessibility of the host chromatin all change during the cell cycle. Cellular co-factors may include components of DNA repair pathways that are thought to participate in the integration process, and may likewise function only during certain phases of the cell cycle. The association of host cell proteins with chromatin determines the accessibility of PICs to chromosomal DNA. Both HIV and MLV integrate preferentially into histone-bound DNA rather than into naked DNA (Pryciak *et al.*, 1992a, 1992b, 1992c) because the histones bend or kink the DNA and thereby expose the phosphodiester bonds in the major groove that IN targets. Similarly, the highly condensed chromatin in mitotic cells may not be compatible with retroviral integration. Other DNA binding proteins block integration into both the host chromatin (Bushman, 1994; Pryciak and Varmus, 1992, 1994; Muller and Varmus, 1994; Muller *et al.*, 1993) and viral cDNA (Chen and Engelman, 1998).

Recent work has provided new and more refined insights into viral integration patterns within cells (Schroder *et al.*, 2002; Wu *et al.*, 2003). First, Schroder mapped the integration sites of over 500 proviruses and found that most were in genes actively expressed upon HIV-1 infection (Schroder *et al.*, 2002). MLV-based vectors also preferentially targeted actively transcribed genes, integrating at or near the promoter (Wu *et al.*, 2003). This integration pattern increases the likelihood of gene dysregulation and therefore the possibility of integration-associated pathology. The differences between HIV-1 and murine leukaemia virus are too numerous to permit speculation about the factors responsible for this subtle yet important difference in integration site selection.

Integration consists of several coordinated steps and does not require exogenous energy to complete (Katzman *et al.*, 1989). The sequences that are recognized in the viral DNA are minimal. The most constant sequence is a cytidine-adenosine (CA) dinucleotide attachment site that is generally two residues from the 3′ end of the reverse transcript (Figure 4.5). Additionally, IN also recognizes 11–20 bp sequences upstream of the CA dinucleotide (Roth *et al.*, 1989; Katzman *et al.*, 1989; Sherman and Fyfe, 1990). The IN must process the termini of the newly synthesized viral DNA. The processing most frequently consists of removing two nucleotides from the 3′ termini by nucleophilic attack with a water molecule (Engelman *et al.*, 1991; Vink *et al.*, 1991). This leaves a reactive 3′-terminal hydroxyl group for the subsequent joining reaction. The oxygen from the hydroxyl group acts as a nucleophile to attack the phosphodiester bond in the acceptor DNA strand (Engelman *et al.*, 1991; Vink *et al.*, 1991) (Figure 4.5(c)). The 3′ ends of both viral cDNA strands are thus attached to the 5′ termini of the acceptor DNA strands, typically four to six bases apart (Figure 4.5(d)). This distance typically corresponds to an insertion in the major groove on the same face of the helix (Brown, 1997). At this stage of integration, the 5′ overhang of the viral DNA is likely to be mismatched with the acceptor DNA, leaving a gap between the 3′ end of the acceptor DNA and the viral DNA. This gap is probably filled by polymerization from the 3′ terminus of the host DNA, displacement of the mismatch and nick repair; the precise mechanism for this step remains to be defined. This integrated viral DNA is termed the provirus.

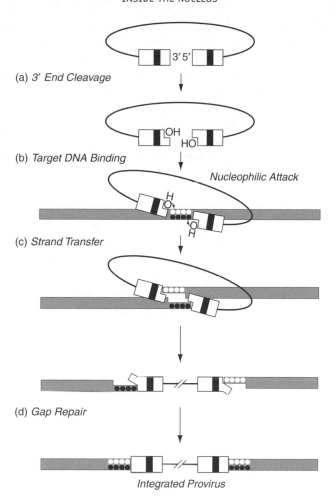

(a) *3′ End Cleavage*

(b) *Target DNA Binding*

Nucleophilic Attack

(c) *Strand Transfer*

(d) *Gap Repair*

Integrated Provirus

Figure 4.5 Integration. IN recognizes sequences 13–22 bp from the end of the completed reverse transcription product. The termini are processed by IN to remove two nucleotides from the 3′ termini by nucleophilic attack with a water molecule (a). A reactive 3′ hydroxyl group then carries out a nucleophilic attack on the host DNA internucleotide phosphate (b), (c). These reactions are usually four to six nucleotides apart and result in mismatched sequences. Cellular enzymes probably repair both the mismatch and the gap (d)

IN is a cleavage product of the Gag–Pol polyprotein. A stop codon between the Gag and Pol reading frames assures that each virion contains more of the structural Gag proteins than of the enzymatic Pol proteins. Pol is translated only when the stop codon function is suppressed, either by a suppressor tRNA or by a frame-shifting reaction mediated by sliding of the ribosome complex, an event that takes place once for every 20–50 polyproteins produced. IN is found at the carboxyl terminus of the polyprotein in every virus characterized except avian sarcoma/leucosis virus (ASLV), and is released by proteolytic cleavage during viral maturation.

IN can be divided into three domains: a 50 amino acid amino-terminal domain, the core domain consisting of the next 185 amino acids, and the carboxy-terminal domain. The amino-terminal domain contains an HHCC motif that is characteristic of a zinc-binding domain similar to that in zinc-finger structures. Mutation of the histidines or cysteines interferes with the proper function of IN (Engelman and Craigie, 1992; Engelman *et al.*, 1995) but this region is not involved in DNA binding (Roth *et al.*, 1990; Khan *et al.*, 1991; Schauer and Billich, 1992; Bushman *et al.*, 1993; Vincent *et al.*, 1993; Vink *et al.*, 1993; Woerner *et al.*, 1992; Mumm and Grandgenett, 1991) and does not contain the catalytic function of the enzyme. Instead, the amino-terminal domain appears to be critical for IN multimerization (Zheng *et al.*, 1996). The IN core domain is characterized by a $D–X_{(39–58)}–D–X_{35}–E$ motif. X-ray crystal-lography studies of HIV-1 and avian sarcoma virus core subunits indicate that these residues are all found in close proximity (Dyda *et al.*, 1994; Bujacz *et al.*, 1995, 1996) and that the two aspartic acids coordinate a divalent cation (Bujacz *et al.*, 1996). The structure of this domain resembles that present in a number of nucleases (Bujacz *et al.*, 1995; Rice *et al.*, 1996; Yang and Steitz, 1995).

The carboxyl domain shows the most diversity among the INs characterized (Johnson *et al.*, 1986; Lutzke *et al.*, 1994). It appears to contribute to DNA binding (Woerner *et al.*, 1992; Mumm and Grandgenett, 1991; Lutzke *et al.*, 1994; Engelman *et al.*, 1994; Woerner and Marcus-Sekura, 1998), although IN mutants lacking this domain continue to associate with DNA (Engelman *et al.*, 1994).

4.4.2 Transcription and Tat

The full-length viral RNA that is packaged as the viral genome contains promoter and enhancer sequences exclusively within the U3 region at the 3′ end (Figure 4.4(b), (j)). Depending upon the type of virus, the polyadenylation signal is either in the U3 region at the 3′ end (HTLV-1, MMTV and ASLV) or in the R region at both ends (HIV-1, Mo-MLV and SNV). Transcription begins at the start of the 5′ R region (Figure 4.4(a), (i), ↦). As described above, the sequences unique to one end of the viral RNA are duplicated onto both ends of the reverse transcription products (Figure 4.4(h), (p)).

Retroviral transcription depends on RNA polymerase II and therefore centres on a TATA box (Figure 4.6(a)). This DNA sequence is the core promoter element around which the transcription complex is assembled. The TATA box is recognized and bound by the 38 kDa TATA-binding protein (TBP), which then assembles with a collection of other proteins known as TBP-associated factors or TAFs. The resulting complex is termed transcription factor IID (TFIID) which, together with RNA polymerase II, constitutes the preinitiation complex, which, with the addition of ATP as an energy source and ribonucleotide triphosphates, can initiate RNA synthesis. Elongation of the RNA requires phosphorylation of the RNA polymerase large subunit carboxy-terminal domain.

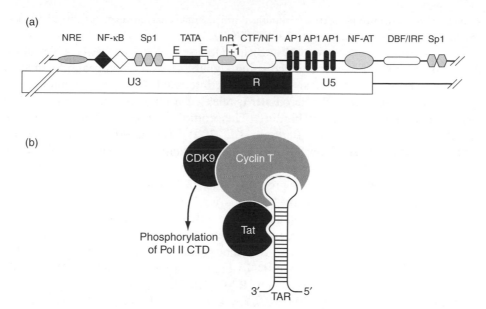

Figure 4.6 Regulation of HIV-1 expression. The 5′ LTR of HIV-1 is shown here to illustrate the transcription regulatory elements of a well-characterized complex retrovirus (a). The positions of regulatory protein binding sites are indicated relative to the three regions of the LTR. The RNA transcribed from this promoter is further involved in regulating RNA expression. The first 59 residues of the transcribed RNA form a stem–loop structure, to which the viral Tat protein binds. Tat recruits the kinase regulator cyclin T, which in turn recruits the Cdk9 kinase (b). The Cdk9–cyclin T complex phosphorylates the C-terminal domain of RNA Pol II and thereby makes it processive

The HIV-1 core promoter, which includes both a TATA box and an initiator region, spans from 30 bp before (−30) to 50 bp after (+50) the transcription initiation site. A pair of binding sites for basic helix–loop–helix proteins (E boxes) flank the TATA box and aid in the transition from basal to activated transcription (Ou *et al.*, 1994).

The rate of transcription is influenced by both nearby promoter elements and more distant sequences termed enhancer and negative regulatory elements. These sequences bind factors that alter the physical configuration of DNA and thereby affect the assembly of other protein complexes, influence chromatin remodelling, help to recruit and/or stabilize the initiation complex or the RNA polymerase, influence promoter clearance and enhance polymerase processivity.

The overall size of retroviral genomes, commonly around 10 kb, constrains the arrangement of the promoters and enhancers, which are contained largely within the 1 kb U3 region. Some additional regulatory elements have been defined within the R and U5 downstream regions. HIV-1 promoter sequences between −46 and −78 provide a GC-rich sequence, characteristic of Sp1 binding sites (Jones *et al.*, 1986), which work in concert with the HIV-1 Tat protein to enhance transcription (Sune and Garcia-Blanco,

1995). Vpr also binds Sp1 to enhance immediate-early transcription of HIV-1 (Sawaya *et al.*, 1998; Wang *et al.*, 1995, 1996).

5′ of the Sp1 sites are a pair of NF-κB sites (Jones and Peterlin, 1994), which allow the virus to respond to immune activating signals as well as to pro-inflammatory signals, such as tumour necrosis factor-α and interleukin-1. Upon stimulation, the cytoplasmic p50/p65 NF-κB heterodimer is released from IκBα, allowing its nuclear import and subsequent DNA binding. This complex enhances both the rate of initiation and RNA elongation (Barboric *et al.*, 2001). The same enhancer region also contains recognition motifs for the nuclear factor of activated T cells (NFAT), which can assemble with AP-1 to form a fully functional transcription complex (Crabtree, 1999). NFAT is activated and proceeds into the cell nucleus in response to dephosphorylation by calcineurin.

The 5′-end of the U3 region can also downregulate viral transcription. Removal of the region, termed the negative regulatory element, can enhance transcription both *in vivo* (Lu *et al.*, 1989) and *in vitro* (Okamoto *et al.*, 1990). Since this element has recognition sites for a wide variety of DNA-binding proteins, including TCF-1, USF, NF-AT, ILF1, GRE, AP-1, COUP-TF, RAR, NRT1/2, T-cell factor B, myb and GATA3 (Atchison and Perry, 1987; Cooney *et al.*, 1991; Franza *et al.*, 1988; Ivanov *et al.*, 1989; Lu *et al.*, 1990; Poeschla and Looney, 1998; Reddy and Dasgupta, 1992; Waterman and Jones, 1990), it has been difficult to identify the specific origin of the transcriptional downmodulation. The cell type-specific availability and interplay among these factors may contribute to this negative effect.

Assembly of the RNA Pol II preinitiation complex, as described above, results in the production of short non-polyadenylated RNA transcripts. Phosphorylation of the RNA polymerase II large subunit carboxy-terminal domain at a repeated Tyr–Ser–Pro–Thr–Ser–Pro–Ser sequence (Shilatifard, 1998) is required for the efficient production of full-length, polyadenylated RNAs. This modification, which alters the preinitiation complex into an effective elongation complex, takes place efficiently in lentiviruses only in the presence of Tat (Rittner *et al.*, 1995; Zhou and Sharp, 1995). Tat is a 14 kDa, 86 amino acid protein that is encoded by a highly spliced viral mRNA. The coding sequence consists of two exons that flank the *env* coding sequence.

How does Tat function? The HIV LTR binds so-called negative transcription elongation factors (Price, 2000) that inhibit activation of the RNA Pol II elongation complex, such as negative elongation factor (Yamaguchi *et al.*, 1999) and the DRB sensitivity-inducing factor (Wada *et al.*, 1998). The function of Tat, which is essential for viral replication (Emerman and Malim, 1998; Cullen, 1998; Karu, 1999; Taube *et al.*, 1999; Adams *et al.*, 1994; Finzi *et al.*, 1999; Peng *et al.*, 1998), counteracts the negative factors, not to increase RNA Pol II binding but rather but to increase elongation activity of this complex.

Unlike many factors that enhance gene expression, Tat binds to viral RNA rather than to proviral DNA. The binding target is an RNA stem–loop structure that is formed by the first 59 residues of the transcribed viral genome (Rosen *et al.*, 1985; Muesing *et al.*, 1987; Hauber and Cullen, 1988). Notable features of the HIV-1 Tat response region (TAR) are a three-nucleotide, uracil-rich bulge within the stem and a hexanucleotide central loop (Jaeger and Tinoco, 1993). Tat, which binds the TAR bulge with an arginine-rich motif, recruits cyclin T1 with its activation domain. Through another arginine-rich motif, cyclin T1, a regulatory subunit for the kinase Cdk9, then associates with the TAR loop (Wei *et al.*, 1998). The association of cyclin T1 with the central TAR bulge further enhances the affinity of this interaction. These proteins recruit Cdk9 to complete the positive transcription elongation factor b complex (Peng *et al.*, 1998; Wei *et al.*, 1998). The Cdk9 kinase then mediates hyperphosphorylation of the C-terminal domain of RNA Pol II and leads to increased elongation of the Pol II complex (Figure 4.6B).

Transcriptional activators like Tat are common among the complex retroviruses; they may facilitate replication in quiescent cells and broaden the host range of these viruses. The simple retroviruses must rely exclusively on specific host factors for transcriptional activation.

4.4.3 RNA processing and Rev

Retroviruses use genetic material very efficiently. In simple retroviruses, this economy is reflected not only in the strategy for reverse transcription, where sequences unique to each terminus of the genomic RNA are duplicated to make fully functional LTRs, but also in the use of the resulting 10 kb provirus to make the proper ratio of viral components, including full-length genomic RNAs for packaging into new virions. The oncoretroviruses generally have a single splice donor and a single splice acceptor that are used to generate the mRNA encoding the Env products, while full-length, unspliced mRNAs are used both as translation templates for Gag and Gag–Pol polyproteins and for viral genomic RNA.

The complex retroviruses, in addition to expressing the Gag, Gag–Pol and Env polyproteins, also express additional gene products, including Tat, Rev, Vif, Vpu, Vpr and Nef, from proviruses that are of comparable size. The additional economy derives from the use of both overlapping reading frames and a much more extensive set of splice donors and acceptors. The use of these splice sites is temporally controlled by a strategy that involves the host RNA splicing machinery, viral factors and nuclear export.

For the RNA processing scheme of even the simple retroviruses to succeed, viral RNA splicing must be suboptimal. Suboptimal splicing allows the nuclear export of sufficient unspliced RNAs for the translation of Gag and Gag–Pol polyproteins and for the production of viral genomes. The spliced products become Env templates.

Complex retroviruses, such as HIV-1, produce numerous RNA products. These fall into three categories: (a) the multiply spliced transcripts, including those for Tat, Rev and Nef; (b) the singly spliced transcripts, including those for Vif, Vpr, Vpu and Env; and (c) the unspliced transcripts for Gag and Gag–Pol and for the viral genomes. The splicing of the complex retrovirus RNAs is determined in part by the suboptimal nature of splice sites, but is also controlled by the viral Rev protein.

Tat and Nef are expressed early after infection, in some instances before integration (Wu and Marsh, 2001). While the function of Nef in early infection remains unclear, Tat, as described above, promotes RNA elongation, thereby increasing production of Tat, Nef and Rev, all from multiply spliced mRNAs. While the viral splice sites are inefficient, the likelihood of splicing increases upon prolonged exposure to nuclear splice factors. HIV-1 benefits from the time-dependence of inefficient splice-site selection to control the temporal expression of gene products through the interplay between Rev, a specific RNA exporter; the Rev response elements, a highly structured RNA element in the *env* gene (Malim *et al.*, 1989), and nuclear RNA retention signals (Schwattz *et al.*, 1992a, 1992b).

The efficiency of splicing may also be influenced by a direct interaction between Rev and the cellular splicing machinery. Rev bound to the Rev response element further inhibits splicing through its physical association with alternative splicing factor/splicing factor 2 (ASF/SF2) (Powell *et al.*, 1997) and its associated p32 protein (Luo *et al.*, 1994). Rev interferes with the incorporation of small nuclear ribonucleoprotein complexes into spliceosomes (Chang and Sharp, 1989; Bohnlein *et al.*, 1991).

Rev is a 116 amino acid, 18 kDa phosphoprotein that, like Tat, is encoded by Env-flanking exons. The amino-terminus contains a basic domain that functions in binding to the Rev response element (Bohnlein *et al.*, 1991; Daly *et al.*, 1989; Hope *et al.*, 1990; Malim and Cullen, 1991; Olsen *et al.*, 1990; Zapp *et al.*, 1989, 1991) and contains the Rev nuclear import signal (Kubota *et al.*, 1989; Malim *et al.*, 1989; Perkins *et al.*, 1989). Interestingly, the arginine-rich import signal functions, like that of Tat (Truant and Cullen, 1999), by binding directly to importin-β, rather than first binding to importin-α (Truant and Cullen, 1999; Henderson and Percipalle, 1997). The overlap between the NLS and the Rev response element binding domain (Henderson and Percipalle, 1997) results in masking of that signal when RNA is bound. This ensures that Rev is not returned into the nucleus before the RNA cargo has been released.

The carboxyl terminus, formerly referred to as the activation domain, contains a leucine-rich sequence characteristic of a nuclear export signal (Fischer *et al.*, 1995; Meyer *et al.*, 1996; Wen *et al.*, 1995). In fact, the Rev nuclear export signal provided a paradigm for the identification of leucine-rich nuclear export signal motifs in other proteins and helped lead to the identification of the importin-β-like, leptomycin B-sensitive protein exportin-1/Crm1 as a major mediator of nuclear export (Stade *et al.*, 1997).

D-type retroviruses, including the Mason–Pfizer Monkey virus (MPMV) and Simian retrovirus type 1, encode a *cis*-acting RNA element (CTE) that serves a Rev-like function (Zolotukhin *et al.*, 1994; Bray *et al.*, 1994). These highly structured RNA segments (Tabernero *et al.*, 1996; Ernst *et al.*, 1997a, 1997b) directly engage the RNA export pathway through TAP, the human homologue of the yeast Mex67 RNA export protein (Segref *et al.*, 1997; Gruter *et al.*, 1998). Another cellular factor, RNA helicase A, also appears to mediate CTE-dependent RNA export from the nucleus (Tang *et al.*, 1997).

4.4.4 Intranuclear Vpr function: induction of cell-cycle arrest

Vpr also induces cell cycle arrest. This function was discovered as a result of the observation that *vpr*, which is among the most highly conserved genes in primary HIV-1 isolates (Meyers *et al.*, 1994), almost invariably acquired stop codons after viruses were propagated in tissue culture (Rogel *et al.*, 1995; Nakaya *et al.*, 1994). Work by several laboratories revealed that viruses with intact Vpr caused cells to stop growing, leading to their eventual death (Rogel *et al.*, 1995; Nakaya *et al.*, 1994; Planelles *et al.*, 1995). Additional investigation revealed that the cells did not die from generalized cytopathicity but rather were blocked in the G_2 phase of the cell cycle, after replication of most or all cellular genomic DNA. Further, Vpr was able to cause cell-cycle arrest in the absence of other viral gene products (Jowett *et al.*, 1995; Planelles *et al.*, 1995).

The biological importance of G_2 arrest in the viral life cycle is reflected in *in vivo* studies. Cells from acutely HIV-infected patients exhibit G_2 arrest (Sherman *et al.*, unpublished data), indicating that this effect of Vpr is not restricted to *in vitro* infections. Further, while viral pathogenesis is maintained in monkeys infected with Vpr-deleted virus (Gibbs *et al.*, 1995), the reopening of closed Vpr reading frames in other simian studies (Hoch *et al.*, 1995), and in a laboratory worker accidentally infected with HIV (Goh *et al.*, 1998) indicates the importance of Vpr and its G_2 arrest functions to the virus.

4.4.5 Regulation of the G_2 to mitosis transition

Gap1 and Gap2 (G_1 and G_2) provide important checkpoints in the cell cycle. These pauses in the cellular replication process provide an opportunity to confirm that critical functions, such as mitosis (M-phase) and host chromatin replication (synthesis or S-phase) have been completed satisfactorily before the next phase of the cell cycle is begun. Defects detected at these checkpoints can be repaired, if they are not too extensive, and the cell cycle can then resume.

A bimolecular complex of cyclin-dependent kinase Cdc2 and its regulator cyclin B1 centrally control the target of Vpr, the G_2-to-mitosis transition. Because the proper completion of events preceding mitosis is critical for cell survival and genetic stability, a number of safeguards have evolved to protect against premature activation of the cyclin B1/Cdc2 complex. Once the cells are prepared to progress from G_2, however, vigorous, irreversible activation of the complex is required (reviewed in O'Farrell, 2001).

Cdc2 pairs with cyclins A and E during early interphase and with cyclin B1 during G_2 and M (reviewed in Smits and Medema, 2001). Cdc2 in the inactive cyclin B1–Cdc2 complex is phosphorylated at residues Thr14, Tyr15 and Thr161 (reviewed in Takizawa and Morgan, 2000). The active complex is dephosphorylated at Thr14 and Tyr15 but phosphorylated at Thr161. The phosphorylation status of Cdc2 is determined by at least three kinases and a phosphatase. The CAK complex is responsible for phosphorylation of Thr161, while the Wee1 and/or Myt1 kinases are responsible for the phosphorylations of Thr14 and Tyr15. These phosphorylations hold the kinase complex in an inactive state. Activation results from dephosphorylation of Thr14 and Tyr15, which is mediated by the phosphatase Cdc25C. Wee1 and Cdc25C are also regulated by phosphorylation, Wee1 being inactivated and Cdc25C being activated. Both proteins are targets for the cyclin B1–Cdc2 kinase complex, which, when activated, results in a positive feedback loop that propagates Cdc2 dephosphorylation. This activation and feedback constitute a positive, irreversible switch for the initiation of mitosis (Figure 4.7).

The cell cycle regulators are physically segregated within cells to provide yet another level of control (reviewed in Takizawa and Morgan, 2000). In mammalian cells, Wee1 is in the nucleus throughout the interphase, while Cdc25C and Myt1 are

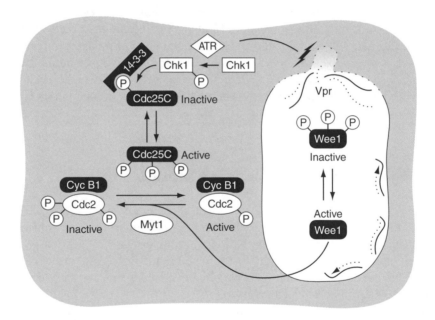

Figure 4.7 Regulation of the G_2-to-mitosis transition and a model for Vpr-mediated G_2 arrest. Cyclin B1 and Cdc2 form a kinase complex that is central to the progression from G_2 to M. In its inactive state, Cdc2 is phosphorylated on residues 14, 15 and 161. Dephosphorylation at residues 14 and 15 is required for activation. Both Wee1 and Myt1 kinases are responsible for maintaining the inactivating phosphorylations, while Cdc25C, a phosphatase, opposes this effect. Both Wee1 and Cdc25C are regulated, as depicted, by phosphorylation. Defects in the nuclear envelope, promoted by expression of the HIV-1 protein Vpr, allow escape from the nucleus of Wee1. This mixing of cell cycle regulatory components could serve to block progression into mitosis. Alternatively, damage to the nuclear structure could interfere with DNA synthesis and thereby trigger activation of ATR, subsequent phosphorylation and activation of Chk1 kinase. Chk1, in turn, phosphorylates Cdc25C on serine 216, which allows its association with a 14–3–3 protein and retention in the cytoplasm, where it is inactive

in the cytoplasm. Cyclin B1 is also in the cytoplasm, associated with microtubules (Jackman *et al.*, 1995). Cdc2, on the other hand, is in the nucleus, paired with other cyclins for at least a portion of the interphase (reviewed in Smits and Medema, 2001). Cdc25C has characteristics of a shuttling protein (Dalal *et al.*, 1999; Graves *et al.*, 2001), but shifts in subcellular localization during the interphase are not readily apparent in time-lapse microscopy experiments (de Noronha, unpublished data). Both Cdc2 and Cdc25C pair with cytoplasmic 14–3–3 proteins upon phosphorylation induced by DNA damage (Chan *et al.*, 1999; Peng *et al.*, 1997).

The subcellular distribution of the key cell cycle regulators at the time of cyclin B1–Cdc2 activation has yet to be determined. A plausible scenario, however, is that Cdc2, hyperphosphorylated by Wee1 in the nucleus, is exported to the cytoplasm, where it joins cyclin B1 on the microtubules. A less likely alternative is that Cdc2 is exported from the nucleus in a hypophosphorylated state to be phosphorylated by Myt1 in the cytoplasm. There the complex becomes a target for dephosphorylation by Cdc25C before entering the nucleus attached to the mitotic spindle.

While the enzymes most proximal to the mitotic switch have been well characterized, the upstream regulators of their functional states are only now being identified. These include sensors of DNA damage, of DNA replication, and of other factors that determine whether cells are prepared to initiate mitosis.

4.4.6   Potential mechanisms of Vpr-mediated arrest

Understanding the mechanism by which Vpr induces cell cycle arrest is crucial for determining the specific role of Vpr in the viral life cycle. G_2 arrest may be important for efficient viral replication under certain conditions, regardless of the mechanism by which it is induced, or it may come about as a byproduct of an unrelated process that arrests actively dividing cells. The central question is whether Vpr directly engages regulators of the cell cycle or whether it otherwise alters the cell and thus triggers a cell cycle-arresting damage response.

The first characterizations of Vpr-mediated cell-cycle arrest demonstrated that cells expressing Vpr have a 4N DNA content, hypophosphorylated Cdc25C and hyperphosphorylated Cdc2 (Jowitt *et al.*, 1995; Re *et al.*, 1995; He *et al.*, 1995). Thus, Vpr expression permits the completion, or near-completion, of cellular DNA synthesis but prevents activation of the two most proximal regulators of progression from G_2 to mitosis: the master regulator, Cdc2, and the phosphatase responsible for its activation, Cdc25C. Mutant strains of *Schizosaccharomyces pombe* have been used to further characterize Vpr-mediated cell-cycle arrest. These studies have revealed the involvement of additional G_2–M transition regulators, including Wee1, PP2A, Cdc25C and a 14–3–3 protein, in promoting Vpr-mediated arrest (Masuda *et al.*, 2000; Elder *et al.*, 2000, 2001, 2002).

Efforts to identify mammalian host cell molecules that are specifically engaged by Vpr to promote cell-cycle arrest have yielded numerous candidates. HHR23A, the human homologue of the yeast DNA repair protein RAD23, interacted with Vpr in yeast two-hybrid and yeast interaction trap screens (de Noronha, unpublished data; Gragerov *et al.*, 1998) and its overexpression partially blocked Vpr-mediated cell-cycle arrest (Gragerov *et al.*, 1998). The human homologue of MOV34, a proteasomal subunit, was similarly identified as a Vpr binding protein (Mahalingam *et al.*, 1998). Upon expression of Vpr, hMOV23 was redistributed from the nucleus to the perinuclear region (Mahalingam *et al.*, 1998). The DNA repair enzyme uracil DNA glycosylase (UNG) also interacted with Vpr in both the yeast two-hybrid system (Bouhamdan *et al.*, 1996) and a yeast interaction trap screen (de Noronha, unpublished data). UNG is an intriguing interaction candidate because none of the lentiviruses that contain Vpr encodes a dUTPase, another enzyme that guards against misincorporation of uracil. Finally, Zhao *et al.* used biochemical means to identify VprBP as an interaction partner for Vpr (Zhao *et al.*, 1994). This protein is thought to interfere with Vpr-mediated arrest by sequestering Vpr in the cytoplasm (Zhang *et al.*, 2001).

Time-lapse video fluorescence microscopy has been used to determine whether the expression of Vpr alters the subcellular distribution of key cell cycle regulators, including Wee1, Cdc25C, cyclin B1 and Cdc2. Rather than causing a stable redistribution of one or more of these regulatory proteins, Vpr weakened the nuclear envelope, causing the formation of nuclear herniations that intermittently ruptured and produced transient admixing of the otherwise segregated nuclear and cytoplasmic components (de Noronha *et al.*, 2001). These herniations were most prominent at the apices of the oblong nuclei. The herniations emanated from defects in the nuclear lamina. Cell-cycle arrest could thus result from the unscheduled intermingling of cell-cycle regulators after nuclear rupture or from interference with DNA synthesis resulting from disruptions in the lamin structure (Moir *et al.*, 2000). Alternatively, other damage signals resulting from the nuclear envelope abnormalities may halt the progression from late S phase or G_2 into mitosis. Efforts to characterize signals upstream of Cdc25C that are responsible for triggering Vpr-mediated G_2 arrest (Roshal *et al.*, 2003) have shown that the DNA damage signalling protein ATR is activated in response to Vpr expression, consistent with S-phase DNA replication problems, such as stalled replication forks and loss of topoisomerase function (Lupardus *et al.*, 2002; Cliby *et al.*, 2002).

Our model of Vpr-mediated G_2 arrest is outlined in Figure 4.7. Briefly, Vpr destabilizes the nuclear envelope, causing either temporally inappropriate mixing of the nuclear and cytoplasmic regulators of the G_2–M transition, interference with host chromatin replication, or both. In the first scenario, Wee1 function likely predominates to assure that the cyclin B1–Cdc2 complex remains inactive. On the other hand, if damage to the nuclear envelope interferes with DNA replication, ATR is activated, leading to the phosphorylation and activation of Chk1. This kinase phosphorylates Cdc25C serine 216, which promotes its association with a 14–3–3 protein and consequent retention in the cytoplasm, where it cannot promote cyclin B1–Cdc2 activation.

Although the mechanism by which Vpr weakens the nuclear lamin structure has not been resolved, a large fraction of nuclear Vpr is tightly associated with the nucleoskeleton (Lu *et al.*, 1993; de Noronha, unpublished data). The nucleoskeleton is a highly stable structure that remains after contents of the nucleus have been exposed to detergent treatment, DNAse and RNAse digestion, and high salt extraction. Nuclear lamins are components of, and may be stabilized by, this structure. While the purpose of the Vpr association with the nucleoskeleton remains unclear, a number of interesting possibilities exist. Vpr could function as an adaptor to anchor other molecules to the nucleoskeleton, including proteins or nucleic acids. In this capacity, Vpr could help guide the PIC to specific areas within the nucleus or modify transcription (Hoover *et al.*, 1996). In these scenarios, destabilization of the nucleoskeleton may be a secondary effect. Alternatively, Vpr could specifically weaken components of the nucleoskeleton, such as the lamin structure or the interface between the lamin and the nuclear pores (Smythe *et al.*, 2000) to allow efficient passage of the large PIC into the nucleus or to cause cell-cycle arrest. Finally, since HIV and other viruses use the nuclear envelope component barrier-to-autointegration factor as an inhibitor of autointegration, Vpr, which can shuttle into and out of the nucleus (Sherman *et al.*, 2001; Jenkins *et al.*, 2001), could subvert this protein and thereby weaken the nucleus.

4.4.7 Benefits of G_2 arrest

The specific role of G_2 arrest in the viral life cycle has yet to be determined, but a few intriguing theories are emerging. Goh *et al.* showed that the G_2 phase provides a favourable milieu for viral production, enhancing transcription from the viral LTR regardless of whether the arrest was induced by Vpr or by other means (Goh *et al.*, 1998). As the authors indicate, the benefits of increased viral production may well outweigh the loss of cell proliferation due to G_2 arrest.

Another attractive theory, which was recently discounted, posited that Vpr inhibits cytotoxic T cell-mediated killing. Granzyme B-mediated apoptosis causes the rapid and temporally inappropriate activation of Cdc2 (Shi *et al.*, 1994). Premature activation of Cdc2, either directly or through its upstream activator Cdc25C, leads to inappropriate entry into mitosis and results in mitotic catastrophe, a process that leads to apoptosis (Heald *et al.*, 1993; Krek and Nigg, 1991). Recently, Lewinsohn *et al.* (2002) investigated whether Vpr, by preventing activation of Cdc25C and Cdc2, protects cells from granzyme B-mediated cytotoxic T lymphocyte killing. In this study, cells infected with virus, with or without Vpr, were equally susceptible to CTL-mediated killing.

4.5 Leaving the nucleus and the cell

After the viral RNA is exported from the cell nucleus, numerous events take place to complete the viral life cycle that are beyond the scope of this chapter. These include translation and, in the case of the envelope glycoproteins, extensive protein modification, assembly at specific locations at the cytoplasmic membrane and budding from the cell. A more comprehensive description of these events is available elsewhere (Coffin *et al.*, 1997).

4.6 Future perspectives

Because of their replication strategies, retroviruses are dependent on the cell nucleus for survival. Integration of the viral genome into host chromatin assures that it is propagated with that of the host and that the virus cannot be destroyed alone. Highly active antiretroviral therapy (HAART) can hinder viral replication to such a level that the supply of infected cells is not replenished but cannot reach all cellular compartments and cannot effectively eradicate long-lived, latently infected cells.

HAART targets viral enzymes that are similar to host enzymes in both structure and function. These similarities make it difficult to design specific, minimally toxic drugs. Our understanding of the function of other viral proteins that are less similar to cellular proteins will therefore be critical to expanding the number of specific drug targets. Some of these potential targets are active in the nucleus, including Tat, Rev and Vpr. Small chemical inhibitors have already been identified for the two better-characterized proteins, Tat (Chao *et al.*, 2000) and Rev (Lind *et al.*, 2002). As we gain a better understanding of the function of Vpr and its interaction with nuclear components, it may well become a therapeutic target as well.

References

1. Bowerman, B. *et al.* (1989). A nucleoprotein complex mediates the integration of retroviral DNA. *Genes Dev* **3**(4), 469–478.
2. Bushman, F. D. (1994). Tethering human immunodeficiency virus 1 integrase to a DNA site directs integration to nearby sequences. *Proc Natl Acad Sci USA* **91**(20), 9233–9237.
3. Lee, M. S. and Craigie, R. (1994). Protection of retroviral DNA from autointegration: involvement of a cellular factor. *Proc Natl Acad Sci USA* **91**(21), 9823–9827.
4. Farnet, C. M. and Haseltine, W. A. (1991). Determination of viral proteins present in the human immunodeficiency virus type 1 preintegration complex. *J Virol* **65**(4), 1910–1915.
5. Karageorgos, L., Li, P. and Burrell, C. (1993). Characterization of HIV replication complexes early after cell-to-cell infection. *AIDS Res Hum Retroviruses* **9**(9), 817–823.
6. Bukrinsky, M. I. *et al.* (1993a). Association of integrase, matrix, and reverse transcriptase antigens of human immunodeficiency virus type 1 with viral nucleic acids following acute infection. *Proc Natl Acad Sci USA* **90**(13), 6125–6129.
7. Farnet, C. M. and Bushman, F. D. (1997). HIV-1 cDNA integration: requirement of HMG I(Y) protein for function of preintegration complexes *in vitro*. *Cell* **88**(4), 483–492.
8. Lori, F. *et al.* (1992). Viral DNA carried by human immunodeficiency virus type 1 virions. *J Virol* **66**(8), 5067–5074.

9. Trono, D. (1992). Partial reverse transcripts in virions from human immunodeficiency and murine leukemia viruses. *J Virol* **66**(8), 4893–4900.

10. Zack, J. A. *et al.* (1992). Incompletely reverse-transcribed human immunodeficiency virus type 1 genomes in quiescent cells can function as intermediates in the retroviral life cycle. *J Virol* **66**(3), 1717–1725.

11. Zhu, J. and Cunningham, J. M. (1993). Minus-strand DNA is present within murine type C ecotropic retroviruses prior to infection. *J Virol* **67**(4), 2385–2388.

12. Varmus, H. E. *et al.* (1978). Kinetics of synthesis, structure and purification of avian sarcoma virus-specific DNA made in the cytoplasm of acutely infected cells. *J Mol Biol* **120**(1), 55–82.

13. Lee, Y. M. and Coffin, J. M. (1991). Relationship of avian retrovirus DNA synthesis to integration *in vitro*. *Mol Cell Biol* **11**(3), 1419–1430.

14. Harris, R. S., Petersen-Mahrt, S. K. and Neuberger, M. S. (2002). RNA editing enzyme APOBEC1 and some of its homologs can act as DNA mutators. *Mol Cell* **10**(5), 1247–1253.

15. Mariani, R. *et al.* (2003). Species-specific exclusion of APOBEC3G from HIV-1 virions by Vif. *Cell* **114**(1), 21–31.

16. Stopak, K. *et al.* (2003). HIV-1 Vif blocks the antiviral activity of APOBEC3G by impairing both its translation and intracellular stability. *Mol Cell* **12**(3), 591–601.

17. Harris, R. S. *et al.* (2003). DNA deamination mediates innate immunity to retroviral infection. *Cell* **113**(6), 803–809.

18. Mangeat, B. *et al.* (2003). Broad antiretroviral defence by human APOBEC3G through lethal editing of nascent reverse transcripts. *Nature* **424**(6944), 99–103.

19. Zhang, H. *et al.* (2003). The cytidine deaminase CEM15 induces hypermutation in newly synthesized HIV-1 DNA. *Nature* **424**(6944), 94–98.

20. Lecossier, D. *et al.* (2003). Hypermutation of HIV-1 DNA in the absence of the Vif protein. *Science* **300**(5622), 1112.

21. Bukrinskaya, A. *et al.* (1998). Establishment of a functional human immunodeficiency virus type 1 (HIV-1) reverse transcription complex involves the cytoskeleton. *J Exp Med* **188**(11), 2113–2125.

22. Smith, G. A., Gross, S. P. and Enquist, L. W. (2001). Herpesviruses use bidirectional fast axonal transport to spread in sensory neurons. *Proc Natl Acad Sci USA* **98**(6), 3466–3470.

23. Suomalainen, M. *et al.* (2001). Adenovirus-activated PKA and p38/MAPK pathways boost microtubule-mediated nuclear targeting of virus. *EMBO J* **20**(6), 1310–1319.

24. McDonald, D. *et al.* (2002). Visualization of the intracellular behavior of HIV in living cells. *J Cell Biol* **159**(3), 441–452.

25. Roe, T. *et al.* (1993). Integration of murine leukemia virus DNA depends on mitosis. *EMBO J* **12**(5), 2099–2108.

26. Lewis, P. F. and Emerman, M. (1994). Passage through mitosis is required for oncoretroviruses but not for the human immunodeficiency virus. *J Virol* **68**(1), 510–516.

27. Weinberg, J. B. *et al.* (1991). Productive human immunodeficiency virus type 1 (HIV-1) infection of nonproliferating human monocytes. *J Exp Med* **174**(6), 1477–1482.

28. Zhang, Z. *et al.* (1999). Sexual transmission and propagation of SIV and HIV in resting and activated CD4$^+$ T cells. *Science* **286**(5443), 1353–1357.

29. Ostrowski, M. A. *et al.* (1999). Both memory and CD45RA$^+$/CD62L$^+$ naive CD4(+) T cells are infected in human immunodeficiency virus type 1-infected individuals. *J Virol* **73**(8), 6430–6435.

30. Blaak, H. *et al.* (2000). *In vivo* HIV-1 infection of CD45RA(+)CD4(+) T cells is estab-lished primarily by syncytium-inducing variants and correlates with the rate of CD4(+) T cell decline. *Proc Natl Acad Sci USA* **97**(3), 1269–1274.

31. Amado, R. G. and Chen, I. S. (1999). Lentiviral vectors – the promise of gene therapy within reach? *Science* **285**(5428), 674–676.

32. Mattaj, I. W. and Englmeier, L. (1998). Nucleocytoplasmic transport: the soluble phase. *Annu Rev Biochem* **67**, 265–306.

33. Weis, K. (1998). Importins and exportins: how to get in and out of the nucleus [published erratum appears in *Trends Biochem Sci* 1998 Jul;**23**(7):235]. *Trends Biochem Sci* **23**(5), 185–189.

34. Kalderon, D. *et al.* (1984). A short amino acid sequence able to specify nuclear location. *Cell* **39**(3 Pt 2), 499–509.

35. Robbins, J. *et al.* (1991). Two interdependent basic domains in nucleoplasmin nuclear targeting sequence: identification of a class of bipartite nuclear targeting sequence. *Cell* **64**(3), 615–623.

36. Gorlich, D. *et al.* (1996). A 41 amino acid motif in importin-α confers binding to importin-β and hence transit into the nucleus. *EMBO J* **15**(8), 1810–1817.

37. Weis, K., Ryder, U. and Lamond, A. I. (1996). The conserved amino-terminal domain of hSRP1α is essential for nuclear protein import. *EMBO J* **15**(8), 1818–1825.

38. Moroianu, J., Blobel, G. and Radu, A. (1995). Previously identified protein of uncer-tain function is karyopherin-α and together with karyopherin-β docks import substrate at nuclear pore complexes. *Proc Natl Acad Sci USA* **92**(6), 2008–2011.

39. Gorlich, D. *et al.* (1995). Two different subunits of importin cooperate to recognize nuclear localization signals and bind them to the nuclear envelope. *Curr Biol* **5**(4), 383–392.

40. Imamoto, N. *et al.* (1995). The nuclear pore-targeting complex binds to nuclear pores after association with a karyophile. *FEBS Lett* **368**(3), 415–419.

41. Radu, A., Blobel, G. and Moore, M. S. (1995). Identification of a protein complex that is required for nuclear protein import and mediates docking of import substrate to distinct nucleoporins. *Proc Natl Acad Sci USA* **92**(5), 1769–1773.

42. Ohno, M., Fornerod, M. and Mattaj, I. W. (1998). Nucleocytoplasmic transport: the last 200 nanometers. *Cell* **92**(3), 327–336.

43. Gorlich, D. and Kutay, U. (1999). Transport between the cell nucleus and the cytoplasm. *Annu Rev Cell Dev Biol* **15**, 607–660.

44. Yoneda, Y. (2000). New steps toward the nucleocytoplasmic traffic of macromole-cules. *Cell Struct Funct* **25**(4), 205–206.

45. Chi, N. C., Adam, E. J. and Adam, S. A. (1995). Sequence and characterization of cyto-plasmic nuclear protein import factor p97. *J Cell Biol* **130**(2), 265–274.

46. Bischoff, F. R. and Gorlich, D. (1997). RanBP1 is crucial for the release of RanGTP from importin β-related nuclear transport factors. *FEBS Lett* **419**(2–3), 249–254.

47. Izaurralde, E. *et al.* (1997). The asymmetric distribution of the constituents of the Ran system is essential for transport into and out of the nucleus. *EMBO J* **16**(21), 6535–6547.

48. Ossareh-Nazari, B., Bachelerie, F. and Dargemont, C. (1997). Evidence for a role of *CRM1* in signal-mediated nuclear protein export. *Science* **278**(5335), 141–144.

49. Richards, S. A., Carey, K. L. and Macara, I. G. (1997). Requirement of guanosine triphosphate-bound ran for signal-mediated nuclear protein export. *Science* **276**(5320), 1842–1844.

50. Fornerod, M. *et al.* (1997). *CRM1* is an export receptor for leucine-rich nuclear export signals [see comments]. *Cell* **90**(6), 1051–1060.

51. Kehlenbach, R. H. *et al.* (1999). A role for RanBP1 in the release of *CRM1* from the nuclear pore complex in a terminal step of nuclear export. *J Cell Biol* **145**(4), 645–657.

52. Fouchier, R. A. and Malim, M. H. (1999). Nuclear import of human immunodeficiency virus type-1 preintegration complexes. *Adv Virus Res* **52**, 275–299.

53. Chen, H. and Engelman, A. (1998). The barrier-to-autointegration protein is a host factor for HIV type 1 integration. *Proc Natl Acad Sci USA* **95**(26), 15270–15274.

54. Bukrinsky, M. I. *et al.* (1993b). A nuclear localization signal within HIV-1 matrix protein that governs infection of non-dividing cells. *Nature* **365**(6447), 666–669.

55. Heinzinger, N. K. *et al.* (1994). The Vpr protein of human immunodeficiency virus type 1 influences nuclear localization of viral nucleic acids in nondividing host cells. *Proc Natl Acad Sci USA* **91**(15), 7311–7315.

56. Gallay, P. *et al.* (1997). HIV-1 infection of nondividing cells through the recognition of integrase by the importin/karyopherin pathway. *Proc Natl Acad Sci USA* **94**(18), 9825–9830.

57. Gallay, P. *et al.* (1995). HIV nuclear import is governed by the phosphotyrosine-mediated binding of matrix to the core domain of integrase. *Cell* **83**(4), 569–576.

58. Fouchier, R. A. *et al.* (1998). Interaction of the human immunodeficiency virus type 1 Vpr protein with the nuclear pore complex. *J Virol* **72**(7), 6004–6013.

59. Pluymers, W. *et al.* (1999). Nuclear localization of human immunodeficiency virus type 1 integrase expressed as a fusion protein with green fluorescent protein. *Virology* **258**(2), 327–332.

60. Vodicka, M. A. *et al.* (1998). HIV-1 Vpr interacts with the nuclear transport pathway to promote macrophage infection. *Genes Dev* **12**(2), 175–185.

61. Zennou, V. *et al.* (2000). HIV-1 genome nuclear import is mediated by a central DNA flap. *Cell* **101**(2), 173–185.

62. Krausslich, H. G. and Welker, R. (1996). Intracellular transport of retroviral capsid components. *Curr Top Microbiol Immunol* **214**, 25–63.

63. Yu, X. *et al.* (1992). The matrix protein of human immunodeficiency virus type 1 is required for incorporation of viral envelope protein into mature virions. *J Virol* **66**(8), 4966–4971.

64. Dorfman, T. *et al.* (1994). Role of the matrix protein in the virion association of the human immunodeficiency virus type 1 envelope glycoprotein. *J Virol* **68**(3), 1689–1696.

65. Bryant, M. and Ratner, L. Myristoylation-dependent replication and assembly of human immunodeficiency virus 1. *Proc Natl Acad Sci USA* **87**(2), 523–527.

66. Gottlinger, H. G., Sodroski, J. G. and Haseltine, W. A. (1989). Role of capsid precursor processing and myristoylation in morphogenesis and infectivity of human immunodeficiency virus type 1. *Proc Natl Acad Sci USA* **86**(15), 5781–5578.

67. Zhou, W. *et al.* (1994). Identification of a membrane-binding domain within the amino-terminal region of human immunodeficiency virus type 1 Gag protein which interacts with acidic phospholipids. *J Virol* **68**(4) 2556–2569.

68. Gallay, P. *et al.* (1995b). HIV-1 infection of nondividing cells: C-terminal tyrosine phosphorylation of the viral matrix protein is a key regulator. *Cell* **80**(3), 379–388.

69. Freed, E. O. *et al.* (1997). Phosphorylation of residue 131 of HIV-1 matrix is not required for macrophage infection. *Cell* **88**(2), 171–173 (discussion, 173–174).

70. Dupont, S. *et al.* (1999). A novel nuclear export activity in HIV-1 matrix protein required for viral replication. Nature **402**(6762), 681–685.

71. Fouchier, R. A. *et al.* (1997). HIV-1 infection of non-dividing cells: evidence that the amino-terminal basic region of the viral matrix protein is important for Gag processing but not for post-entry nuclear import. *EMBO J* **16**(15), 4531–4539.

72. Reil, H. *et al.* (1998). Efficient HIV-1 replication can occur in the absence of the viral matrix protein. *EMBO J* **17**(9), 2699–2708.

73. Charneau, P. *et al.* (1994). HIV-1 reverse transcription. A termination step at the center of the genome. *J Mol Biol* **241**(5), 651–626.

74. Hungnes O., Tjotta, E. and Grinde, B. (1992). Mutations in the central polypurine tract of HIV-1 result in delayed replication. *Virology* **190**(1) 440–442.

75. Charneau, P., Alizon, M. and Clavel, F. (1992). A second origin of DNA plus-strand synthesis is required for optimal human immunodeficiency virus replication. *J Virol* **66**(5), 2814–2820.

76. Hamm, J. and Mattaj, I. W. (1990). Monomethylated cap structures facilitate RNA export from the nucleus. *Cell* **63**(1), 109–118.

77. Dvorin, J. D. *et al.* (2002). Reassessment of the roles of integrase and the central DNA flap in human immunodeficiency virus type 1 nuclear import. *J Virol* **76**(23), 12087–12096.

78. Demaison, C. *et al.* (2002). High-level transduction and gene expression in hematopoietic repopulating cells using a human immunodeficiency virus type 1-based lentiviral vector containing an internal spleen focus forming virus promoter. *Hum Gene Ther* **13**(7), 803–813.

79. Lewis, B. C. *et al.* (2001). Development of an avian leukosis-sarcoma virus subgroup A pseudotyped lentiviral vector. *J Virol* **75**(19), 9339–9344.

80. Mautino, M. R. *et al.* (2000). Modified human immunodeficiency virus-based lentiviral vectors display decreased sensitivity to *trans*-dominant Rev. *Hum Gene Ther* **11**(6), 895–908.

81. Mautino, M. R. and Morgan, R. A. (2002). Enhanced inhibition of human immunodeficiency virus type 1 replication by novel lentiviral vectors expressing human immunodeficiency virus type 1 envelope antisense RNA. *Hum Gene Ther* **13**(9), 1027–1037.

82. Follenzi, A. *et al.* (2000). Gene transfer by lentiviral vectors is limited by nuclear translocation and rescued by *HIV-1 pol* sequences. *Nat Genet* **25**(2), 217–222.

83. Lu, Y. L. *et al.* (1995). A leucine triplet repeat sequence (LXX)4 in p6gag is important for Vpr incorporation into human immunodeficiency virus type 1 particles. *J Virol* **69**(11), 6873–6879.

84. Singh, S. P. *et al.* (2001). Virion-associated HIV-1 Vpr: variable amount in virus particles derived from cells upon virus infection or proviral DNA transfection. *Virology* **283**(1), 78–83.

85. von Schwedler, U., Kornbluth, R. S. and Trono, D. (1994). The nuclear localization signal of the matrix protein of human immunodeficiency virus type 1 allows the establishment of infection in macrophages and quiescent T lymphocytes. *Proc Natl Acad Sci USA* **91**(15), 6992–6996.

86. Gallay, P. *et al.* (1996). Role of the karyopherin pathway in human immunodeficiency virus type 1 nuclear import. *J Virol* **70**(2), 1027–1032.

87. Bouyac-Bertoia, M. *et al.* (2001). HIV-1 infection requires a functional integrase NLS. *Mol Cell* **7**(5) 1025–1035.

88. Jowett, J. B. *et al.* (1995). The human immunodeficiency virus type 1 *vpr* gene arrests infected T cells in the $G_2 + M$ phase of the cell cycle. *J Virol* **69**(10), 6304–6313.

89. Rogel, M. E., Wu, L. I. and Emerman, M. (1995). The human immunodeficiency virus type 1 *vpr* gene prevents cell proliferation during chronic infection. *J Virol* **69**(2), 882–888.

90. Fletcher, T. M. III *et al.* (1996). Nuclear import and cell cycle arrest functions of the HIV-1 Vpr protein are encoded by two separate genes in HIV-2/SIV(SM). *EMBO J* **15**(22) 6155–6165.

91. Di Marzio, P. *et al.* (1995). Mutational analysis of cell cycle arrest, nuclear localization and virion packaging of human immunodeficiency virus type 1 Vpr. *J Virol* **69**(12), 7909–7916.

92. Lu, Y. L., Spearman, P. and Ratner, L. (1993). Human immunodeficiency virus type 1 viral protein R localization in infected cells and virions. *J Virol* **67**(11), 6542–6550.

93. Mahalingam, S. *et al.* (1997). Nuclear import, virion incorporation, and cell cycle arrest/differentiation are mediated by distinct functional domains of human immunodeficiency virus type 1 Vpr. *J Virol* **71**(9), 6339–6347.

94. Mahalingam, S. *et al.* (1995). Functional analysis of HIV-1 Vpr: identification of determinants essential for subcellular localization. *Virology* **212**(2), 331–339.

95. Zhou, Y., Lu, Y. and Ratner, L. (1998). Arginine residues in the C-terminus of HIV-1 Vpr are important for nuclear localization and cell cycle arrest. *Virology* **242**(2), 414–424.

96. Yao, X. J. *et al.* (1995). Mutagenic analysis of human immunodeficiency virus type 1 Vpr: role of a predicted N-terminal α-helical structure in Vpr nuclear localization and virion incorporation. J Virol **69**(11), 7032–7044.

97. Jenkins, Y. *et al.* (1998). Characterization of HIV-1 vpr nuclear import: analysis of signals and pathways. *J Cell Biol* **143**(4), 875–885.

98. Sherman, M. P. *et al.* (2001). Nucleocytoplasmic shuttling by human immunodeficiency virus type 1 Vpr. *J Virol* **75**(3), 1522–1532.

99. Popov, S. *et al.* (1998a). Viral protein R regulates docking of the HIV-1 preintegration complex to the nuclear pore complex. *J Biol Chem* **273**(21), 13347–13352.

100. Popov, S. *et al.* (1998b). Viral protein R regulates nuclear import of the HIV-1 preintegration complex. *EMBO J* **17**(4), 909–917.

101. Bukrinsky, M. I. *et al.* (1992). Active nuclear import of human immunodeficiency virus type 1 preintegration complexes. *Proc Natl Acad Sci USA* **89**(14), 6580–6584.

102. Connor, R. I. *et al.* (1995). Vpr is required for efficient replication of human immunodeficiency virus type-1 in mononuclear phagocytes. *Virology* **206**(2), 935–944.

103. Subbramanian, R. A. *et al.* (1998). Human immunodeficiency virus type 1 Vpr is a positive regulator of viral transcription and infectivity in primary human macrophages. *J Exp Med* **187**(7), 1103–1111.

104. Eckstein, D. A. *et al.* (2001). HIV-1 actively replicates in naive CD4(+) T cells residing within human lymphoid tissues. *Immunity* **15**(4), 671–682.

105. Tsurutani, N. *et al.* (2000). Identification of critical amino acid residues in human immunodeficiency virus type 1 IN required for efficient proviral DNA formation at steps prior to integration in dividing and nondividing cells. *J Virol* **74**(10), 4795–4806.

106. Kukolj, G., Jones, K. S. and Skalka, A. M. (1997). Subcellular localization of avian sarcoma virus and human immunodeficiency virus type 1 integrases. *J Virol* **71**(1), 843–847.

107. Kukolj, G., Katz, R. A. and Skalka, A. M. (1998). Characterization of the nuclear localization signal in the avian sarcoma virus integrase. *Gene* **223**(1–2), 157–163.

108. Hatziioannou, T. and Goff, S. P. (2001). Infection of nondividing cells by Rous sarcoma virus. *J Virol* **75**(19), 9526–9531.

109. Seamon, J. A. *et al.* (2002). Inserting a nuclear targeting signal into a replication-competent Moloney murine leukemia virus affects viral export and is not sufficient for cell cycle-independent infection. *J Virol* **76**(16), 8475–8484.

110. Fassati, A. *et al.* (2003). Nuclear import of HIV-1 intracellular reverse transcription complexes is mediated by importin 7. *EMBO J* **22**(14), 3675–3685.

111. Korin, Y. D. and Zack, J. A. (1998). Progression to the G_{1b} phase of the cell cycle is required for completion of human immunodeficiency virus type 1 reverse transcription in T cells. *J Virol* **72**(4), 3161–3168.

112. Miller, M. D., Wang, B. and Bushman, F. D. (1995). Human immunodeficiency virus type 1 preintegration complexes containing discontinuous plus strands are competent to integrate *in vitro. J Virol* **69**(6), 3938–3944.

113. Brown, P. O. *et al.* (1987). Correct integration of retroviral DNA *in vitro. Cell* **49**(3), 347–356.

114. Brown, P. O. *et al.* (1989). Retroviral integration: structure of the initial covalent product and its precursor, and a role for the viral IN protein. *Proc Natl Acad Sci USA* **86**(8), 2525–2529.

115. Fujiwara, T. and Mizuuchi, K. (1988). Retroviral DNA integration: structure of an integration intermediate. *Cell* **54**(4), 497–504.

116. Roth, M. J., Schwartzberg, P. L. and Goff, S. P. (1989). Structure of the termini of DNA intermediates in the integration of retroviral DNA: dependence on IN function and terminal DNA sequence. *Cell* **58**(1), 47–54.

117. Roe, T., Chow, S. A. and Brown, P. O. (1997). 3′-End processing and kinetics of 5′-end joining during retroviral integration *in vivo. J Virol* **71**(2), 1334–1340.

118. Varmus, H. E. *et al.* (1974). Synthesis of viral DNA in the cytoplasm of duck embryo fibroblasts and in enucleated cells after infection by avian sarcoma virus. *Proc Natl Acad Sci USA* **71**(10), 3874–3878.

119. Pryciak, P. M., Muller, H. P. and Varmus, H. E. (1992a). Simian virus 40 minichromosomes as targets for retroviral integration *in vivo. Proc Natl Acad Sci USA* **89**(19), 9237–9241.

120. Pryciak, P. M., Sil, A. and Varmus, H. E. (1992b). Retroviral integration into minichromosomes *in vitro. EMBO J* **11**(1), 291–303.

121. Pryciak, P. M. and Varmus, H. E. (1992c). Nucleosomes, DNA-binding proteins, and DNA sequence modulate retroviral integration target site selection. *Cell* **69**(5), 769–780.

122. Muller, H. P. and Varmus, H. E. (1994). DNA bending creates favored sites for retroviral integration: an explanation for preferred insertion sites in nucleosomes. *EMBO J* **13**(19), 4704–4714.

123. Muller, H. P., Pryciak, P. M. and Varmus, H. E. (1993). Retroviral integration machinery as a probe for DNA structure and associated proteins. *Cold Spring Harb Symp Quant Biol* **58** 533–541.

124. Schroder, A. R. *et al.* (2002). HIV-1 integration in the human genome favors active genes and local hotspots. *Cell* **110**(4), 521–529.

125. Wu, X. *et al.* (2003). Transcription start regions in the human genome are favored targets for MLV integration. *Science* **300**(5626), 1749–1751.

126. Katzman, M. *et al.* (1989). The avian retroviral integration protein cleaves the terminal sequences of linear viral DNA at the *in vivo* sites of integration. *J Virol* **63**(12), 5319–5327.

127. Sherman, P. A. and Fyfe, J. A. (1990). Human immunodeficiency virus integration protein expressed in *Escherichia coli* possesses selective DNA cleaving activity. *Proc Natl Acad Sci USA* **87**(13), 5119–5123.

128. Engelman, A., Mizuuchi, K. and Craigie, R. (1991). HIV-1 DNA integration: mechanism of viral DNA cleavage and DNA strand transfer. *Cell* **67**(6), 1211–1221.

129. Vink, C. *et al.* (1991). Site-specific hydrolysis and alcoholysis of human immunodeficiency virus DNA termini mediated by the viral integrase protein. *Nucleic Acids Res* **19**(24), 6691–6698.

130. Brown, P. O. (1997). Integration. In *Retroviruses*, pp. 161–203, H. E. Varmus (ed.). Cold Spring Harbor Laboratory Press: Cold Spring Harbor, NY.

131. Engelman, A. and Craigie, R. (1992). Identification of conserved amino acid residues critical for human immunodeficiency virus type 1 integrase function *in vitro*. *J Virol* **66**(11), 6361–6369.

132. Engelman, A. *et al.* (1995). Multiple effects of mutations in human immunodeficiency virus type 1 integrase on viral replication. *J Virol* **69**(5), 2729–2736.

133. Roth, M. J. *et al.* (1990). Analysis of mutations in the integration function of Moloney murine leukemia virus: effects on DNA binding and cutting. *J Virol* **64**(10), 4709–4717.

134. Khan, E. *et al.* (1991). Retroviral integrase domains: DNA binding and the recognition of LTR sequences. *Nucleic Acids Res* **19**(4), 851–860.

135. Schauer, M. and Billich, A. (1992). The N-terminal region of HIV-1 integrase is required for integration activity, but not for DNA-binding. *Biochem Biophys Res Commun* **185**(3), 874–880.

136. Bushman, F. D. *et al.* (1993). Domains of the integrase protein of human immunodeficiency virus type 1 responsible for polynucleotidyl transfer and zinc binding. *Proc Natl Acad Sci USA* **90**(8), 3428–3432.

137. Vincent, K. A. *et al.* (1993). Characterization of human immunodeficiency virus type 1 integrase expressed in *Escherichia coli* and analysis of variants with amino-terminal mutations. *J Virol* **67**(1), 425–437.

138. Vink, C., Oude Groeneger, A. M. and Plasterk, R. H. (1993). Identification of the catalytic and DNA-binding region of the human immunodeficiency virus type I integrase protein. *Nucleic Acids Res* **21**(6), 1419–1425.

139. Woerner, A. M. *et al.* (1992). Localization of DNA binding activity of HIV-1 integrase to the C-terminal half of the protein. *AIDS Res Hum Retrovir* **8**(2), 297–304.

140. Mumm, S. R. and Grandgenett, D. P. (1991). Defining nucleic acid-binding properties of avian retrovirus integrase by deletion analysis. *J Virol* **65**(3) 1160–1167.

141. Zheng, R., Jenkins, T. M. and Craigie, R. (1996). Zinc folds the N-terminal domain of HIV-1 integrase, promotes multimerization, and enhances catalytic activity. *Proc Natl Acad Sci USA* **93**(24) 13659–13664.

142. Dyda, F. *et al.* (1994). Crystal structure of the catalytic domain of HIV-1 integrase: similarity to other polynucleotidyl transferases. *Science* **266**(5193), 1981–1986.

143. Bujacz, G. *et al.* (1996). The catalytic domain of avian sarcoma virus integrase: conformation of the active-site residues in the presence of divalent cations. *Structure* **4**(1) 89–96.

144. Bujacz, G. *et al.* (1995). High-resolution structure of the catalytic domain of avian sarcoma virus integrase. *J Mol Biol* **253**(2) 333–346.

145. Rice, P., Craigie, R. and Davies, D. R. (1996). Retroviral integrases and their cousins. *Curr Opin Struct Biol* **6**(1) 76–83.

146. Yang, W. and Steitz, T. A. (1995). Crystal structure of the site-specific recombinase gamma delta resolvase complexed with a 34 bp cleavage site. *Cell* **82**(2) 193–207.

147. Johnson, M. S. *et al.* (1986). Computer analysis of retroviral pol genes: assignment of enzymatic functions to specific sequences and homologies with nonviral enzymes. *Proc Natl Acad Sci USA* **83**(20), 7648–7652.

148. Lutzke, R. A., Vink, C. and Plasterk, R. H. (1994). Characterization of the minimal DNA-binding domain of the HIV integrase protein. *Nucleic Acids Res* **22**(20), 4125–4131.

149. Engelman, A., Hickman, A. B. and Craigie, R. (1994). The core and carboxyl-terminal domains of the integrase protein of human immunodeficiency virus type 1 each contribute to nonspecific DNA binding. *J Virol* **68**(9) 5911–5917.

150. Woerner, A. M. and Marcus-Sekura, C. J. (1993). Characterization of a DNA binding domain in the C-terminus of HIV-1 integrase by deletion mutagenesis. *Nucleic Acids Res* **21**(15), 3507–3511.

151. Ou, S. H. *et al.* (1994). Role of flanking E box motifs in human immunodeficiency virus type 1 TATA element function. *J Virol* **68**(11), 7188–7199.

152. Jones, K. A. *et al.* (1986). Activation of the AIDS retrovirus promoter by the cellular transcription factor, Sp1. *Science* **232**(4751), 755–759.

153. Sune, C. and Garcia-Blanco, M. A. (1995). Sp1 transcription factor is required for *in vitro* basal and Tat-activated transcription from the human immunodeficiency virus type 1 long terminal repeat. *J Virol* **69**(10), 6572–6576.

154. Sawaya, B. E. *et al.* (1998). Cooperative actions of HIV-1 Vpr and p53 modulate viral gene transcription. *J Biol Chem* **273**(32), 20052–20057.

155. Wang, L. *et al.* (1995). Interaction of virion protein Vpr of human immunodeficiency virus type 1 with cellular transcription factor Sp1 and transactivation of viral long terminal repeat. *J Biol Chem* **270**(43), 25564–25569.

156. Wang, L. *et al.* (1996). Characterization of a leucine-zipper-like domain in Vpr protein of human immunodeficiency virus type 1. *Gene* **178**(1–2) 7–13.

157. Jones, K. A. and Peterlin, B. M. (1994). Control of RNA initiation and elongation at the HIV-1 promoter. *Annu Rev Biochem* **63** 717–743.

158. Barboric, M. *et al.* (2001). NF-kappaB binds P-TEFb to stimulate transcriptional elongation by RNA polymerase II. *Mol Cell* **8**(2), 327–337.

159. Crabtree, G. R. (1999). Generic signals and specific outcomes: signaling through Ca2$^+$, calcineurin, and NF-AT. *Cell* **96**(5), 611–614.

160. Lu, Y. *et al.* (1989). Effects of long terminal repeat mutations on human immunodeficiency virus type 1 replication. *J Virol* **63**(9), 4115–4119.

161. Okamoto, T. *et al.* (1990). Transcriptional activation from the long-terminal repeat of human immunodeficiency virus *in vitro*. *Virology* **177**(2) 606–614.

162. Atchison, M. L. and Perry, R. P. (1987). The role of the kappa enhancer and its binding factor NF-kappa B in the developmental regulation of kappa gene transcription. *Cell* **48**(1), 121–128.

163. Cooney, A. J. *et al.* (1991). Chicken ovalbumin upstream promoter transcription factor binds to a negative regulatory region in the human immunodeficiency virus type 1 long terminal repeat. *J Virol* **65**(6), 2853–2860.

164. Franza, B. R. Jr *et al.* (1988). The Fos complex and Fos-related antigens recognize sequence elements that contain AP-1 binding sites. *Science* **239**(4844), 1150–1153.

165. Ivanov, V. *et al.* (1989). Infection with the intracellular protozoan parasite *Theileria parva* induces constitutively high levels of NF-kappa B in bovine T lymphocytes. *Mol Cell Biol* **9**(11), 4677–4686.

166. Lu, Y. C. *et al.* (1990). Identification of *cis*-acting repressive sequences within the negative regulatory element of human immunodeficiency virus type 1. *J Virol* **64**(10), 5226–5229.

167. Poeschla, E. M. and Looney, D. J. (1998). CXCR4 is required by a nonprimate lentivirus: heterologous expression of feline immunodeficiency virus in human, rodent, and feline cells. *J Virol* **72**(8), 6858–6866.

168. Reddy, E. P. and Dasgupta, P. (1992). Regulation of HIV-1 gene expression by cellular transcription factors. *Pathobiology* **60**(4), 219–224.

169. Waterman, M. L. and Jones, K. A. (1990). Purification of TCF-1 alpha, a T-cell-specific transcription factor that activates the T-cell receptor C alpha gene enhancer in a context-dependent manner. *New Biol* **2**(7), 621–636.

170. Shilatifard, A. (1998). The RNA polymerase II general elongation complex. *Biol Chem* **379**(1), 27–31.

171. Rittner, K. *et al.* (1995). The human immunodeficiency virus long terminal repeat includes a specialised initiator element which is required for Tat-responsive transcription. *J Mol Biol* **248**(3), 562–580.

172. Zhou, Q. and Sharp, P. A. (1995). Novel mechanism and factor for regulation by HIV-1 Tat. *EMBO J* **14**(2), 321–328.

173. Price, D. H. (2000). P-TEFb, a cyclin-dependent kinase controlling elongation by RNA polymerase II. *Mol Cell Biol* **20**(8), 2629–2634.

174. Yamaguchi, Y. *et al.* (1999). NELF, a multisubunit complex containing RD, cooperates with DSIF to repress RNA polymerase II elongation. *Cell* **97**(1), 41–51.

175. Wada, T. *et al.* (1998). DSIF, a novel transcription elongation factor that regulates RNA polymerase II processivity, is composed of human Spt4 and Spt5 homologs. *Genes Dev* **12**(3), 343–356.

176. Emerman, M. and Malim, M. H. (1998) HIV-1 regulatory/accessory genes: keys to unraveling viral and host cell biology. *Science* **280**(5371), 1880–1884.

177. Cullen, B. R. (1998). HIV-1 auxiliary proteins: making connections in a dying cell. *Cell* **93**(5), 685–692.

178. Karn, J. (1999). Tackling Tat. *J Mol Biol* **293**(2), 235–254.

179. Taube, R. *et al.* (1999). Tat *trans*-activation: a model for the regulation of eukaryotic transcriptional elongation. *Virology* **264**(2), 245–253.

180. Adams, M. *et al.* (1994). Cellular latency in human immunodeficiency virus-infected individuals with high CD4 levels can be detected by the presence of promoter-proximal transcripts. *Proc Natl Acad Sci USA* **91**(9), 3862–3866.

181. Finzi, D. *et al.* (1999). Latent infection of CD4$^+$ T cells provides a mechanism for lifelong persistence of HIV-1, even in patients on effective combination therapy. *Nat Med* **5**(5), 512–517.

182. Peng, J. *et al.* (1998). Identification of multiple cyclin subunits of human P-TEFb. *Genes Dev* **12**(5), 755–762.

183. Rosen, C. A., Sodroski, J. G. and Haseltine, W. A. (1985). The location of *cis*-acting regulatory sequences in the human T cell lymphotropic virus type III (HTLV-III/LAV) long terminal repeat. *Cell* **41**(3) 813–823.

184. Muesing, M. A., Smith, D. H. and Capon, D. J. (1987). Regulation of mRNA accumulation by a human immunodeficiency virus *trans*-activator protein. *Cell* **48**(4), 691–701.

185. Hauber, J. and Cullen, B. R. (1988). Mutational analysis of the *trans*-activation-responsive region of the human immunodeficiency virus type I long terminal repeat. *J Virol* **62**(3), 673–679.

186. Jaeger, J. A. and Tinoco, I. Jr. (1993). An NMR study of the HIV-1 TAR element hairpin. *Biochemistry* **32**(46), 12522–12530.

187. Wei, P. *et al.* (1998). A novel CDK9-associated C-type cyclin interacts directly with HIV-1 Tat and mediates its high-affinity, loop-specific binding to TAR RNA. *Cell* **92**(4), 451–462.

188. Wu, Y. and Marsh, J. W. (2001). Selective transcription and modulation of resting T cell activity by preintegrated HIV DNA. *Science* **293**(5534), 1503–1506.

189. Malim, M. H. *et al.* (1989). The HIV-1 rev transactivator acts through a structured target sequence to activate nuclear export of unspliced viral mRNA. *Nature* **338**(6212), 254–257.

190. Schwartz, S. *et al.* (1992). Mutational inactivation of an inhibitory sequence in human immunodeficiency virus type 1 results in Rev-independent *gag* expression. *J Virol* **66**(12), 7176–7182.

191. Schwartz, S., Felber, B. K. and Pavlakis, G. N. (1992b). Distinct RNA sequences in the *gag* region of human immunodeficiency virus type 1 decrease RNA stability and inhibit expression in the absence of Rev protein. *J Virol* **66**(1), 150–159.

192. Powell, D. M. *et al.* (1997). HIV Rev-dependent binding of SF2/ASF to the Rev response element: possible role in Rev-mediated inhibition of HIV RNA splicing. *Proc Natl Acad Sci USA* **94**(3), 973–978.

193. Luo, Y., Yu, H. and Peterlin, B. M. (1994). Cellular protein modulates effects of human immunodeficiency virus type 1 Rev. *J Virol* **68**(6), 3850–3556.

194. Chang, D. D. and Sharp, P. A. (1989). Regulation by HIV Rev depends upon recognition of splice sites. *Cell* **59**(5), 789–795.

195. Bohnlein, E., Berger, J. and Hauber, J. (1991). Functional mapping of the human immunodeficiency virus type 1 Rev RNA binding domain: new insights into the domain structure of Rev and Rex. *J Virol* **65**(12), 7051–7055.

196. Daly, T. J. *et al.* (1989). Specific binding of HIV-1 recombinant Rev protein to the Rev-responsive element *in vitro*. *Nature* **342**(6251), 816–819.

197. Hope, T. J. *et al.* (1990). Steroid-receptor fusion of the human immunodeficiency virus type 1 Rev *trans*-activator: mapping cryptic functions of the arginine-rich motif. *Proc Natl Acad Sci USA* **87**(19), 7787–7791.

198. Malim, M. H. and Cullen, B. R. (1991). *HIV-1* structural gene expression requires the binding of multiple Rev monomers to the viral RRE: implications for HIV-1 latency. *Cell* **65**(2), 241–248.

199. Olsen, H. S. *et al.* (1990). Interaction of the human immunodeficiency virus type 1 Rev protein with a structured region in *env* mRNA is dependent on multimer formation mediated through a basic stretch of amino acids. *Genes Dev* **4**(8), 1357–1364.

200. Zapp, M. L. and Green, M. R. (1989). Sequence-specific RNA binding by the HIV-1 Rev protein. *Nature* **342**(6250), 714–716.

201. Zapp, M. L. *et al.* (1991). Oligomerization and RNA binding domains of the type 1 human immunodeficiency virus Rev protein: a dual function for an arginine-rich binding motif. *Proc Natl Acad Sci USA* **88**(17), 7734–7738.

202. Kubota, S. *et al.* (1989). Functional similarity of HIV-1 *rev* and HTLV-1 *rex* proteins: identification of a new nucleolar-targeting signal in *rev* protein. *Biochem Biophys Res Commun* **162**(3), 963–970.

203. Malim, M. H. *et al.* (1989). Functional dissection of the HIV-1 Rev *trans*-activator – derivation of a *trans*-dominant repressor of Rev function. *Cell* **58**(1): 205–214.

204. Perkins, A. *et al.* (1989). Structural and functional characterization of the human immunodeficiency virus *rev* protein. *J Acqu Immune Defic Syndr* **2**(3), 256–263.

205. Truant, R. and Cullen, B. R. (1999). The arginine-rich domains present in human immunodeficiency virus type 1 Tat and Rev function as direct importin beta-dependent nuclear localization signals. *Mol Cell Biol* **19**(2), 1210–1217.

206. Henderson, B. R. and Percipalle, P. (1997). Interactions between HIV Rev and nuclear import and export factors: the Rev nuclear localisation signal mediates specific binding to human importin-β. *J Mol Biol* **274**(5): 693–707.

207. Fischer, U. *et al.* (1995). The HIV-1 Rev activation domain is a nuclear export signal that accesses an export pathway used by specific cellular RNAs. *Cell* **82**(3), 475–483.

208. Meyer, B. E., Meinkoth, J. L. and Malim, M. H. (1996). Nuclear transport of human immunodeficiency virus type 1, visna virus, and equine infectious anemia virus Rev proteins: identification of a family of transferable nuclear export signals. *J Virol* **70**(4), 2350–2359.

209. Wen, W. *et al.* (1995). Identification of a signal for rapid export of proteins from the nucleus. *Cell* **82**(3), 463–473.

210. Stade, K. *et al.* (1997). Exportin 1 (Crm1p) is an essential nuclear export factor. *Cell* **90**(6), 1041–1050.

211. Zolotukhin, A. S. *et al.* (1994). Continuous propagation of RRE(−) and Rev(−)RRE(−) human immunodeficiency virus type 1 molecular clones containing a *cis*-acting element of simian retrovirus type 1 in human peripheral blood lymphocytes. *J Virol* **68**(12), 7944–7952.

212. Bray, M. *et al.* (1994). A small element from the Mason–Pfizer monkey virus genome makes human immunodeficiency virus type 1 expression and replication Rev-independent. *Proc Natl Acad Sci USA* **91**(4), 1256–1260.

213. Tabernero, C. *et al.* (1996). The posttranscriptional control element of the simian retrovirus type 1 forms an extensive RNA secondary structure necessary for its function. *J Virol* **70**(9), 5998–6011.

214. Ernst, R. K. *et al.* (1997a). Secondary structure and mutational analysis of the Mason–Pfizer monkey virus RNA constitutive transport element. *RNA* **3**(2), 210–222.

215. Ernst, R. K. *et al.* (1997b). A structured retroviral RNA element that mediates nucleocytoplasmic export of intron-containing RNA. *Mol Cell Biol* **17**(1), 135–144.

216. Segref, A. *et al.* (1997). Mex67p, a novel factor for nuclear mRNA export, binds to both poly(A)+ RNA and nuclear pores. *EMBO J* **16**(11), 3256–3271.

217. Gruter, P. *et al.* (1998). TAP, the human homolog of Mex67p, mediates CTE-dependent RNA export from the nucleus. *Mol Cell* **1**(5), 649–659.

218. Tang, H. *et al.* (1997). A cellular cofactor for the constitutive transport element of type D retrovirus. *Science* **276**(5317), 1412–1415.

219. Meyers, G. *et al.* (eds) (1994). *Human Retroviruses and AIDS*. Los Alamos National Laboratory: Los Alamos, CA.

220. Nakaya, T. *et al.* (1994). Nonsense mutations in the *vpr* gene of HIV-1 during *in vitro* virus passage and in HIV-1 carrier-derived peripheral blood mononuclear cells. *FEBS Lett* **354**(1), 17–22.

221. Planelles, V. *et al.* (1995). Fate of the human immunodeficiency virus type 1 provirus in infected cells: a role for *vpr*. *J Virol* **69**(9), 5883–5889.

222. Sherman, M. P. *et al.* (unpublished data).

223. Gibbs, J. S. *et al.* (1995). Progression to AIDS in the absence of a gene for *vpr* or *vpx*. *J Virol* **69**(4), 2378–2383.

224. Hoch, J. *et al.* (1995). *vpr* Deletion mutant of simian immunodeficiency virus induces AIDS in rhesus monkeys. *J Virol* **69**(8), 4807–4813.

225. Goh, W. C. *et al.* (1998). HIV-1 Vpr increases viral expression by manipulation of the cell cycle: a mechanism for selection of Vpr *in vivo*. *Nat Med* **4**(1), 65–71.

226. O'Farrell, P. H. (2001). Triggering the all-or-nothing switch into mitosis. *Trends Cell Biol* **11**(12), 512–519.

227. Smits, V. A. and Medema, R. H. (2001). Checking out the G(2)/M transition. *Biochim Biophys Acta* **1519**(1–2), 1–12.

228. Takizawa, C. G. and Morgan, D. O. (2000). Control of mitosis by changes in the subcellular location of cyclin-B1-Cdk1 and Cdc25C. *Curr Opin Cell Biol* **12**(6), 658–665.

229. Jackman, M., Firth, M. and Pines, J. (1995). Human cyclins B1 and B2 are localized to strikingly different structures: B1 to microtubules, B2 primarily to the Golgi apparatus. *EMBO J* **14**(8), 1646–1654.

230. Dalal, S. N. *et al.* (1999). Cytoplasmic localization of human cdc25C during interphase requires an intact 14-3-3 binding site. *Mol Cell Biol* **19**(6), 4465–4479.

231. Graves, P. R. *et al.* (2001). Localization of human Cdc25C is regulated both by nuclear export and 14-3-3 protein binding. *Oncogene* **20**(15), 1839–1851.

232. de Noronha, C. M. (unpublished data).

233. Chan, T. A. *et al.* (1999). 14-3-3 Sigma is required to prevent mitotic catastrophe after DNA damage. *Nature* **401**(6753), 616–620.

234. Peng, C. Y. *et al.* (1997). Mitotic and G_2 checkpoint control: regulation of 14-3-3 protein binding by phosphorylation of Cdc25C on serine-216. *Science* **277**(5331), 1501–1505.

235. Re, F. *et al.* (1995). Human immunodeficiency virus type 1 Vpr arrests the cell cycle in G_2 by inhibiting the activation of p34cdc2-cyclin B. *J Virol* **69**(11) 6859–6864.

236. He, J. *et al.* (1995). Human immunodeficiency virus type 1 viral protein R (Vpr) arrests cells in the G_2 phase of the cell cycle by inhibiting p34cdc2 activity. *J Virol* **69**(11), 6705–6711.

237. Masuda, M. *et al.* (2000). Genetic studies with the fission yeast *Schizosaccharomyces pombe* suggest involvement of *wee1, ppa2*, and *rad24* in induction of cell cycle arrest by human immunodeficiency virus type 1 Vpr. *J Virol* **74**(6) 2636–2646.

238. Elder, R. T. *et al.* (2000). Cell cycle G_2 arrest induced by HIV-1 Vpr in fission yeast (*Schizosaccharomyces pombe*) is independent of cell death and early genes in the DNA damage checkpoint. *Virus Res* **68**(2), 161–173.

239. Elder, R. T. *et al.* (2001). HIV-1 Vpr induces cell cycle G_2 arrest in fission yeast (*Schizosaccharomyces pombe*) through a pathway involving regulatory and catalytic subunits of PP2A and acting on both Wee1 and Cdc25. *Virology* **287**(2), 359–370.

240. Elder, R. T., Benko, Z. and Zhao, Y. (2002). HIV-1 VPR modulates cell cycle G_2/M transition through an alternative cellular mechanism other than the classic mitotic checkpoints. *Front Biosci* **7**, D349–357.

241. Gragerov, A. *et al.* (1998). HHR23A, the human homologue of the yeast repair protein RAD23, interacts specifically with Vpr protein and prevents cell cycle arrest but not the transcriptional effects of Vpr. *Virology* **245**(2), 323–330.

242. Mahalingam, S. *et al.* (1998). HIV-1 Vpr interacts with a human 34-kDa mov34 homologue, a cellular factor linked to the G_2/M phase transition of the mammalian cell cycle. *Proc Natl Acad Sci USA* **95**(7), 3419–3424.

243. Bouhamdan, M. *et al.* (1996). Human immunodeficiency virus type 1 Vpr protein binds to the uracil DNA glycosylase DNA repair enzyme. *J Virol* **70**(2), 697–704.

244. Zhao, L. J., Mukherjee, S. and Narayan, O. (1994). Biochemical mechanism of HIV-1 Vpr function. Specific interaction with a cellular protein. *J Biol Chem* **269**(22), 15577–15582.

245. Zhang, S. G. *et al.* (2001). Cytoplasmic retention of HIV-1 regulatory protein Vpr by protein–protein interaction with a novel human cytoplasmic protein VprBP. *Gene* **V263**(NI–2), 131–140.

246. de Noronha, C. M. *et al.* (2001). Dynamic disruptions in nuclear envelope architecture and integrity induced by HIV-1 Vpr. *Science* **294**(5544), 1105–1108.

247. Moir, R. D. *et al.* (2000). Disruption of nuclear lamin organization blocks the elongation phase of DNA replication. *J Cell Biol* **149**(6), 1179–1192.

248. Roshal, M. *et al.* (2003). Activation of the ATR-mediated DNA damage response by the HIV-1 viral protein R. *J Biol Chem* **278**(28), 25879–25886.

249. Lupardus, P. J. *et al.* (2002). A requirement for replication in activation of the ATR-dependent DNA damage checkpoint. *Genes Dev* **16**(18), 2327–2332.

250. Cliby, W. A. *et al.* (2002). S phase and G_2 arrests induced by topoisomerase I poisons are dependent on ATR kinase function. *J Biol Chem* **277**(2), 1599–1606.

251. Hoover, T. *et al.* (1996). A nuclear matrix-specific factor that binds a specific segment of the negative regulatory element (NRE) of HIV-1 LTR and inhibits NF-kappa(B) activity. *Nucleic Acids Res* **24**(10), 1895–1900.

252. Smythe, C., Jenkins, H. E. and Hutchison, C. J. (2000). Incorporation of the nuclear pore basket protein Nup 153 into nuclear pore structures is dependent upon lamina assembly: evidence from cell-free extracts of *Xenopus* eggs. *EMBO J* **19**(15), 3918–3931.

253. Jenkins, Y. *et al.* (2001). Nuclear export of human immunodeficiency virus type 1 Vpr is not required for virion packaging. *J Virol* **75**(17), 8348–8352.

254. Shi, L. *et al.* (1994). Premature p34cdc2 activation required for apoptosis. *Science* **263**(5150), 1143–1145.

255. Heald, R., McLoughlin, M. and McKeon, F. (1993). Human *Wee1* maintains mitotic timing by protecting the nucleus from cytoplasmically activated Cdc2 kinase. *Cell* **74**(3), 463–474.

256. Krek, W. and Nigg, E. A. (1991). Mutations of p34cdc2 phosphorylation sites induce premature mitotic events in HeLa cells: evidence for a double block to p34cdc2 kinase activation in vertebrates. *EMBO J* **10**(11), 3331–3341.

257. Lewinsohn, D. A. *et al.* (2002). HIV-1 Vpr does not inhibit CTL-mediated apoptosis of HIV-1 infected cells. *Virology* **294**(1), 13–21.

258. Coffin, J. M. Hughes, S. H. and Varmus, H. E. (eds) (1997). *Retroviruses*, p. 843. Cold Spring Harbor Laboratory Press: Cold Spring Harbor, NY.

259. Chao, S. H. *et al.* (2000). Flavopiridol inhibits P-TEFb and blocks HIV-1 replication. *J Biol Chem* **275**(37), 28345–28348.

260. Lind, K. E. *et al.* (2002). Structure-based computational database screening, *in vitro* assay, and NMR assessment of compounds that target TAR RNA. *Chem Biol* **9**(2), 185–193.

5 Negative-sense RNA Viruses and the Nucleus

Debra Elton and Paul Digard

Division of Virology, Department of Pathology, University of Cambridge, Tennis Court Road, Cambridge, UK

5.1 Introduction

Viruses with negative-sense single-strand RNA genomes form a broad group that includes several major genera (Tables 5.1, 5.2). These viruses can infect hosts throughout the animal and plant kingdoms and some are of great economic and/or medical importance. In terms of genome structure, phylogeny and replicative mechanisms, the negative-sense viruses fall into two broad categories: those with one strand of genomic RNA (order *Mononegavirales*) and those with multiple segments (Families *Arenaviridae, Bunyaviridae* and *Orthomyxoviridae*). There are clear functional and sequence homologies within the segmented and non-segmented families that strongly suggest that the current members of each group share a common ancestor (Tordo *et al.*, 1992; Pringle and Easton, 1997). In addition, there are enough similarities between the two orders to make it plausible that the segmented viruses are descended from a distant mononegavirus ancestor (Tordo *et al.*, 1992). Despite these evolutionary relationships, negative-sense RNA viruses show some surprising differences in their sites of genome transcription and replication. Like most RNA viruses, the majority of negative-sense viruses replicate in the cytoplasm. Of the segmented viruses, the families *Arenaviridae* and *Bunyaviridae* are all known to use the cytoplasm (Table 5.2), but the *Orthomyxoviridae* replicate in the nucleus. Within the order *Mononegavirales*, the *Filoviridae* and *Paramyxoviridae* replicate in the cytoplasm, the *Bornaviridae* in the nucleus, whereas the various members of the *Rhabdoviridae*, use either compartment (Table 5.1).

Viruses and the Nucleus   Edited by Julian A. Hiscox
© 2006 John Wiley & Sons, Ltd

Table 5.1   Replication sites for the order *Mononegavirales*

Family Subfamily	Genus	Type species	Replication site
Bornaviridae			
	Bornavirus	*Borna disease virus*	Nucleus
Filoviridae			
	'Marburg-like viruses'	*Marburg virus*	Cytoplasm
	'Ebola-like viruses'	Zaire *Ebola virus*	Cytoplasm
Paramyxoviridae *Paramyxovirinae*			
	Respirovirus	*Sendai virus*	Cytoplasm
	Rubulavirus	*Mumps virus*	Cytoplasm
	Morbiliivirus	*Measles virus*	Cytoplasm
Pneumovirinae			
	Pneumovirus	*Human respiratory syncytial virus*	Cytoplasm
	Metapneumovirus	*Turkey rhinotracheitis virus*	Cytoplasm
Rhabdoviridae			
	Vesiculovirus	*Vesicular stomatitis Indiana virus*	Cytoplasm
	Lyssavirus	*Rabies virus*	Cytoplasm
	Ephemerovirus	*Bovine ephemeral fever virus*	Cytoplasm
	Novirhabdovirus	*Infectious haematopoietic necrosis virus*	Cytoplasm
	Cytorhabdovirus	*Lettuce necrotic yellows virus*	Cytoplasm
	Nucleorhabdovirus	*Potato yellow dwarf virus*	Nucleus

The differing sites of replication used by the negative-sense viruses lead to fundamental differences in how they interact with their host cell nuclei. Before considering these interactions, a brief overview of their replication cycles follows. In all cases, the genome structure consists of one or more pieces of single-strand RNA encapsidated into ribonucleoprotein (RNP) structures. The major protein component is a single-strand RNA binding nucleoprotein (N or NP) which binds to RNA in a stoichiometric manner; the other essential component is the RNA-dependent RNA polymerase. In all animal negative-sense viruses, the genomic RNPs are packaged into enveloped virus particles which enter the cell by receptor-mediated fusion. For some plant-infecting species, however, it is not

Table 5.2 Viruses with segmented genomes

Family Subfamily	Genus	Type species	Replication site	Genome segments
Orthomyxoviridae			Nucleus	
Influenzavirus A		*Influenza A virus*		8
Influenzavirus B		*Influenza B virus*		8
Influenzavirus C		*Influenza C virus*		7
Thogotovirus		*Thogoto virus*		6
Bunyaviridae			Cytoplasm	3
	Bunyavirus	*Bunyamwera virus*		
	Hantavirus	*Hantaan virus*		
	Nairovirus	*Dugbe virus*		
	Phlebovirus	*Rift Valley fever virus*		
	Tospovirus	*Tomato spotted wilt virus*		
	Tenuivirus	*Rice stripe virus*	Cytoplasm	4–6
Arenaviridae	*Arenavirus*	*Lymphocytic choriomeningitis virus*	Cytoplasm	2

clear whether virus particles are ever formed and the infectious unit may be the RNPs themselves, which are transmitted by mechanical damage. Regardless of their route of entry, after their initial delivery to the cytoplasm the RNPs migrate to their site of replication. In all cases, the obligatory first transcriptional event is the partial transcription of the input genome to produce capped mRNAs (Figure 5.1; transcription). Once these mRNAs have been translated to produce viral proteins, the genome undergoes replicative transcription to produce a full-length uncapped molecule, which is also encapsidated to form an RNP. This positive-sense replicative intermediate is repeatedly transcribed to produce new unit-length negative-sense RNPs (Figure 5.1; replication). This pool of amplified vRNPs serve as transcriptional templates for secondary mRNA transcription and also as genomes for packaging into new virus particles. This basic scheme of genome transcription and replication is used by the orthomyxoviruses, where it occurs for each separate RNA segment (Figure 5.1A). The non-segmented viruses elaborate the process by introducing a decreasing gradient of mRNA transcription for the tandem array of virus genes contained on the genome, with the mRNA initiating at the 3′ end of vRNA transcribed at the highest level (Figure 5.1B). Bunyaviruses and arenaviruses blur the distinction between negative- and positive-stranded RNA viruses by possessing ambisense segments with non-overlapping genes encoded in both senses. Here, the cRNA replicative intermediate is transcribed into mRNA as well as replicated back to vRNA. After genome replication, events centre around virion assembly. The envelope is acquired by budding through a cellular membrane containing the viral glycoprotein(s). In most cases a viral matrix protein is thought to be integral

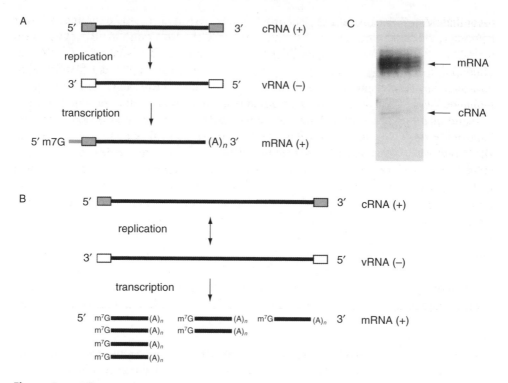

Figure 5.1 RNA transcription and replication strategies of negative-sense viruses. Basic life cycles of (A) segmented and (B) non-segmented viruses are depicted. 5′-mRNA cap-structures are indicated by m⁷G, poly(A) tails by $(A)_n$. Boxes represent RNA polymerase promoter sequences. (C) Primer extension analysis of the 5′-ends of positive-sense influenza A virus RNA. Note that mRNA species are heterogeneous and 10–15 nucleotides larger than cRNA because of the presence of host-derived capped primers

to the process, to drive membrane curvature and to link the RNPs to the virus envelope. However, a matrix protein has not yet been defined for the bunyaviruses.

5.2   The necessity of a nucleus

A theme of early work studying virus replication in tissue culture was to establish whether the host cell nucleus was essential for virus replication by examining the growth of the virus in enucleated cells. This type of experiment established that there were two classes of negative-sense virus: those that could replicate at least to some extent in the absence of a nucleus, and those that could not (Follet *et al.*, 1974, 1975, 1976; Kelly *et al.*, 1974; Banerjee *et al.*, 1976). The former class, such as the rhabdovirus vesicular stomatitis virus (VSV), were presumed to be viruses which replicated in the cytoplasm. The latter group, including influenza A virus, could represent an RNA virus which either replicated in the nucleus or a cytoplasmic replicating virus which required a host function dependent on the nucleus. Although earlier experiments on influenza virus had provided evidence that host-cell DNA was somehow involved in virus replication (Barry *et al.*, 1962), and even earlier studies had found viral antigens in the nucleus (Breitenfeld and Schafer, 1957), the actual site of virus transcription remained controversial. For over a decade numerous investigations produced conflicting results and it was not until the early 1980s that the question was regarded as finally solved. Studies using non-aqueous cell-fractionation (Herz *et al.*, 1981) and *in situ* analysis of the localization of viral RNA products (Jackson *et al.*, 1982) provided independent convincing evidence that influenza virus transcription occurred in the nucleus. Thus, the influenza viruses replicate in the nucleus. For many years this was regarded as a peculiarity of the orthomyxoviruses. However, in more recent years, non-segmented viruses have been identified that also synthesize their RNA in the nucleus. Initial characterization of Borna disease virus (BDV) suggested that it was an RNA virus with nuclear involvement in its life cycle but which did not need host cell transcription (Danner and Mayr, 1979). Subsequent work showed that BDV possessed a single segment of negative-stranded RNA that was not only associated with the nucleus of infected cells (de la Torre *et al.*, 1990; Carbone *et al.*, 1991) but was also transcribed and replicated in that compartment (Briese *et al.*, 1992; Cubitt and de la Torre, 1994). Similarly, work in the 1960s and 1970s on plant rhabdoviruses hinted at nuclear involvement (MacLeod *et al.*, 1966; Sinha, 1971; Christie *et al.*, 1974), such was subsequently confirmed by electron microscopic studies of virus morphogenesis and cell fractionation-based analysis of transcription (van Beek *et al.*, 1985; Ismail *et al.*, 1987; Wagner *et al.*, 1996).

Thus, individual families of segmented and non-segmented negative-sense viruses have evolved that use either the nucleus or the cytoplasm for RNA synthesis. The nuclear-replicating viruses interact with host cell nuclear function on many levels; both positively, in using cellular functions for their own benefit, and negatively, to prevent unwanted interference from the cell. The cytoplasmic replicating viruses do not necessarily need the host nucleus, but theirs is not a relationship of total indifference. At least in some cases, it is largely an antagonistic one in which the virus seeks to effectively enucleate the cell. The following sections will discuss these various virus–host interactions according to the class of cellular function that is being considered.

5.3 Nuclear import machinery

On infection of the cell, the nuclear-transcribing viruses must first deliver their genomes to that compartment. As the negative-sense genomes are always found as RNPs, this entails the transport of a mega-Dalton molecular weight assembly through the nuclear pore complex. An active transport mechanism is therefore necessary and in all cases so far studied the viruses have evolved to use the classical importin-α-mediated host cell nuclear import machinery (Gorlich and Kutay, 1999). Nuclear localization signals (NLSs) have been identified in the nucleoproteins of most orthomyxoviruses (reviewed in Portela and Digard, 2002), BDV (Pyper and Gartner, 1997; Kobayashi *et al.*, 1998) and the nucleorhabdovirus sonchus yellow net virus (SYNV) (Goodin *et al.*, 2001). In influenza A virus, the best-studied case, NP has been shown to be sufficient to mediate nuclear import of the virus genome (O'Neill *et al.*, 1995). However, additional NLSs have been shown to be present in many of the other protein components of the various viral RNPs, such as the polymerase and accessory proteins (Nath and Nayak, 1990; Mukaigawa and Nayak, 1991; Nieto *et al.*, 1994; Shoya *et al.*, 1998; Goodin *et al.*, 2001; Walker and Lipkin, 2002; Fodor and Smith, 2004). These may provide a degree of redundancy to the nuclear targeting of the infecting RNPs, but are most likely also essential for the individual nuclear import of newly translated RNP components during genome replication. Similarly, other viral structural and non-structural components also contain active NLSs, such as the influenza virus matrix (M1) and NS1 polypeptides (Greenspan *et al.*, 1988; Ye *et al.*, 1995) and the BDV p10 protein (Wolff *et al.*, 2002).

As often as not, the NLSs in the nuclear-targeted negative-sense virus polypeptides are not easily identifiable by comparison with the known consensus sequences of cellular and simian virus 40 type NLSs. Although bipartite NLSs (Robbins *et al.*, 1991) have been identified in orthomyxovirus and nucleorhabdovirus NP and polymerase subunits (Weber *et al.*, 1998; Goodin *et al.*, 2001; Nieto *et al.*, 1994) and the BDV phospho (P) protein (Shoya *et al.*, 1998; Schwemmle *et al.*, 1999), many other characterized signals bear no strong resemblance to previously characterized NLSs (Mukaigawa and Nayak, 1991; Nieto *et al.*, 1994; Wang *et al.*, 1997; Wolff *et al.*, 2002; Fodor and Smith, 2004). However, where these atypical signals have been fully characterized (influenza A NP and BDV p10), they have been shown nevertheless to function through binding cellular importin-α (O'Neill and Palese, 1995; Wang *et al.*, 1997; Wolff *et al.*, 2002). Another striking feature of the nuclear import of negative-sense virus proteins is that there are often multiple NLSs per polypeptide (Greenspan *et al.*, 1988; Pyper and Gartner, 1997; Wang *et al.*, 1997; Martins *et al.*, 1998; Shoya *et al.*, 1998; Weber *et al.*, 1998). This could represent an adaptation for efficient utilization of the different isotypes of importin-α (Kohler *et al.*, 1999). However, it may reflect nothing more than the fact that NLSs generally contain basic amino acids and that many of the proteins in question are highly basic RNA-binding proteins.

Nuclear import of orthomyxovirus RNPs at the start of infection is also targeted by cellular innate antiviral mechanisms. The human MxA protein has been shown to inhibit nuclear entry of Thogoto virus RNPs (Kochs and Haller, 1999b). MxA also inhibits multiplication of influenza A virus, but probably at a later, transcriptional, stage (Pavlovic *et al.*, 1992). Nevertheless, both inhibitory mechanisms result from an interaction of MxA with the RNP, most likely with NP itself (Kochs and Haller, 1999a; Weber *et al.*, 2000; Turan *et al.*, 2004).

Perhaps unexpectedly, negative-sense viruses that replicate in the cytoplasm also interact with the nuclear import machinery. Although the formation of cytoplasmic inclusion bodies (such as Negri bodies by rabies virus) is diagnostic for these viruses, many induce nuclear inclusions too. These could represent abnormal accumulations of cellular proteins, but specific viral proteins are also found in the nucleus. In the case of rabies virus, the P gene produces 5 carboxy-co-terminal polypeptides by a process of leaky ribosomal scanning. The P and P2 polypeptides are cytoplasmic, but the shorter P3, P4 and P5 proteins are mainly nuclear (Chenik *et al.*, 1995). The small non-structural protein (NSs) of the bunyavirus Rift Valley fever virus forms fibrillar structures in the nucleus (Struthers and Swanepoel, 1982; Yadani *et al.*, 1999). The possible function of these nuclear proteins will be discussed later. A 28 kDa fragment of the Pichinde arenavirus NP accumulates in the nuclei of infected cells (Young *et al.*, 1987), but the significance of this is unknown. However, a well-studied example of a cytoplasmically replicating negative-sense virus that interacts with the host nuclear import machinery is VSV. Initial observations suggested that the matrix (M) protein entered the nucleus of infected cells (Lyles *et al.*, 1988) and subsequent studies showed that a significant fraction is associated with the nuclear pore complex (Petersen *et al.*, 2000; von Kobbe *et al.*, 2000). Nuclear localization of M is achieved through signal-mediated active transport, using multiple sequences that again are not obviously related to cellular NLSs (Glodowski *et al.*, 2002). A number of functions have been proposed for the nuclear localization of VSV M. As well as effects on cellular transcription, discussed below, several studies have shown that the matrix proteins from VSV and other vesiculoviruses are effective inhibitors of nucleocyto-plasmic transport (Her *et al.*, 1997; von Kobbe *et al.*, 2000; Petersen *et al.*, 2000, 2001). In microinjected *Xenopus* oocytes, M inhibited multiple transport pathways, including importin-α-mediated import and snRNP import (Her *et al.*, 1997), although this finding was not replicated in mammalian cells (von Kobbe *et al.*, 2000). In addition to M, the first VSV transcription product, the 49 nucleotide leader RNA, transiently localizes to the nucleus early in infection (Kurilla *et al.*, 1982). The mechanisms responsible for this temporally regulated import of the RNA are unknown but its function will be returned to later. In addition, fluorescently labelled glycoprotein (G) from infecting virions accumulates at the nucleus and although the significance of this remains to be determined, it was suggested that it results from RNP uncoating occurring in the perinuclear region (Da Poian *et al.*, 1996).

5.4   The cellular transcription machinery

5.4.1   RNA pol II transcription

In many ways, influenza A virus provided an early paradigm for the unexpected means by which an RNA virus might subvert normal host processes. Initial observations that cellular DNA function was required for influenza virus RNA synthesis (Barry *et al.*, 1962; Mahy *et al.*, 1977) were refined by the use of the specific host RNA polymerase II inhibitor, α-amanitin. If added early enough, this drug inhibited all detectable viral RNA synthesis unless the cells carried a mutant polymerase that was resistant to the drug (Rott and Scholtissek, 1970; Mahy *et al.*, 1972; Spooner and Barry, 1977; Mark *et al.*, 1979). With no evidence for reverse transcription in the influenza virus life-cycle, these results seemed inexplicable. However, like cellular mRNA, viral mRNA synthesized *in vivo* contained 5′ cap-structures (Krug *et al.*, 1976). In contrast, the positive-sense RNA synthesized *in vitro* by RNPs packaged in virions was not capped or methylated, although it was polyadenylated (Plotch and Krug, 1977; Plotch *et al.*, 1978). Thus, the seemingly paradoxical requirement for pol II transcription *in vivo* was explained by the finding that cap structures from cellular mRNAs are recycled as primers for influenza virus mRNA synthesis (Plotch *et al.*, 1979, 1981). In a reaction confirmed by sequence analysis of *in vivo* produced mRNA (Dhar *et al.*, 1980) and now extensively characterized at the molecular level (reviewed by Elton *et al.*, 2002), the influenza virus RNA polymerase binds to methylated cap structures on host pre-mRNAs. It then cleaves the cellular RNA around 10–15 bases 3′ to the cap structure and uses the capped RNA fragment as a primer for transcription initiation (Figure 5.1A). Influenza virus mRNAs thus contain a host-derived sequence at the 5′-end that is heterogeneous in sequence and length (Figure 5.1C). A similar 'cap-snatching' mechanism is used by all other negative-sense segmented virus families so far studied, including arenaviruses, bunyaviruses and tenuiviruses (Bishop *et al.*, 1983; Patterson *et al.*, 1984; Raju *et al.*, 1990; Huiet *et al.*, 1993), even though these viruses replicate in the cytoplasm (Rossier *et al.*, 1986; Southern, 1996; Estabrook *et al.*, 1998). However, it is less clear whether these viruses require post-infection pol II transcription for their replication. Studies on bunyaviruses suggest that ongoing cellular transcription is not needed (Bouloy and Hannoun, 1973; Vezza *et al.*, 1979; Raju and Kolakofsky, 1988), possibly reflecting a greater steady-state pool of capped RNAs in the cytoplasm. Indeed, more recent studies suggest that they actively inhibit host cell transcription (Billecocq *et al.*, 2004, Le May *et al.*, 2004, Thomas *et al.*, 2004). The NSs protein of bunyamwera virus inhibits the specific phosphorylation of the C-terminal domain of RNA Pol II required for the transition from initiation to elongation modes of transcription (Thomas *et al.*, 2004). The NSs polypeptide of Rift Valley fever virus causes the same ultimate effect, but by reducing the functional levels of the essential transcription factor TFIIH, possibly by sequestering it in the nuclear fibrillar structures referred to earlier (Le May *et al.*, 2004). However, studies on the sensitivity of arenaviruses to α-amanitin and actinomycin D have generally (but not always) shown a sensitivity to transcription inhibitors and removal of the nucleus (Banerjee *et al.*, 1976; Rawls *et al.*, 1976; Mersich *et al.*, 1981; Lopez *et al.*, 1986). The reason for the difference between the two virus families is not known. Many arenavirus infections are not particularly cytopathic, but persistent infections *in vitro* and *in vivo* with the arenavirus lymphocytic choriomeningitis virus (LCMV) show subtle effects on host-cell transcription, with specific downregulation of growth hormone gene transcription (Oldstone *et al.*, 1982; de la Torre and Oldstone, 1992). This effect maps to the glycoprotein (G) gene of the virus (Teng *et al.*, 1996) and is mediated through the cellular growth hormone gene-specific transcription factor Pit 1 (de la Torre and Oldstone, 1992), presumably via an as-yet uncharacterized signalling pathway.

Mononegaviruses that replicate in the cytoplasm, such as VSV or rabies virus (Wagner *et al.*, 1972), do not require host transcription to support their replication (Black and Brown, 1968; Villarreal and Holland, 1976). The lack of any requirement for host cell transcription can be explained by the fact that, in contrast to the segmented viruses, the mononegavirus polymerase proteins possess RNA capping and methylation activities (Abraham *et al.*, 1975). Moreover, at least some members of the group actively inhibit cellular transcription (Weck and Wagner, 1978). In the case of VSV, this inhibition is multifactorial. The matrix (M) protein is sufficient to cause a potent inhibition of host cell pol I and pol II transcription (Black and Lyles, 1992; Ahmed and Lyles, 1998; Chiou *et al.*, 2000). The mechanism of this downregulation is not fully understood but involves an indirect inhibitory effect of M on the pol II transcription factor TFIID (Yuan *et al.*, 1998, 2001), that may arise from the M-induced block in nuclear import/export (Her *et al.*, 1997; von Kobbe *et al.*, 2000). In addition, there is also evidence that the viral leader transcript can downregulate cellular transcription (Grinnell and Wagner, 1983). The precise mechanism here is not known, but the leader RNA inhibits *in vitro* transcription reactions carried out by cell extracts (McGowan *et al.*, 1982) and it has been hypothesized that this may occur through its binding to cellular La protein and/or heterogeneous nuclear ribonucleoprotein U (Kurilla and Keene, 1983; Gupta *et al.*, 1998).

Mononegaviruses which replicate in the nucleus apparently do not require host cell transcription (Danner and Mayr, 1979; Pyper *et al.*, 1998; Mizutani *et al.*, 1999). Sequence analyses of the polymerase genes of BDV and the nucleorhabdovirus Sonchus yellow net virus suggest that these enzymes also possess RNA capping and methylation activities (Choi *et al.*, 1992; Cubitt *et al.*, 1994a). Consistent with this, analysis of the 5′-end of BDV mRNAs has excluded the possibility that they are formed from host-cell donors (Schneeman *et al.*, 1994).

5.4.2 Post-transcriptional processing

Several of the negative-sense viruses that replicate in the nucleus have evolved strategies to take advantage of other host nuclear functions and again, influenza A virus serves as the first and best-studied example. One such function is the process of host cell RNA methylation. This is crucial for virus RNA synthesis, as recognition of donor host-cell mRNA cap structures by the influenza virus polymerase requires that the cap is fully methylated (Bouloy *et al.*, 1980). In addition, viral mRNA itself is methylated on internal adenine residues (Krug *et al.*, 1976; Narayan *et al.*, 1987), although the significance of this is not known. However, various methylation inhibitors interfere with the synthesis and transport of viral mRNA (Ransohoff *et al.*, 1987; Vogel *et al.*, 1994). This may simply reflect a reduction in functional cap-donor molecules for synthesis of mature viral mRNA, or it may be that methylation of viral RNA itself plays a role in regulation of transport.

An interaction of unquestionable importance for influenza and Borna disease viruses is that with the cellular RNA splicing machinery. BDV and all orthomyxoviruses so far studied access alternative reading frames in their genome through cell-mediated differential splicing of primary viral transcripts. This first became apparent for influenza A virus, when a ninth viral polypeptide was identified as the product of the eighth genome segment (Lamb *et al.*, 1978; Inglis *et al.*, 1979; reviewed by Krug *et al.*, 1989). It was subsequently shown that mRNA transcripts from the two smallest segments of the influenza A genome undergo alternative splicing reactions to generate five different mRNA species (Inglis *et al.*, 1979; Lamb and Lai, 1980, 1982, 1984; Lamb *et al.*, 1981). Similar analysis of BDV has shown that the genome contains at least three introns that are processed by the cellular splicing machinery (Cubitt *et al.*, 1994b, 2001; Schneider *et al.*, 1994; Tomonaga *et al.*, 2000). Curiously, however, there is no evidence so far for splicing in the nucleorhabdoviruses.

The use of alternative splicing strategies increases the coding capacity of relatively small RNA genomes and presents opportunities for regulating viral gene expression. However, unlike the majority of cellular pre-mRNAs, the primary unspliced viral transcripts have crucial coding functions. In the case of influenza A virus, the unspliced transcript from segment 7 encodes the M1 polypeptide, the major structural component of the virion (Lamb *et al.*, 1981), while the unspliced transcript from segment 8 is translated to produce the equally abundant NS1 protein (Lamb and Lai, 1980). The spliced products of segment 7 produce either the minor virion component M2 (an ion channel) or a hypothetical nine amino acid peptide that has never been detected in infected cells. The spliced mRNA2 from segment 8 encodes the minor virion component NS2. In the case of BDV, the two major introns occur in the matrix (M) and glycoprotein (G) coding regions (Cubitt *et al.*, 1994b; Schneider *et al.*, 1994), while splicing of the third intron is predicted to lead to the production of two proteins of unknown function (Cubitt *et al.*, 2001).

In these viruses, efficient replication presumably depends on inefficient and/or regulated use of the splice sites present in the viral mRNAs. Splicing of influenza A virus segment 7 in infected cells is inefficient, with the majority of mRNAs remaining intact to produce the M1 polypeptide. However, pol II-directed expression of segment 7 cDNA in the absence of other influenza virus proteins produces transcripts which are largely spliced down to the vestigial nine amino-acid M3 ORF (Lamb and Lai, 1982; Valcarcel *et al.*, 1993). Comparison of the influenza splice sites with cellular sequences shows that the 5′ proximal site used to produce the M3 transcript is a close match to the cellular consensus. However, the 5′-distal site used to produce the M2 mRNA is a relatively poor match. Clearly, therefore, a mechanism exists in infected cells to downregulate use of the strong 5′-splice site and this is thought to be mediated by the viral polymerase. All influenza virus mRNAs contain a copy of the polymerase-binding site downstream of the host-derived primer (Figure 5.1A), and binding of the polymerase to this sequence in segment 7 transcripts blocks the strong 5′-proximal splice site. In the presence of excess polymerase, the system is biased towards use of the inefficient downstream 5′-splice site to produce low levels of the M2 mRNA and a majority of the unspliced M1 mRNA (Shih *et al.*, 1995). In addition, host cell RNA pol II transcription and pre-mRNA processing events occur in a multiprotein complex and thus are linked (Hirose and Manley, 2000; Hammell *et al.*, 2002).

Thus, influenza virus transcripts may be intrinsically poor substrates for splicing, simply because they are produced by the viral polymerase rather than the RNA pol II complex (Li *et al.*, 2001). Influenza A virus segment 8 mRNA appears to be an inefficient substrate for splicing (Lamb and Lai, 1984; Alonso-Caplen and Krug, 1991), but expression of the unspliced NS1 and spliced NS2 ORFs is still modulated in infected cells. This may involve autoregulation through NS1 itself (Smith and Inglis, 1985; Alonso-Caplen and Krug, 1991; Alonso-Caplen *et al.*, 1992). There is much evidence to show that NS1 is a general inhibitor of splicing that functions by binding to U6 snRNAs and also causes the redistribution of splicing factors within the nucleus (Fortes *et al.*, 1994, 1995; Lu *et al.*, 1994; Qiu *et al.*, 1995; Wang and Krug, 1998; Wolff *et al.*, 1998). Nevertheless, a clear explanation of how these general negative effects on splicing would benefit the virus is lacking. However, NS1 also affects the nuclear export of mRNA and it is possible that this indirectly affects splicing of viral transcripts (Alonso-Caplen and Krug, 1991; Alonso-Caplen *et al.*, 1992), as discussed below.

In the case of BDV, inefficient splicing of mRNA also occurs in infected cells but not in cells transfected with viral cDNA under the control of a pol II promoter (Jehle *et al.*, 2000; Tomonaga *et al.*, 2000). Splicing of the plasmid-derived BDV transcript was not altered by co-expression of individual virus polypeptides or, indeed, by concomitant virus infection (Jehle *et al.*, 2000; Tomonaga *et al.*, 2000). This supports the hypothesis that transcripts synthesized by a viral polymerase are inherently poor substrates for splicing because they are not associated with the host pol II transcriptosome.

In addition to its involvement with early post-transcription events, influenza virus also interferes with the final processing step of pre-mRNA maturation; 3'-end formation. Normally, this occurs through two reactions: endonucleolytic cleavage of the primary transcript and polyadenylation of the new 3'-end (Colgan and Manley, 1997). Without these processing events, most mRNAs are not efficiently exported to the cytoplasm (Huang and Carmichael, 1996; Hammell *et al.*, 2002). Influenza A virus infection results in the inhibition of host gene expression (Katze and Krug, 1984) and this has been attributed in part to selective inhibition of the nuclear export of cellular, but not viral, mRNAs by the NS1 protein (Fortes *et al.*, 1994; Qian *et al.*, 1994; Qiu and Krug, 1994). The current hypothesis is that NS1 inhibits both the RNA cleavage event and the subsequent polyadenylation reaction (Nemeroff *et al.*, 1998; Chen *et al.*, 1999; Shimizu *et al.*, 1999). Inhibition of the initial cleavage step is not total (Chen *et al.*, 1999) but the additional inhibition of the polyadenylation reaction achieves a very efficient block to 3'-end processing and thus RNA export (reviewed by Chen and Krug, 2000). At the molecular level, NS1 binds to the 30 kDA subunit of the cellular cleavage and polyadenylation stimulation factor and inhibits binding of this multiprotein complex to the RNA substrate (Nemeroff *et al.*, 1998). NS1 also binds to poly(A)-binding protein II and prevents it from stimulating poly(A) polymerase (Chen *et al.*, 1999). Selective inhibition of cellular RNA processing occurs because viral mRNAs are polyadenylated through a totally separate reaction carried out by the viral polymerase (reviewed by Elton *et al.*, 2002). In addition, as alluded to above, the ability of NS1 to inhibit 3'-end formation may contribute to its negative effects on splicing (Li *et al.*, 2001) because these processes are normally linked in cells through the integrated function of the pol II 'transcriptosome' (Hirose and Manley, 2000).

5.5   Nuclear architecture

In many cell types the nucleus is the largest single organelle, often with a greater volume than the remainder of the cell. Although this large compartment is not further subdivided by membranes, it does contain multiple organizational domains (see Chapter 1). In the herpesvirus field in particular, there is much data showing that viruses take advantage of specific features of the nuclear architecture (see Chapter 3). The study of nuclear-replicating RNA viruses is less advanced in this regard but there is evidence that they also do not use the nucleus as a large undifferentiated space.

In the case of the nucleorhabdoviruses, the gross morphological changes in the nuclei of infected cells provided the initial evidence that they replicate in this compartment. The nuclei of infected cells swell to two or three times their original volume and large intranuclear inclusion bodies form that are visible by light microscopy (Sinha, 1971; Christie *et al.*, 1974; van Beek *et al.*, 1985; Ismail *et al.*, 1987; Martins *et al.*, 1998). These inclusion bodies are of two types: amorphous 'viroplasms' and aggregations of virus particles, often surrounded by a single membrane (Sinha, 1971; van Beek *et al.*, 1985; Ismail *et al.*, 1987). Immuno-electron microscopy has shown that the viroplasms contain the protein components of the RNP and that expression of the nucleoprotein alone leads to formation of a similar structure (Martins *et al.*, 1998). There is also evidence that the viroplasms exclude host-cell chromatin (Christie *et al.*, 1974; Martins *et al.*, 1998), which provides an interesting parallel with the way herpesvirus DNA replication compartments reorganize the host DNA (Monier *et al.*, 2000; Chapter 3). Formation of these viroplasms is the earliest event that can be detected after infection, preceding the appearance of large numbers of RNP particles (van Beek *et al.*, 1985). RNP cores acquire a membrane by budding through the inner nuclear membrane into the perinuclear space (see following section), but if this process is blocked by the glycosylation inhibitor tunicamycin, the viroplasms swell to fill almost the whole volume of the nucleus (van Beek *et al.*, 1985). Therefore, it seems likely that these are classic viral inclusion bodies in which the components required for new RNPs are concentrated prior to their assembly.

The novel membranous inclusion bodies formed within the nucleus of cells infected by nucleorhabdoviruses contain enveloped virus particles and most likely result from invagination of the inner nuclear membrane (Sinha, 1971; van Beek *et al.*, 1985). Strikingly, the amount of membrane required to surround the intracellular virus particles was estimated to be ten times that of the normal inner nuclear envelope, suggesting that the virus induces considerable membrane synthesis (van Beek *et al.*, 1985). Presumably these morphological changes in nuclear structure involve the specific intervention of viral proteins in normal cellular processes, but the molecular details are currently unknown.

Although BDV tends to establish persistent infections and is not particularly cytopathic, it also induces nuclear inclusion bodies that contain the viral N, phosphoprotein (P) and p10 polypeptides (Herzog and Rott, 1980; Thierer *et al.*, 1992; Wehner *et al.*, 1997). These inclusion bodies might represent sites of RNP assembly, similar to those of the nucleorhabdoviruses. In addition, it has been suggested that BDV uses the nucleolus as the site for RNA synthesis (Pyper *et al.*, 1998). In some but not all infected rat brain cells at early times post-infection, *in situ* RNA hybridization experiments showed positive-sense RNA in the nucleolus and negative-sense (genomic) RNA around its periphery. Cell fractionation experiments also provided evidence for the association of N and P polypeptides with nucleolar fractions (Pyper *et al.*, 1998). However, immunofluorescence images of the nuclear inclusion bodies in infected cells are not characteristic of nucleoli (Herzog and Rott, 1980; Thierer *et al.*, 1992; Wehner *et al.*, 1997) and no co-localization studies of viral and the appropriate cellular antigens have yet been reported. A recent study demonstrated a modest increase in the length of the cell cycle of BDV-infected fibroblasts that resulted from an increase in the length of the G_2 phase (Planz *et al.*, 2003). This delay in progression across the G_2–M checkpoint was linked to an interaction between the viral nucleoprotein and the cellular cdc2–cyclin B1 complex. The function of this phenomenon in the virus life cycle is not clear but it has been suggested that it may reflect the fact that the virus normally replicates in non-dividing neuronal cells. (Planz *et al.*, 2003).

In contrast to the nucleorhabdoviruses, the orthomyxoviruses do not cause gross structural changes to the nucleus. However, there is evidence that they do interact with structural as well as enzymatic elements of the compartment. Biochemical fractionation and *in situ* autoradiography of pulse-labelled RNA indicate that a substantial fraction of RNPs are bound tightly to insoluble 'nuclear matrix' or chromatin components (Bukrinskaya *et al.*, 1979; Jackson *et al.*, 1982; Lopez-Turiso *et al.*, 1990; Bui *et al.*, 2000). Immunofluorescence analysis of the intracellular distribution of NP in infected cells during the first half or so of the infectious cycle generally shows predominantly nuclear staining (Breitenfeld and Schafer, 1957; Martin and Helenius, 1991), consistent with the known location of viral RNA synthesis. In recent years, the advent of techniques for higher-resolution optical microscopy are permitting this general picture to be refined. Single optical sections taken through the nuclei of infected fibroblast cells usually show NP distributed in a speckled pattern throughout the nucleus, with the exception of nucleoli (Martin and Helenius, 1991; Elton *et al.*, 2001, 2005; Ma *et al.*, 2001) (Figure 5.2A). However, when nuclear export of RNPs is blocked by treatment of the cells with the toxin leptomycin B (LMB; see below), then the RNPs redistribute to the nuclear periphery (Elton *et al.*, 2001; Ma *et al.*, 2001). More recent analysis has shown that this change in intranuclear distribution of NP is not limited to drug-treated cells but is a normal feature of infection, best seen in cells of epithelial origin. In this type of cell, NP is distributed throughout the nucleus for the first 2 h or so of infection (using reasonably high multiplicities of infection), but then redistributes to the nuclear periphery over the next few hours before the onset of nuclear export (Figure 5.2A; Elton *et al.*, 2005). The molecular mechanisms that control and mediate this specific intranuclear distribution of the protein have yet to be

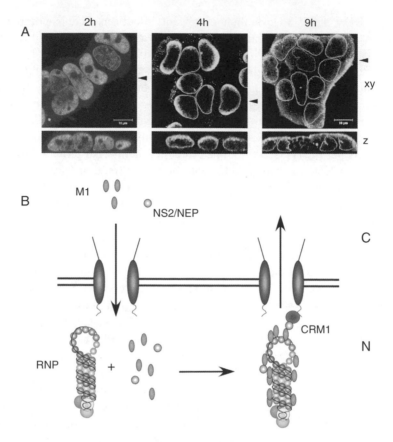

Figure 5.2   Intracellular trafficking of influenza virus RNPs. (A) Immunofluorescent staining of NP and nuclear lamina-associated polypeptide 2 of 293-T cells infected with influenza virus A/PR/8/34 at the indicated times post-infection. Adapted from Elton *et al.* (2005). (B) Diagrammatic model of influenza virus RNP nuclear export. Viral M1 and NS2/NEP polypeptides enter the nucleus, bind sequentially to the RNPs and are then exported through the nuclear pore complex because of interactions between NS2/NEP and cellular CRM1

elucidated, but it is worth noting that the viral M1 and NEP polypeptides do not show a similar redistribution in response to LMB treatment (Elton *et al.*, 2001; Ma *et al.*, 2001) or during normal infection of epithelial cells (Elton *et al.*, 2005). Furthermore, NP expressed in the absence of other viral proteins shows a similar ability to localize to the periphery of the nucleus, suggesting that it is interacting with a cellular structure (Elton *et al.*, 2005). The bias NP exhibits towards regions of the nucleus facing the apical plasma membrane prior to RNP nuclear export correlates with the apical distribution of the viral membrane proteins and, indeed, the site of virus budding (Barman *et al.*, 2001). Thus, polarized intranuclear trafficking of the RNPs may provide a functional advantage to the virus by directing the virus genome towards the correct surface of the cell for virion assembly before it leaves the nucleus. The temporally distinct patterns of RNP localization within the nucleus might also have functional consequences for viral RNA synthesis. One study showed a correlation between attachment to the insoluble nuclear matrix and the ability to synthesize negative-sense genomic RNA, with the less tightly bound RNPs biased towards mRNA synthesis (Lopez-Turiso *et al.*, 1990). The factors that control the balance between transcription and replication of the influenza virus genome are poorly understood (Elton *et al.*, 2002) and it is a fascinating possibility that the virus might use the spatial organization of the nucleus to differentiate the two processes. Although it is not known to what cellular structure(s) the RNPs bind, the interaction must be regulated, as late in infection the RNPs exit the nucleus. Nuclear export of RNPs is discussed in detail in the following section, but it is worth noting that there is *in vitro* evidence that the M1 protein binds to histones (Zhirnov and Klenk, 1997) and these authors proposed the interesting hypothesis that this activity of M1 serves to displace RNPs from chromatin prior to their nuclear export.

Cytoplasmic replicating negative-sense viruses have also been shown to interact with a nuclear structure. Infection with the arenavirus LCMV results in the dispersal of the cellular promyelocytic leukaemia protein (PML) from its characteristic punctate nuclear distribution to the cytoplasm (Borden *et al.*, 1997). Transfection experiments showed that this effect was due to the viral Z protein, and an interaction between the LCMV and Lassa fever virus Z proteins and PML was demonstrated (Borden *et al.*, 1998b). The same laboratory has also shown that both the PML and LCMV Z proteins interact with the ribosomal P proteins in the nucleus but that virus infection does not alter their distribution (Borden *et al.*, 1998a). Similarly, there is evidence that rabies virus modifies PML bodies. Virus infection increases the size of PML domains and alters the ultrastructural distribution of the PML protein within them (Blondel *et al.*, 2002). Transfection experiments indicate that this is a function of the P family of products, most likely the nuclear resident P3 protein, and is mediated through binding to PML (Blondel *et al.*, 2002).

5.6 Nuclear export

The eukaryotic nucleus has evolved numerous specific export pathways to direct traffic of RNA and proteins to the cytoplasm. Negative-strand viruses have also produced a variety of mechanisms that interfere with and subvert these cellular processes. Those viruses that replicate in the nucleus must export first their mRNAs and second their genomic RNPs. The cytoplasmic viruses do not have the same requirements, but in at least two instances they and their nuclear cousins benefit from preventing the nuclear export of cellular RNA. The mechanism by which the influenza A virus NS1 protein inhibits export of cellular mRNAs has already been described. The M protein of VSV (and related vesiculoviruses) also blocks nuclear export of cellular RNA, but by a different mechanism. Rather than interfering with the maturation of host cell mRNAs, VSV M binds to the Nup98 component of the nuclear pore complex (Petersen *et al.*, 2000; von Kobbe *et al.*, 2000). This leads to the inhibition of nuclear export pathways that depend on cellular CRM1, CAS and the relatively poorly characterized factor(s) responsible for mRNA export (von Kobbe *et al.*, 2000; Petersen *et al.*, 2000) and thus produces a wider spectrum of export inhibition than influenza NS1. As well as mRNA, ribosomal RNA, snRNPs and some proteins are retained by VSV M (Her *et al.*, 1997; von Kobbe *et al.*, 2000; Petersen *et al.*, 2000). This general export inhibition may be the primary cause of all other effects on host gene expression for which VSV M is responsible (von Kobbe *et al.*, 2000).

The means by which the nuclear replicating viruses export their mRNAs is very poorly understood. In the case of influenza A virus, export of viral mRNA is not blocked by NS1, because the viral polymerase polyadenylates its own messages in a reaction independent of the cellular cleavage and polyadenylation complex inhibited by NS1 (Chen and Krug, 1999). However, this does not provide an explanation for what promotes nuclear export of the predominantly unspliced viral mRNAs. No influenza virus counterpart of the retroviral Rev protein or *cis*-acting constitutive transport RNA element has been identified, and viral protein synthesis (and therefore presumably, mRNA export) is largely insensitive to the CRM1 inhibitor leptomycin B (Elton *et al.*, 2001). It is possible that viral transcripts do not undergo the same checks that a cellular mRNA undergoes before nuclear export because they are not synthesized by the pol II transcriptosome. On the other hand, there is evidence suggestive of regulated export of viral mRNA (Alonso-Caplen and Krug, 1991; Alonso-Caplen *et al.*, 1992; Vogel *et al.*, 1994). On a speculative note, the influenza A NP has been shown to interact with the cellular polypeptide UAP56/BAT1 (Momose *et al.*, 2001). Although originally identified as a cellular splicing factor (Fleckner *et al.*, 1997) and proposed to modulate influenza virus transcription (Shimizu *et al.*, 1994; Momose *et al.*, 2001), more recent work has implicated UAP56 as a linker molecule between the processes of mRNA splicing and export (Luo *et al.*, 2001; Kiesler *et al.*, 2002). It is widely accepted that NP does not encapsidate viral mRNAs as it does v- and cRNA, but the available evidence does not rule out some degree of association between the two molecules. The hypothesis that the NP-UAP56 interaction is involved in viral mRNA export would be an interesting one to test.

Non-segmented viruses that replicate in the nucleus face the same problem as the orthomyxoviruses: how to achieve nuclear export of mRNAs that are totally (the nucleorhabdoviruses) or largely (BDV) unspliced. Little is known of the events involved in nucleorhabdovirus transcription. However, a *cis*-acting RNA element has been identified in the unspliced transcripts of BDV that directs their cytoplasmic accumulation (Schneider *et al.*, 1997). Function of this sequence element did not depend on the presence of BDV polypeptides, suggesting that it interacts with an as-yet unidentified cellular transport factor (Schneider *et al.*,1997).

Towards the end of their replication cycle, nuclear-replicating viruses face the problem of how to get very large RNP particles across a nuclear envelope with relatively small gated channels to the cytoplasm. In addition, the process must be carefully regulated, as the negative-sense RNPs have two functions: to serve as templates for transcription in the nucleus and to be packaged into progeny virions for release from the cell. Therefore, inappropriately early export of newly assembled RNPs would adversely affect viral gene expression. The plant nucleorhabdoviruses apparently side-step the nuclear pore complex and achieve nuclear exit by budding their RNP cores through the inner nuclear envelope (Sinha, 1971; van Beek *et al.*, 1985; Ismail *et al.*, 1987). The process is not understood in any molecular detail and has largely been defined by electron microscopy. Accumulation of enveloped virus particles in the perinuclear space is a commonly observed feature, as is the presence of budding intermediates at the inner nuclear membrane (Figure 5.3). Inhibition of glycosylation leads to the failure of this budding event and the accumulation of crystalline arrays of RNPs in the nucleus (van Beek *et al.*, 1985). Virus particles can also be observed in the rough endoplasmic reticulum, implying that they may leave the cell by the exocytic pathway (van Beek *et al.*, 1985). Free RNPs are also seen in the cytoplasm and it has been suggested that these result from fusion of the viral membrane with that of the endoplasmic reticulum (van Beek *et al.*, 1985). This model for viral envelopment has many parallels with that suggested for the herpesviruses, which also bud through the inner nuclear membrane (Mettenleiter, 2002; see Chapter 3).

In contrast, influenza virus uses a conventional cellular nuclear export pathway for RNP trafficking. Nuclear export of influenza A virus RNPs is inhibited by the toxin leptomycin B (Elton *et al.*, 2001; Ma *et al.*, 2001; Watanabe *et al.*, 2001; Iwatsuki-Horimoto *et al.*, 2004), indicating the involvement of the cellular CRM1-dependent nuclear export pathway. It is thought that the viral M1 and NS2 proteins are responsible for targeting the RNPs to this particular export pathway. M1 binds directly to the RNP and silences its transcriptional activity (Zvonarjev and Ghendon, 1980), while NS2

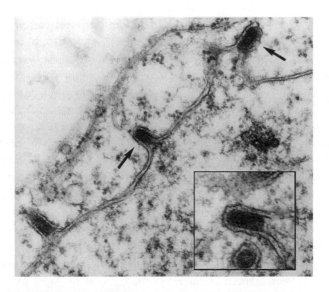

Figure 5.3 Electron micrographs showing nucleorhabdovirus RNPs (arrows, and inset at higher magnification) budding through the inner nuclear membrane of infected protoplasts. Adapted with permission from van Beek *et al.* (1985) Copyright © 1985, Elsevier

binds to M1–RNP complexes (Yasuda *et al.*, 1993). NS2 contains a functional nuclear export signal that interacts with cellular CRM1 and this directs the transport of the entire complex to the cytoplasm (Figure 5.2B; O'Neill *et al.*, 1998; Neumann *et al.*, 2000; Akarsu *et al.*, 2003). The NS2 proteins of influenza B and C viruses have also been shown to have nuclear export activity, suggesting the presence of a similar export mechanism in these orthomyxoviruses (Paragas *et al.*, 2001), although it should be noted that Thogoto virus does not obviously code for an NS2 analogue (Wagner *et al.*, 2001). There is a large body of experimental evidence that supports the M1/NS2 export hypothesis. Evidence for an essential role for M1 comes from microinjection of anti-M1 antibodies, the examination of defective viruses, heat shock and the use of a drug that blocks late viral gene expression; all circumstances which can lead to a block in RNP nuclear export (Martin and Helenius, 1991; Whittaker *et al.*, 1995; Bui *et al.*, 2000; Sakaguchi *et al.*, 2003). Evidence for NS2 involvement comes from microinjection of anti-NS2 IgG (O'Neill *et al.*, 1998) and, most compellingly, from the creation of a virus lacking NS2. This virus grows normally in the presence of complementing NS2 and can initiate infection of normal cells, but RNP export then fails and only a single round of infection is achieved (Neumann *et al.*, 2000). However, some evidence is at variance with this hypothesis. Several studies show RNP export occurring in the presence of much reduced amounts or even the complete absence of NS2 (Wolstenholme *et al.*, 1980; Bui *et al.*, 2000; Elton *et al.*, 2001; Huang *et al.*, 2001). In one case, RNP export occurred with undetectable amounts of M1 (Mahy *et al.*, 1977). Furthermore, NP itself contains a nuclear export signal, interacts with CRM1 and shows leptomycin B-sensitive localization in the absence of other viral proteins (Whittaker *et al.*, 1996; Neumann *et al.*, 1997; Elton *et al.*, 2001).

The M1/NS2 hypothesis goes some way towards explaining the temporal regulation of RNP localization observed in infected cells. The exact timing varies, depending on virus strain and cell type, but in synchronous infections the onset of detectable RNP export occurs at a predictable time. Around an hour later, the nuclei are apparently devoid of NP in the majority of cells (Breitenfeld and Schaffer, 1957). Both the M1 and NS2 polypeptides are considered to be late gene products whose synthesis lags behind the early components of NP and NS1 (Skehel, 1973; Lamb *et al.*, 1978). However, it is not clear whether the delayed synthesis of M1 and NS2 is sufficient to account for the definite switch between nuclear and cytoplasmic accumulation of NP. There is also evidence that a phosphorylation event is needed as well as M1 synthesis, although the target of this modification is not known (Bui *et al.*, 2000; Pleschka *et al.*, 2001). It has also been proposed that cellular caspase activity is needed for efficient nuclear export of RNPs, possibly through modification of the nuclear pore complex (Wurzer *et al.*, 2003).

The BDV N protein also contains nuclear import and export activities and in transient transfection experiments is retained in the nucleus by leptomycin B treatment (Kobayashi *et al.*, 2001). The nuclear export signal has been mapped to a leucine-rich sequence typical of the nuclear export signals (NESs) recognized by CRM1 (Kobayashi *et al.*, 2001). Furthermore, the NES sequence in N coincides with the sequence recognized by the viral P protein (Berg *et al.*, 1998) and co-expression of P prevents cytoplasmic accumulation of N in a titratable manner (Kobayashi *et al.*, 2001). Thus, these authors suggested the hypothesis that BDV RNPs are exported through the CRM1 pathway using the NES in N and that the activity of the latter signal is controlled by the relative concentration of P protein in the nucleus. An alternative hypothesis for RNP export revolves around the viral p10 polypeptide. This protein contains a leucine-rich sequence akin to an NES, binds to P (and thus indirectly to N and RNPs) and colocalizes with them in the nucleus (Schwemmle *et al.*, 1998; Malik *et al.*, 2000; Wolff *et al.*, 2000). Thus, it has been suggested that p10 has an analogous function to the influenza virus NS2 protein (Schwemmle *et al.*, 1998). However, the putative NES sequence is also the binding site for the P protein, so it is not obvious how a p10-P-RNP complex would function for nuclear export (Malik *et al.*, 2000; Wolff *et al.*, 2000). In addition, both the P- and p10-dependent hypotheses for NES-driven nuclear export of RNPs are founded largely on data from transient-transfection studies rather than authentic virus infection, as there is not yet a reverse-genetics system for BDV and, at least in a short treatment, leptomycin B has no effect on RNP localization in infected cells (Bajramovic *et al.*, 2002). The mechanism by which BDV achieves nuclear export of its genome therefore remains obscure.

5.7   Evasion of innate antiviral responses

A vast amount of work investigating the interactions of segmented and non-segmented viruses with the host nucleus has been summarized in the sections above. There are common themes, such as the way in which the orthomyxoviruses and BDV use the cellular splicing machinery. However, possibly most significant is the number of nuclear intervention steps that are concerned with evading innate antiviral responses. Influenza A and VSV block the nuclear export of host mRNA by distinct mechanisms and yet in both cases the effect has been linked with the suppression of an inducible interferon response (Kim *et al.*, 2002; Ahmed *et al.*, 2003; Noah *et al.*, 2003). Similarly, the effects of the arenavirus Z protein on PML bodies has been correlated with interferon-mediated antiviral effects (Djavani *et al.*, 2001; Bonilla *et al.*, 2002). The interaction of the rabies virus P protein with PML has also been suggested to be a countermeasure to innate antiviral effects (Blondel *et al.*, 2002). The NSs polypeptides of the bunyaviruses bunyamwera virus and Rift Valley fever virus have both been implicated as interferon antagonists (Bouloy *et al.*, 2001; Weber *et al.*, 2002). As discussed above, both proteins have been shown to have a general inhibitory effect on host cell transcription, albeit potentially through mechanistically distinct pathways (Thomas *et al.*, 2004; Billecocq *et al.*, 2004). Despite the broad effects on mammalian transcription, the function of both bunyavirus NSs would appear to be suppression of the interferon response, since deletion mutants lacking NSs are more efficient inducers of IFN than wild-type viruses (Bouloy *et al.*, 2001;

Weber *et al.*, 2002) and are able to replicate in IFN-deficient mice but are attenuated in IFN-competent animals (Bouloy *et al.*, 2001; Bridgen *et al.*, 2001; Weber *et al.*, 2002). Although other bunyavirus NSs polypeptides have interferon and apoptosis antagonist functions without obviously entering the nucleus (Simons *et al.*, 1992; Weber *et al.*, 2002; Kohl *et al.*, 2003), it is possible that this may reflect mechanistic differences in activity between related virus families, as with influenza virus A and B NS1 proteins (Wang and Krug, 1996; Kim *et al.*, 2002). Thus, gene products with interferon antagonist function can be found in virtually all negative-sense viruses, although many apparently function in the cytoplasm of infected cells (Basler *et al.*, 2000; Goodbourn *et al.*, 2000; Hagemaier *et al.*, 2003; Park *et al.*, 2003), possibly by preventing nuclear translocation of essential transcription factors. The variety of mechanisms used emphasize both the selection pressure and the evolutionary responses involved in setting the virus–host balance.

Acknowledgements

We thank Anne Mullin for providing Figure 5.1C and Dick Peters for Figure 5.3. Work in the authors' laboratory is supported by grants from the Royal Society, the Wellcome Trust (nos. 059151 and 073126), BBSRC (no. S18874) and MRC (nos. G9901213 and G0300009).

References

Abraham, G., Rhodes, D. P. and Banerjee, A. K. (1975). The 5′ terminal structure of the methylated mRNA synthesized *in vitro* by vesicular stomatitis virus. *Cell* **5**, 51–58.

Ahmed, M. and Lyles, D. S. (1998). Effect of vesicular stomatitis virus matrix protein on transcription directed by host RNA polymerases I, II, and III. *J Virol* **72**, 8413–8419.

Ahmed, M., McKenzie, M. O., Puckett, S., Hojnacki, M., Poliquin, L. and Lyles, D. S. (2003). Ability of the matrix protein of vesicular stomatitis virus to suppress beta interferon gene expression is genetically correlated with the inhibition of host RNA and protein synthesis. *J Virol* **77**, 4646–4657.

Akarsu, H., Burmeister, W. P., Petosa, C., Petit, I., Muller, C. W., Ruigrok, R. W. and Baudin, F. (2003). Crystal structure of the M1 protein-binding domain of the influenza A virus nuclear export protein (NEP/NS2). *EMBO J* **22**, 4646–4655.

Alonso-Caplen, F. V. and Krug, R. M. (1991). Regulation of the exteuenza virus NS1 mRNA: role of the rates of splicing and of the nucleocytoplasmic transport of NS1 mRNA. *Mol Cell Biol* **11**, 1092–1098.

Alonso-Caplen, F. V., Nemeroff, M. E., Qiu, Y. and Krug, R. M. (1992). Nucleocytoplasmic transport: the influenza virus NS1 protein regulates the transport of spliced NS2 mRNA and its precursor NS1 mRNA. *Genes Dev* **6**, 255–267.

Bajramovic, J. J., Syan, S., Brahic, M., de la Torre, J. C. and Gonzalez-Dunia, D. (2002). 1-β-D-arabinofuranosylcytosine inhibits borna disease virus replication and spread. *J Virol* **76**, 6268–6276.

Banerjee, S. N., Buchmeier, M. and Rawls, W. E. (1976). Requirement of cell nucleus for the replication of an arenavirus. *Intervirology* **6**, 190–196.

Barman, S., Ali, A., Hui, E. K., Adhikary, L. and Nayak, D. P. (2001). Transport of viral proteins to the apical membranes and interaction of matrix protein with glycoproteins in the assembly of influenza viruses. *Virus Res* **77**, 61–69.

Barry, R. D., Ives, D. R. and Cruickshank, J. G. (1962). Participation of deoxyribonucleic acid in the multiplication of influenza virus. *Nature* **194**, 1139–1140.

Basler, C. F., Wang, X., Muhlberger, E., Volchkov, V., Paragas, J., Klenk, H. D., Garcia-Sastre, A. and Palese, P. (2000). The Ebola virus VP35 protein functions as a type I IFN antagonist. *Proc Natl Acad Sci USA* **97**, 12289–12294.

Berg, M., Ehrenborg, C., Blomberg, J., Pipkorn, R. and Berg, A. L. (1998). Two domains of the Borna disease virus p40 protein are required for interaction with the p23 protein. *J Gen Virol* **79**, 2957–2963.

Billecocq, A., Spiegel, M., Vialat, P., Kohl, A., Weber, F., Bouloy, M. and Haller, O. (2004). NSs protein of Rift Valley fever virus blocks interferon production by inhibiting host gene transcription. *J Virol* **78**, 9798–9806.

Bishop, D. H., Gay, M. E. and Matsuoko, Y. (1983). Nonviral heterogeneous sequences are present at the 5' ends of one species of snowshoe hare bunyavirus S complementary RNA. *Nucleic Acids Res* **11**, 6409–6418.

Black, D. N. and Brown, F. (1968). The influence of mitomycin C, actinomycin D and ultraviolet light on the replication of the viruses of foot-and-mouth disease and vesicular stomatitis. *J Gen Virol* **3**, 453–457.

Black, B. L. and Lyles, D. S. (1992). Vesicular stomatitis virus matrix protein inhibits host cell-directed transcription of target genes *in vivo*. *J Virol* **66**, 4058–4064.

Blondel, D., Regad, T., Poisson, N., Pavie, B., Harper, F., Pandolfi, P. P., de The, H. and Chelbi-Alix, M. K. (2002). Rabies virus P and small P products interact directly with PML and reorganise PML nuclear bodies. *Oncogene* **21**, 7957–7970.

Bonilla, W. V., Pinschewer, D. D., Klenerman, P., Rousson, V., Gaboli, M., Pandolfi, P. P., Zinkernagel, R. M., Salvato, M. S. and Hengartner, H. (2002). Effects of promyelocytic leukemia protein on virus–host balance. *J Virol* **76**, 3810–3818.

Borden, K. L., Campbell-Dwyer, E. J. and Salvato, M. S. (1997). The promyelocytic leukemia protein PML has a pro-apoptotic activity mediated through its RING domain. *FEBS Lett* **418**, 30–34.

Borden, K. L., Campbelldwyer, E. J., Carlile, G. W., Djavani, M. and Salvato, M. S. (1998a). Two RING finger proteins, the oncoprotein PML and the arenavirus Z protein, colocalize with the nuclear fraction of the ribosomal P proteins. *J Virol* **72**, 3819–3826.

Borden, K. L., Campbell-Dwyer, E. J. and Salvato, M. S. (1998b). An arenavirus RING (zinc-binding) protein binds the oncoprotein promyelocyte leukemia protein (PML) and relocates PML nuclear bodies to the cytoplasm. *J Virol* **72**, 758–766.

Bouloy, M. and Hannoun, C. (1973). Effet de l'actinomycine D sur la multiplication du virus Tahyna. *Ann Microbiol (Inst Pasteur)* **124b**, 547–553.

Bouloy, M., Janzen, C., Vialat, P., Khun, H., Pavlovic, J., Huerre, M. and Haller, O. (2001). Genetic evidence for an interferon-antagonistic function of Rift Valley fever virus nonstructural protein NSs. *J Virol* **75**, 1371–1377.

Bouloy, M., Plotch, S. J., and Krug, R. M. (1980). Both the 7-methyl and the 2'-O-methyl groups in the cap of mRNA strongly influence its ability to act as primer for influenza virus RNA transcription. *Proc Natl Acad Sci USA* **77**, 3952–3956.

Breitenfeld, P. M. and Schafer, W. (1957). The formation of fowl plague antigens in infected cells, as studied with fluorescent antibodies. *Virology* **4**, 328–345.

Bridgen, A., Weber, F., Fazakerley, J. K. and Elliott, R. M. (2001). Bunyamwera bunyavirus nonstructural protein NSs is a nonessential gene product that contributes to viral pathogenesis. *Proc Natl Acad Sci USA* **98**, 664–669.

Briese, T., de la Torre, J. C., Lewis, A., Ludwig, H. and Lipkin, W. I. (1992). Borna disease virus, a negative-strand RNA virus, transcribes in the nucleus of infected cells. *Proc Natl Acad Sci USA* **89**, 11486–11489.

Bui, M., Wills, E. G., Helenius, A. and Whittaker, G. R. (2000). Role of the influenza virus M1 protein in nuclear export of viral ribonucleoproteins. *J Virol* **74**, 1781–1786.

Bukrinskaya, A. G., Vorkunova, G. K. and Vorkunova, N. K. (1979). Cytoplasmic and nuclear input virus RNPs in influenza virus-infected cells. *J Gen Virol* **45**, 557–567.

Carbone, K. M., Moench, T. R. and Lipkin, W. I. (1991). Borna disease virus replicates in astrocytes, Schwann cells and ependymal cells in persistently infected rats: location of viral genomic and messenger RNAs by *in situ* hybridization. *J Neuropathol Exp Neuro* **50**, 205–214.

Chen, Z. and Krug, R. M. (2000). Selective nuclear export of viral mRNAs in influenza-virus-infected cells. *Trends Micro* **8**, 376–383.

Chen, Z., Li, Y. and Krug, R. M. (1999). Influenza A virus NS1 protein targets poly(A)-binding protein II of the cellular 3′-end processing machinery. *EMBO J* **18**, 2273–2283.

Chenik, M., Chebli, K. and Blondel, D. (1995). Translation initiation at alternate in-frame AUG codons in the rabies virus phosphoprotein mRNA is mediated by a ribosomal leaky scanning mechanism. *J Virol* **69**, 707–712.

Chiou, P. P., Kim, C. H., Ormonde, P. and Leong, J.-A. C. (2000). Infectious hematopoetic necrosis virus matrix protein inhibits host-directed gene expression and induces morphological changes of apoptosis in cell cultures. *J Virol* **74**, 7619–7627.

Choi, T. J., Kuwata, S., Koonin, E. V., Heaton, L. A. and Jackson, A. O. (1992). Structure of the L (polymerase) protein gene of Sonchus yellow net virus. *Virology* **189**, 31–39.

Christie, S. R., Christie, R. G. and Edwardson, J. R. (1974). Transmission of a bacilliform virus of sowthistle and *Bidens pilosa*. *Phytopathology* **64**, 840–845.

Colgan, D. F. and Manley, J. L. (1997). Mechanism and regulation of mRNA polyadenylation. *Genes Dev* **11**, 2755–2766.

Cubitt, B. and de la Torre, J. C. (1994). Borna disease virus (BDV), a nonsegmented RNA virus, replicates in the nuclei of infected cells where infectious BDV ribonucleoproteins are present. *J Virol* **68**, 1371–1381.

Cubitt, B., Ly, C. and de la Torre, J. C. (2001). Identification and characterization of a new intron in Borna disease virus. *J Gen Virol* **82**, 641–646.

Cubitt, B., Oldstone, C. and de la Torre, J. C. (1994a). Sequence and genome organization of Borna disease virus. *J Virol* **68**, 1382–1396.

Cubitt, B., Oldstone, C., Valcarcel, J. and de la Torre, J. C. (1994b). RNA splicing contributes to the generation of mature mRNAs of Borna disease virus, a non-segmented negative strand RNA virus. *Virus Res* **34**, 69–79.

Danner, K. and Mayr, A. (1979). *In vitro* studies on Borna virus. II. Properties of the virus. *Arch Virol* **61**, 261–271.

Da Poian, A. T., Gomes, A. M., Oliveira, R. J. and Silva, J. L. (1996). Migration of vesicular stomatitis virus glycoprotein to the nucleus of infected cells. *Proc Natl Acad Sci USA* **93**, 8268–8273.

de la Torre, J. C. and Oldstone, M. B. (1992). Selective disruption of growth hormone transcription machinery by viral infection. *Proc Natl Acad Sci USA* **89**, 9939–9943.

de la Torre, J. C., Carbone, K. M. and Lipkin, W. I. (1990). Molecular characterization of the Borna disease agent. *Virology* **179**, 853–856.

Dhar, R., Chanock, R. M. and Lai, C.-J. (1980). Non viral oligonucleotides at the 5′ terminus of cytoplasmic viral mRNA deduced from cloned complete genome sequences. *Cell* **21**, 495–500.

Djavani, M., Rodas, J., Lukashevich, I. S., Horejsh, D., Pandolfi, P. P., Borden, K. L. and Salvato, M. S. (2001). Role of the promyelocytic leukemia protein PML in the interferon sensitivity of lymphocytic choriomeningitis virus. *J Virol* **75**, 6204–6208.

Elton, D., Amorim, M. J., Medcalf, L. and Digard P. (2005). Genome gating: polarised intranuclear trafficking of influenza virus ribonucleoproteins. *Biol Lett* **1**, 113–117.

Elton, D., Simpson-Holley, M., Archer, K., Medcalf, L., Hallam, R., McCauley, J. and Digard, P. (2001). Interaction of the influenza virus nucleoprotein with the cellular CRM1-mediated nuclear export pathway. *J Virol* **75**, 408–419.

Elton, D., Tiley, L. and Digard, P. (2002). Molecular mechanisms of influenza virus transcription. *Recent Res Devel Virol* **4**, 121–146.

Estabrook, E. M., Tsai, J. and Falk, B. W. (1998). *In vivo* transfer of barley stripe mosaic hordeivirus ribonucleotides to the 5′ terminus of maize stripe tenuivirus RNAs. *Proc Natl Acad Sci USA* **95**, 8304–8309.

Fleckner, J., Zhang, M., Valcarel, J. and Grenn, M. R. (1997). U2AF65 recruits a novel DEAD box protein required for the U2 snRNP-branchpoint interaction. *Genes Dev* **11**, 1864–1872.

Fodor, E. and Smith, M. (2004). The PA subunit is required for efficient nuclear accumulation of the PB1 subunit of the influenza A virus RNA polymerase complex. *J Virol* **78**, 9144–9153.

Follet, E. A. C., Pringle, C. R., Wunner, W. H. and Skehel, J. J. (1974). Virus replication in enucleate cells: vesicular stomatitis virus and influenza virus. *J Virol* **13**, 394–399.

Follett, E. A., Pringle, C. R. and Pennington, T. H. (1975). Virus development in enucleate cells: echovirus, poliovirus, pseudorabies virus, reovirus, respiratory syncytial virus and Semliki Forest virus. *J Gen Virol* **26**, 183–196.

Follett, E. A., Pringle, C. R. and Pennington, T. H. (1976). Events following the infections of enucleate cells with measles virus. *J Gen Virol* **32**, 163–175.

Fortes, P., Beloso, A. and Ortin, J. (1994). Influenza virus NS1 protein inhibits pre mRNA splicing and blocks mRNA nucleocytoplasmic transport. *EMBO J* **13**, 704–712.

Fortes, P., Lamond, A. I. and Ortin, J. (1995). Influenza virus NS1 protein alters the subnuclear localisation of cellular splicing components. *J Gen Virol* **76**, 1001–1007.

Glodowski, D. R., Petersen, J. M. and Dahlberg, J. E. (2002). Complex nuclear localization signals in the matrix protein of vesicular stomatitis virus. *J Biol Chem* **277**, 46864–46870.

Goodbourn, S., Didcock, L. and Randall, R. E. (2000). Interferons: cell signalling, immune modulation, antiviral response and virus countermeasures. *J Gen Virol* **81**, 2341–2364.

Goodin, M. M., Austin, J., Tobias, R., Fujita, M., Morales, C. and Jackson, A. O. (2001). Interactions and nuclear import of the N and P proteins of sonchus yellow net virus, a plant nucleorhabdovirus. *J Virol* **75**, 9393–9406.

Gorlich, D. and Kutay, U. (1999). Transport between the cell nucleus and the cytoplasm. *Annu Rev Cell Dev Biol* **15**, 607–660.

Greenspan, D., Palese, P. and Krystal, M. (1988). Two nuclear location signals in the influenza virus NS1 nonstructural protein. *J Virol* **62**, 3020–3026.

Grinnell, B. W. and Wagner, R. R. (1983). Comparative inhibition of cellular transcription by vesicular stomatitis virus serotypes New Jersey and Indiana: role of each viral leader RNA. *J Virol* **48**, 88–101.

Gupta, A. K., Drazba, J. A. and Banerjee, A. K. (1998). Specific interaction of heterogeneous nuclear ribonucleoprotein particle U with the leader RNA sequence of vesicular stomatitis virus. *J Virol* **72**, 8532–8540.

Hagmaier, K., Jennings, S., Buse, J., Weber, F. and Kochs, G. (2003). Novel gene product of Thogoto virus segment 6 codes for an interferon antagonist. *J Virol* **77**, 2747–2752.

Hammell, C. M., Gross, S., Zenklusen, D., Heath, C. V., Stutz, F., Moore, C. and Cole, C. N. (2002). Coupling of termination, 3′ processing, and mRNA transport. *Mol Cell Biol* **22**, 6441–6457.

Her, L., Lund, E. and Dahlberg, J. E. (1997). Inhibition of Ran guanosine triphosphatase-dependent nuclear transport by the matrix protein of vesicular stomatitis virus. *Science* **276**, 1845–1848.

Herz, C., Stavnezer, E. and Krug, R. M. (1981). Influenza virus, an RNA virus, synthesises its messenger RNA in the nucleus of infected cells. *Cell* **26**, 391–400.

Herzog, S. and Rott, R. (1980). Replication of Borna disease virus in cell cultures. *Med Microbiol Immunol* **168**, 153–158.

Hirose, Y. and Manley, J. L. (2000). RNA polymerase II and the integration of nuclear events. *Genes Dev* **14**, 1415–1429.

Huang, Y. and Carmichael, G. (1996). Role of polyadenylation in nucleocytoplasmic transport of mRNA. *Mol Cell Biol* **16**, 1534–1542.

Huang, X., Liu, T., Muller, J., Levandowski, R. A. and Ye, Z. (2001). Effect of influenza virus matrix protein and viral RNA on ribonucleoprotein formation and nuclear export. *Virology* **287**, 405–416.

Huiet, L., Feldstein, P. A., Tsai, J. H. and Falk, B. W. (1993). The maize stripe virus major noncapsid protein messenger RNA transcripts contain heterogeneous leader sequences at their 5-termini. *Virology* **197**, 808–812.

Inglis, S. C., Barrett, T., Brown, C. M. and Almond, J. W. (1979). The smallest genome RNA segment of influenza virus contains two genes that may overlap. *Proc Natl Acad Sci USA* **76**, 3790–3794.

Ismail, I. D., Hamilton, I. D., Robertson, E. and Milner, J. J. (1987). Movement and intracellular location of Sonchus yellow net virus within infected *Nicotiana edwardsonii*. *J Gen Virol* **68**, 2429–2438.

Iwatsuki-Horimoto, K., Horimoto, T., Fujii, Y. and Kawaoka, Y. (2004). Generation of influenza A virus NS2 (NEP) mutants with an altered nuclear export signal sequence. *J Virol* **78**, 10149–10155.

Jackson, D. A., Caton, A. J., McCready, S. J. and Cook, P. R. (1982). Influenza virus RNA is synthesised at fixed sites in the nucleus. *Nature* **296**, 366–368.

Jehle, C., Lipkin, W. I., Staeheli, P., Marion, R. M. and Schwemmle, M. (2000). Authentic Borna disease virus transcripts are spliced less efficiently than cDNA-derived viral RNAs. *J Gen Virol* **81**, 1947–1954.

Katze, M. G. and Krug, R. M. (1984). Metabolism and expression of RNA polymerase II transcripts in influenza virus-infected cells. *Mol Cell Biol* **4**, 2198–2206.

Kelly, D. C., Avery, R. J. and Dimmock, N. J. (1974). failure of an influenza virus to initiate infection in enucleate BHK cells. *J Virol* **13**, 1156–1161.

Kiesler, E., Miralles, F. and Visa, N. (2002). HEL/UAP56 binds cotranscriptionally to the Balbiani ring pre-mRNA in an intron-independent manner and accompanies the BR mRNP to the nuclear pore. *Curr Biol* **12**, 859–862.

Kim, M. J., Latham, A. G. and Krug, R. M. (2002). Human influenza viruses activate an interferon-independent transcription of cellular antiviral genes: outcome with influenza A virus is unique. *Proc Natl Acad Sci USA* **99**, 10096–10101.

Kobayashi, T., Kamitani, W., Zhang, G., Watanabe, M., Tomonaga, K. and Ikuta, K. (2001). Borna disease virus nucleoprotein requires both nuclear localization and export activities for viral nucleocytoplasmic shuttling. *J Virol* **75**, 3404–3412.

Kobayashi, T., Shoya, Y., Koda, T., Takashima, I., Lai, P. K., Ikuta, K., Kakinuma, M. and Kishi, M. (1998). Nuclear targeting activity associated with the amino terminal region of the Borna disease virus nucleoprotein. *Virology* **243**, 188–197.

Kochs, G. and Haller, O. (1999a). GTP-bound MxA protein interacts with the nucleocapsids of Thogoto virus (Orthomyxoviridae). *J Biol Chem* **274**, 4370–4376.

Kochs, G. and Haller, O. (1999b). Interferon-induced human MxA GTPase blocks nuclear import of Thogoto virus nucleocapsids. *Proc Natl Acad Sci USA* **96**, 2082–2086.

Kohl, A., Clayton, R. F., Weber, F., Bridgen, A., Randall, R. E. and Elliott, R. M. (2003). Bunyamwera virus non-structural protein Nss counteracts interferon regulatory factor 3-mediated induction of early death. *J Virol* **77**, 7999–8008.

Kohler, M., Speck, C., Christiansen, M., Bischoff, F. R., Prehn, S., Haller, H., Gorlich, D. and Hartmann, E. (1999). Evidence for distinct substrate specificities of importin alpha family members in nuclear protein import. *Mol Cell Biol* **19**, 7782–7791.

Kurilla, M. G. and Keene, J. D. (1983). The leader RNA of vesicular stomatitis virus is bound by a cellular protein reactive with anti-La lupus antibodies. *Cell* **34**, 837–845.

Kurilla, M. G., Piwnica-Worms, H. and Keene, J. D. (1982). Rapid and transient localization of the leader RNA of vesicular stomatitis virus in the nuclei of infected cells. *Proc Natl Acad Sci USA* **79**, 5240–5244.

Krug, R. M., Alonso-Caplen, F. V., Julkunen, I. and Katze, M. G. (1989). Expression and replication of the influenza virus genome. In *The Influenza Viruses*, pp. 89–152, R. M. Krug (ed.). Plenum: New York.

Krug, R. M., Morgan, M. A. and Shatkin, A. J. (1976). Influenza viral mRNA contains internal N6-methyladenosine and 5′-terminal 7-methylguanosine in cap structures. *J Virol* **20**, 45–53.

Lamb, R. A. and Lai, C.-J. (1980). Sequence of interrupted and uninterrupted mRNAs and cloned DNA coding for two overlapping nonstructural proteins of influenza virus. *Cell* **21**, 475–485.

Lamb, R. A. and Lai, C.-J. (1982). Spliced and unspliced messenger RNAs synthesised from cloned influenza virus M DNA in an SV40 vector: expression of the influenza virus membrane protein (M1). *Virology* **123**, 237–256.

Lamb, R. A. and Lai, C.-J. (1984). Expression of unspliced NS1 mRNA, spliced NS2 mRNA and a chimaeric mRNA from cloned influenza virus NS DNA in an SV40 vector. *Virology* **135**, 139–147.

Lamb, R. A., Etkind, P. R. and Choppin, P. W. (1978). Evidence for a ninth influenza virus polypeptide. *Virology* **91**, 60–78.

Lamb, R. A., Lai, C.-J. and Choppin, P. W. (1981). Sequences of mRNAs derived from genome RNA segment 7 of influenza virus: colinear and interrupted mRNAs code for overlapping proteins. *Proc Natl Acad Sci USA* **78**, 4170–4174.

Le May, N., Dubaele, S., Proietti de Santis, L., Billecocq, A., Bouloy, M. and Egly, J-M. (2004). TFIIH transcription factor, a target for the Rift Valley hemorrhagic fever virus. *Cell* **116**, 541–550.

Li, Y., Chen, Z.-Y., Wang, W., Baker, C. C. and Krug, R. M. (2001). The 3′-end-processing factor CPSF is required for the splicing of single-intron pre-mRNAs *in vivo*. *RNA* **7**, 920–931.

Lopez, R., Grau, O. and Franze-Fernandez, M. T. (1986). Effect of actinomycin D on arenavirus growth and estimation of the generation time for a virus particle. *Virus Res* **5**, 213–220.

Lopez-Turiso, J. A., Martinez, C., Tanaka, T. and Ortin, J. (1990). The synthesis of influenza virus negative-strand RNA takes place in insoluble complexes present in the nuclear matrix fraction. *Virus Res* **16**, 325–338.

Lu, Y., Qian, X. Y. and Krug, R. M. (1994). The influenza virus NS1 protein: a novel inhibitor of pre-mRNA splicing. *Genes Dev* **8**, 1817–1828.

Luo, M. L., Zhou, Z., Magni, K., Christoforides, C., Rappsilber, J., Mann, M. and Reed, R. (2001). Pre-mRNA splicing and mRNA export linked by direct interactions between UAP56 and Aly. *Nature* **413**, 644–647.

Lyles, D. S., Puddington, L. and McCreedy, B. J. Jr (1988). Vesicular stomatitis virus M protein in the nuclei of infected cells. *J Virol* **62**, 4387–4392.

Ma, K., Roy, A. M. M. and Whittaker, G. R. (2001). Nuclear export of influenza virus ribonucleoproteins: identification of an export intermediate at the nuclear periphery. *Virology* **282**, 215–220.

MacLeod, R., Black, L. M. and Moyer, F. H. (1966). The fine structure and intracellular localization of potato yellow dwarf virus. *Virology* **29**, 540–552.

Mahy, B. W. J., Carroll, A. R., Brownson, J. M. T. and McGeoch, D. J. (1977). Block to influenza virus replication in cells preirradiated with ultraviolet light. *Virology* **83**, 150–162.

Mahy, B. W. J., Hastie, N. D. and Armstrong, S. J. (1972). Inhibition of influenza virus replication by α-amanitin: mode of action. *Proc Natl Acad Sci USA* **69**, 1421–1424.

Malik, T. H., Kishi, M. and Lai, P. K. (2000). Characterization of the P protein-binding domain on the 10 kDa protein of Borna disease virus. *J Virol* **74**, 3413–3417.

Mark, G. E., Taylor, J. M., Broni, B. and Krug, R. M. (1979). Nuclear accumulation of influenza viral RNA transcripts and the effects of cycloheximide, actinomycin D, and α-amanitin. *J Virol* **29**, 744–752.

Martin, K. and Helenius, A. (1991). Nuclear transport of influenza virus ribonucleoproteins: the viral matrix protein (M1) promotes export and inhibits import. *Cell* **67**, 117–130.

Martins, C. R., Johnson, J. A., Lawrence, D. M., Choi, T. J., Pisi, A. M., Tobin, S. L., Lapidus, D., Wagner, J. D., Ruzin, S., McDonald, K. and Jackson, A. O. (1998). Sonchus yellow net rhabdovirus nuclear viroplasms contain polymerase-associated proteins. *J Virol* **72**, 5669–5679.

McGowan, J. J., Emerson, S. U. and Wagner, R. R. (1982). The plus-strand leader RNA of VSV inhibits DNA-dependent transcription of adenovirus and SV40 genes in a soluble whole cell extract. *Cell.* **28**, 325–333.

Mersich, S. E., Damonte, E. B. and Coto, C. E. (1981). Induction of RNA polymerase II activity in Junin virus-infected cells. *Intervirology* **16**, 123–127.

Mettenleiter, T. C. (2002). Herpesvirus assembly and egress. *J Virol* **76**, 1537–1547.

Mizutani, T., Inagaki, H., Hayasaka, D., Shuto, S., Minakawa, N., Matsuda, A., Kariwa, H. and Takashima, I. (1999). Transcriptional control of Borna disease virus (BDV) in persistently BDV-infected cells. *Arch Virol* **144**, 1937–1946.

Momose, F., Basler, C. F., O'Neill, R. E., Iwamatsu, A., Palese, P. and Nagata, K. (2001). Cellular splicing factor RAF-2p48/NPI-5/BAT1/UAP56 interacts with the influenza virus nucleoprotein and enhances viral RNA synthesis. *J Virol* **75**, 1899–1908.

Monier, K., Armas, J. C., Etteldorf, S., Ghazal, P. and Sullivan, K. F. (2000). Annexation of the interchromosomal space during viral infection. *Nat Cell Biol* **2**, 661–665.

Mukaigawa, J. and Nayak, D. P. (1991). Two signals mediate nuclear localization of influenza virus (A/WSN/33) polymerase basic protein-2. *J Virol* **65**, 245–253.

Narayan, P., Ayers, D. F., Rottman, F. M., Maroney, P. A. and Nilsen, T. W. (1987). Unequal distribution of N6-methyladenosine in influenza virus mRNAs. *Mol Cell Biol* **7**, 1572–1575.

Nath, S. T. and Nayak, D. P. (1990). Function of two discrete regions is required for nuclear localization of polymerase basic protein 1 of A/WSN/33 influenza virus (H1 N1). *Mol Cell Biol* **10**, 4139–4145.

Nemeroff, M. E., Barabino, S. M., Li, Y., Keller, W. and Krug, R. M. (1998). Influenza virus NS1 protein interacts with the cellular 30 kDa subunit of CPSF and inhibits 3'-end formation of cellular pre-mRNAs. *Mol Cell* **1**, 991–1000.

Neumann, G., Castrucci, M. R. and Kawaoka, Y. (1997). Nuclear import and export of influenza virus nucleoprotein. *J Virol* **71**, 9690–9700.

Neumann, G., Hughes, M. T. and Kawaoka, Y. (2000). Influenza A virus NS2 protein mediates vRNP nuclear export through NES-independent interaction with hCRM1. *EMBO J* **19**, 6751–6758.

Nieto, A., de la Luna, S., Barcena, J., Portela, A. and Ortin, J. (1994). Complex structure of the nuclear translocation signal of influenza virus polymerase PA subunit. *J Gen Virol* **75**, 29–36.

Noah, D. L., Twu, K. Y. and Krug, R. M. (2003). Cellular antiviral responses against influenza A virus are countered at the posttranscriptional level by the viral NS1A protein via its binding to a cellular protein required for the 3' end processing of cellular pre-mRNAs. *Virology* **307**, 386–395.

Oldstone, M. B., Sinha, Y. N., Blount, P., Tishon, A., Rodriguez, M., von Wedel, R. and Lampert, P. W. (1982). Virus-induced alterations in homeostasis: alteration in differentiated functions of infected cells *in vivo*. *Science* **218**, 1125–1127.

O'Neil, R. E. and Palese, P. (1995). NPI-1, the human homologue of SRP-1, interacts with influenza virus nucleoprotein. *Virology* **206**, 116–125.

O'Neill, R. E., R. Jaskunas, G. Blobel, P. Palese and J. Moroianu. (1995). Nuclear import of influenza virus RNA can be mediated by viral nucleoprotein and transport factors required for protein import. *J Biol Chem* **270**, 22701–22704.

O'Neill, R. E., Talon, J. and Palese, P. (1998). The influenza virus NEP (NS2 protein) mediates the nuclear export of viral ribonucleoproteins. *EMBO J* **17**, 288–296.

Paragas, J., Talon, J., O'Neill, R. E., Anderson, D. K., Garcí a-Sastre, A. and Palese, P. (2001). Influenza B and C Virus NEP (NS2) proteins possess nuclear export activities. *J Virol* **75**, 7375–7383.

Park, M. S., Shaw, M. L., Munoz-Jordan, J., Cros, J. F, Nakaya, T., Bouvier, N., Palese, P., Garcia-Sastre, A. and Basler, C. F. (2003). Newcastle disease virus (NDV)-based assay demonstrates interferon-antagonist activity for the NDV V protein and the Nipah virus V, W, and C proteins. *J Virol* **77**, 1501–1511.

Patterson, J. L., Holloway, B. and Kolakofsky, D. (1984). La Crosse virions contain a primer-stimulated RNA polymerase and a methylated cap-dependent endonuclease. *J Virol* **52**, 215–222.

Pavlovic, J., Haller, O. and Staeheli, P. (1992). Human and mouse Mx proteins inhibit different steps of the influenza virus multiplication cycle. *J Virol* **66**, 2564–2569.

Petersen, J. M., Her, L. S. and Dahlberg, J. E. (2001). Multiple vesiculoviral matrix proteins inhibit both nuclear export and import. *Proc Natl Acad Sci USA* **98**, 8590–8595.

Petersen, J. M., Her, L.-S., Varvel, V., Lund, E. and Dahlberg, J. E. (2000). The matrix protein of vesicular stomatitis virus inhibits nucleocytoplasmic transport when it is in the nucleus and associated with nuclear pore complexes. *Mol Cell Biol* **20**, 8590–8601.

Planz, O., Pleschka, S., Oesterle, K., Berberich-Siebelt, F., Ehrhardt, C. and Stitz, L. (2003). Borna disease virus nucleoprotein interacts with the CDC2–cyclin B1 complex. *J Virol* **77**, 11186–11192.

Pleschka, S., Wolff, T., Ehrhardt, C., Hobom, G., Planz, O., Rapp, U. R. and Ludwig, S. (2001). Influenza virus propagation is impaired by inhibition of the Raf/MEK/ERK signalling cascade. *Nat Cell Biol* **3**, 301–305.

Plotch, S. J. and Krug, R. M. (1977). Influenza virion transcriptase: synthesis *in vitro* of large, polyadenylic acid containing complementary RNA. *J Virol* **21**, 24–34.

Plotch, S. J., Bouloy, M. and Krug, R. M. (1979). Transfer of 5'-terminal cap of globin mRNA to influenza viral complementary RNA during transcription *in vitro*. *Proc Natl Acad Sci USA* **76**, 1618–1622.

Plotch, S. J., Bouloy, M., Ulmanen, I. and Krug, R. M. (1981). A unique cap(m^{7}GpppXm)-dependent influenza virion endonuclease cleaves capped RNAs to generate the primers that initiate viral RNA transcription. *Cell* **23**, 847–858.

Plotch, S. J., Tomasz, J. and Krug, R. M. (1978). Absence of detectable capping and methylating enzymes in influenza virions. *J Virol* **25**, 75–83.

Portela, A. and Digard, P. (2002). The influenza virus nucleoprotein: a multifunctional RNA-binding protein pivotal to virus replication. *J Gen Virol* **83**, 723–734.

Pringle, C. R. and Easton, A. J. (1997). Monopartite negative strand RNA genomes. *Semin Virol* **8**, 49–57.

Pyper, J. M. and Gartner, A. E. (1997). Molecular basis for the differential subcellular localization of the 38 and 39 kDa structural proteins of Borna disease virus. *J Virol* **71**, 5133–5139.

Pyper, J. M., Clements, J. E. and Zink, M. C. (1998). The nucleolus is the site of Borna disease virus RNA transcription and replication. *J Virol* **72**, 7697–7602.

Qian, X.-Y., Alonso-Caplen, F. and Krug, R. M. (1994). Two functional domains of the influenza virus NS1 protein are required for regulation of nuclear export of mRNA. *J Virol* **68**, 2433–2441.

Qiu, Y. and Krug, R. M. (1994). The influenza virus NS1 protein is a poly(A)-binding protein that inhibits nuclear export of mRNAs containing poly(A). *J Virol* **68**, 2425–2432.

Qiu, Y., Nemeroff, M. and Krug, R. M. (1995). The influenza virus NS1 protein binds to a specific region in human U6 snRNA and inhibits U6–U2 and U6–U4 snRNA interactions during splicing. *RNA* **1**, 304–316.

Raju, R. and Kolakofsky, D. (1988). La Crosse virus infection of mammalian cells induces mRNA instability. *J Virol* **62**, 27–32.

Raju, R., Raju, L., Hacker, D., Garcin, D., Compans, R. and Kolakofsky, D. (1990). Nontemplated bases at the 5' ends of Tacaribe virus mRNAs. *Virology* **174**, 53–59.

Ransohoff, R. M., Narayan, P., Ayers, D. F., Rottman, F. M. and Nilsen, T. W. (1987). Priming of influenza mRNA transcription is inhibited in CHO cells treated with the methylation inhibitor, neplanocin A. *Antiviral Res* **7**, 317–327.

Rawls, W. E., Banerjee, S. N., McMillan, C. A. and Buchmeier, M. J. (1976). Inhibition of Pichinde virus replication by actinomycin D. *J Gen Virol* **33**, 421–434.

Robbins, J., Dilworth, S. M., Laskey, R. A. and Dingwall, C. (1991). Two interdependent basic domains in nucleoplasmin nuclear targeting sequence: identification of a class of bipartite nuclear targeting sequence. *Cell* **64**, 615–623.

Rossier, C., Patterson, J. and Kolakofsky. D. (1986). La Crosse virus small genome mRNA is made in the cytoplasm. *J Virol* **58**, 647–650.

Rott, R. and Scholtissek, C. (1970). Specific inhibition of influenza virus replication by α-amanitin. *Nature* **228**, 56.

Sakaguchi, A., Hirayama, E., Hiraki, A., Ishida, Y. and Kim, J. (2003). Nuclear export of influenza viral ribonucleoprotein is temperature-dependently inhibited by dissociation of viral matrix protein. *Virology* **306**, 244–253.

Schneeman, A., Schneider, P. A., Kim, S. and Lipkin, W. I. (1994). Identification of signal sequences that control transcription of Borna disease virus, a nonsegmented, negative-strand RNA virus. *J Virol* **68**, 6514–6522.

Schneider, P. A., Schneemann, A. and Lipkin, W. I. (1994). RNA splicing in Borna disease virus, a nonsegmented, negative-strand RNA virus. *J Virol* **68**, 5007–5012.

Schneider, P. A., Schwemmle, M. and Lipkin, W. I. (1997). Implication of a cis-acting element in the cytoplasmic accumulation of unspliced Borna disease virus RNAs. *J Virol* **71**, 8940–8945.

Schwemmle, M., Jehle, C., Shoemaker, T. and Lipkin, W. I. (1999). Characterization of the major nuclear localization signal of the Borna disease virus phosphoprotein. *J Gen Virol* **80**, 97–100.

Schwemmle, M., Salvatore, M., Shi, L., Richt, J., Lee, C. H. and Lipkin, W. I. (1998). Interactions of the borna disease virus P, N, and X proteins and their functional implications. *J Biol Chem* **273**, 9007–9012.

Shih, S.-R., Nemeroff, M. E. and Krug, R. M. (1995). The choice of alternative 5′ splice sites in influenza virus M1 mRNA is regulated by the viral polymerase complex. *Proc Natl Acad Sci USA* **92**, 6324–6328.

Shimizu, K., Handa, H., Nakada, S. and Nagata, K. (1994). Regulation of influenza virus RNA polymerase activity by cellular and viral factors. *Nucleic Acids Res* **22**, 5047–5053.

Shimizu, K., Iguchi, A., Gomyou, R. and Ono, Y. (1999). Influenza virus inhibits cleavage of the Hsp70 pre-mRNAs at the polyadenylation site. *Virology* **254**, 213–219.

Shoya, Y., Kobayashi, T., Koda, T., Ikuta, K., Kakinuma, M. and Kishi, M. (1998). Two proline-rich nuclear localization signals in the amino- and carboxy-terminal regions of the Borna Disease virus phosphoprotein. *J Virol* **72**, 9755–9762.

Simons, J. F., Persson, R. and Pettersson, R. F. (1992). Association of the nonstructural protein NS(S) of Uukuniemi virus with the 40S ribosomal-subunit. *J Virol* **66**, 4233–4241.

Sinha, R. C. (1971). Distribution of wheat striate mosaic virus in infected plants and some morphological features of the virus *in situ* and *in vitro*. *Virology* **44**, 342–351.

Skehel, J. J. (1973). Early polypeptide synthesis in influenza virus-infected cells. *Virology* **56**, 394–399.

Smith, D. B. and Inglis, S. C. (1985). Regulated production of an influenza virus spliced mRNA mediated by virus-specific products. *EMBO J* **4**, 2313–2319.

Southern, P. J. (1996). *Arenaviridae*: the viruses and their replication. In *Field's Virology*, 3rd edn, pp. 1505–1519. B. N. Fields, D. M. Knipe and P. Howley (eds). Lippincott-Raven: Philadelphia, PA.

Spooner, L. L. R. and Barry, R. D. (1977). Participation of DNA-dependent RNA polymerase II in replication of influenza viruses. *Nature* **268**, 650–652.

Struthers, J. K. and Swanepoel, R. (1982). Identification of a major non-structural protein in the nuclei of Rift Valley fever virus-infected cells. *J Gen Virol* **60**, 381–384.

Teng, M. N., Borrow, P., Oldstone, M. B. and de la Torre, J. C. (1996). A single amino acid change in the glycoprotein of lymphocytic choriomeningitis virus is associated with the ability to cause growth hormone deficiency syndrome. *J. Virol.* **70**, 8438–8443.

Thierer, J., Riehle, H., Grebenstein, O., Binz, T., Herzog, S., Thiedemann, N., Stitz, L., Rott, R., Lottspeich, F. and Niemann, H. (1992). The 24 K protein of Borna disease virus. *J Gen Virol* **73**, 413–416.

Thomas, D., Blakqori, G., Wagner, V., Banholzer, M., Kessler, N., Elliott, R. M., Haller, O. and Weber, F. (2004). Inhibition of RNA polymerase II phosphorylation by a viral interferon antagonist. *J Biol Chem* **279**, 31471–31477.

Tomonaga, K., Kobayashi, T., Lee, B.-J., Watanabe, M., Kamitani, W. and Ikuta, K. (2000). Identification of alternative splicing and negative splicing activity of a non-segmented negative-strand RNA virus, Borna disease virus. *Proc Natl Acad Sci USA* **97**, 12788–12793.

Tordo, N., De Haan, P., Goldbach, R. and Poch, O. (1992). Evolution of negative-stranded RNA genomes. *Semin Virol* **3**, 341–357.

Turan, K., Mibayashi, M., Sugiyama, K., Saito, S., Numajiri, A. and Nagata, K. (2004). Nuclear MxA proteins form a complex with influenza virus NP and inhibit the transcription of the engineered influenza virus genome. *Nucleic Acids Res* **32**, 643–652.

Valcarcel, J., Fortes, P. and Ortin J. (1993). Splicing of influenza virus matrix protein mRNA expressed from a simian virus 40 recombinant. *J Gen Virol* **74**, 1317–1326.

van Beek, N. A. M., Lohuis, D., Dijkstra, J. and Peters, D. (1985). Morphogenesis of Sonchus yellow net virus in cowpea protoplasts. *J Ultrastruct Res* **90**, 294–303.

Vezza, A. C., Repik, P. M., Cash, P. and Bishop, D. H. L. (1979). *In vivo* transcription and protein synthesis capabilities of bunyaviruses: wild-type snowshoe hare virus and its temperature-sensitive group I, group II and group I/II mutants. *J Virol* **31**, 426–436.

Villarreal, L. P. and Holland, J. J. (1976). RNA synthesis in BHK 21 cells persistently infected with vesicular stomatitis virus and rabies virus. *J Gen Virol* **33**, 213–224.

Vogel, U., Kunerl, M. and Scholtissek, C. (1994). Influenza A virus late mRNAs are specifically retained in the nucleus in the presence of a methyltransferase or protein kinase inhibitor. *Virology* **198**, 227–233.

von Kobbe, C., van Deursen, J. M., Rodrigues, J. P., Sitterlin, D., Bachi, A., Wu, X., Wilm, M., Carmo-Fonseca, M. and Izaurralde, E. (2000). Vesicular stomatitis virus matrix protein inhibits host cell gene expression by targeting the nucleoporin Nup98. *Mol Cell* **6**, 1243–1252.

Wagner, R. R., Kiley, M. P., Snyder, R. M. and Schnaitman, C. A. (1972). Cytoplasmic compartmentalization of the protein and ribonucleic acid species of vesicular stomatitis virus. *J Virol* **9**, 672–683.

Wagner, J. D. O., Choi, T.-J. and Jackson, A. O (1996). Extraction of nuclei from Sonchus yellow net rhabdovirus-infected plants yields a polymerase that synthesizes viral mRNAs and polyadenylated plus-strand leader RNA. *J Virol* **70**, 468–477.

Wagner, E., Engelhardt, O. G., Gruber, S., Haller, O. and Kochs, G. (2001). Rescue of recombinant Thogoto virus from cloned cDNA. *J Virol* **75**, 9282–9286.

Walker, M. P. and Lipkin, W. I. (2002). Characterization of the nuclear localisation signal of the Borna Disease virus polymerase. *J Virol* **76**, 8460–8467.

Wang, P., Palese, P., and O'Neill, R. E. (1997). The NPI-1/NPI-3 (karyopherin α) binding site on the influenza A virus nucleoprotein NP is a nonconventional nuclear localisation signal. *J Virol* **71**, 1850–1856.

Wang, W. and Krug, R. W. (1996). The RNA-binding and effector domains are conserved to different extents among influenza A and B viruses. *Virology* **223**, 41–50.

Wang, W. and Krug, R. M. (1998). U6atac snRNA, the highly divergent counterpart of U6 snRNA, is the specific target that mediates inhibition of AT-AC splicing by the influenza virus NS1 protein. *RNA* **4**, 55–64.

Watanabe, K., Takizawa, N., Katoh, M., Hoshida, K., Kobayashi, N. and Nagata, K. (2001). Inhibition of nuclear export of ribonucleoprotein complexes of influenza virus by leptomycin B. *Virus Res* **77**, 31–42.

Weber, F., Bridgen, A., Fazakerley, J. K., Streitenfeld, H., Kessler, N., Randall, R. E. and Elliott, R. M. (2002). Bunyamwera bunyavirus nonstructural protein NSs counteracts the induction of alpha/beta interferon. *J Virol* **76**, 7949–7955.

Weber, F., Haller, O. and Kochs, G. (2000). MxA GTPase blocks reporter gene expression of reconstituted Thogoto virus ribonucleoprotein complexes. *J Virol* **74**, 560–563.

Weber, F., Kochs, G., Gruber, S. and Haller, O. (1998). A classical bipartite nuclear localization signal on Thogoto and influenza A virus nucleoproteins. *Virology* **250**, 9–18.

Weck, P. K. and Wagner, R. R. (1978). Inhibition of RNA synthesis in mouse myeloma cells infected with vesicular stomatitis virus. *J Virol* **25**, 770–780.

Wehner, T., Ruppert, A., Herden, C., Frese, K., Becht, H. and Richt, J. A. (1997). Detection of a novel Borna disease virus-encoded 10 kDa protein in infected cells and tissues. *J Gen Virol* **78**, 2459–2466.

Whittaker, G., Bui, M. and Helenius, A. (1996). Nuclear trafficking of influenza virus ribonucleoproteins in heterokaryons. *J Virol* **70**, 2743–22756.

Whittaker, G., Kemler, I. and Helenius, A. (1995). Hyperphosphorylation of mutant influenza virus matrix protein, M1, causes its retention in the nucleus. *J Virol* **69**, 439–445.

Wolff, T., O'Neill, R. E. and Palese, P. (1998). NS1-binidng protein (NS1-BP): a novel human protein that interacts with the influenza A virus nonstructural NS1 protein is relocalized in the nuclei of infected cells. *J Virol* **72**, 7170–7180.

Wolff, T., Pfleger, R., Wehner, T., Reinhardt, J. and Richt, J. A. (2000). A short leucine-rich sequence in the Borna disease virus p10 protein mediates association with the viral phospho- and nucleoproteins. *J Gen Virol* **81**, 939–947.

Wolff, T., Unterstab, G., Heins, G., Richt, J. A. and Kann, M. (2002). Characterization of an unusual importin alpha binding motif in the Borna disease virus p10 protein that directs nuclear import. *J Biol Chem* **277**, 12151–12157.

Wolstenholme, A. J., Barrett, T., Nichol, S. T. and Mahy, B. W. (1980). Influenza virus-specific RNA and protein syntheses in cells infected with temperature-sensitive mutants defective in the genome segment encoding nonstructural proteins. *J Virol* **35**, 1–7.

Wurzer, W. J., Planz, O., Ehrhardt, C., Giner, M., Silberzahn, T., Pleshka, S. and Ludwig, S. (2003). Caspase 3 activation is essential for efficient influenza virus propagation. *EMBO J* **22**, 2717–2718.

Yadani, F. Z., Kohl, A., Prehaud, C., Billecocq, A. and Bouloy, M. (1999). The carboxy-terminal acidic domain of Rift Valley fever virus NSs protein is essential for the formation of filamentous structures but not for the nuclear localization of the protein. *J Virol* **73**, 5018–5025.

Yasuda, J., Nakada, S., Kato, A., Toyoda, T. and Ishihama, A. (1993). Molecular assembly of influenza virus: association of the NS2 protein with virion matrix. *Virology* **196**, 249–255.

Ye, Z., Robinson, D. and Wagner, R. R. (1995). Nucleus-targeting domain of the matrix protein (M1) of influenza virus. *J Virol* **69**, 1964–1970.

Young, P. R., Chanas, A. C., Lee, S. R., Gould, E. A. and Howard, C. R. (1987). Localization of an arenavirus protein in the nuclei of infected cells. *J Gen Virol* **68**, 2465–2470.

Yuan, H., Puckett, S. and Lyles, D. S. (2001). Inhibition of host transcription by vesicular stomatitis virus involves a novel mechanism that is independent of phosphorylation of TATA-binding protein (TBP) or association of TBP with TBP-associated factor subunits. *J Virol* **75**, 4453–4458.

Yuan, H., Yoza, B. K. and Lyles, D. S. (1998). Inhibition of host RNA polymerase II-dependent transcription by vesicular stomatitis virus results from inactivation of TFIID. *Virology* **251**, 383–392.

Zhirnov, O. P. and Klenk, H. D. (1997). Histones as a target for influenza virus matrix protein M1. *Virology* **235**, 302–310.

Zvonarjev, A. Y. and Ghendon, Y. Z. (1980). Influence of membrane (M) protein on influenza A virus virion transcriptase activity in vitro and its susceptibility to rimantidine. *J Virol* **33**, 583–586.

6 Positive-strand RNA Viruses and the Nucleus

Kurt E. Gustin[1] and Peter Sarnow[2]

[1]Department of Microbiology, Molecular Biology and Biochemistry, University of Idaho, Moscow, USA
[2]Department of Microbiology and Immunology, Stanford University School of Medicine, Stanford, USA

6.1 Introduction

The positive-strand RNA viruses include a diverse array of pathogens that infect bacteria, yeast, plants, insects and animals (for review, see van Regenmortel *et al.*, 2000). The variety within this class of viruses is truly impressive. Virions can be enveloped or non-enveloped and nucleocapsids can be icosahedral, rod-shaped or filamentous. The RNA genomes can be segmented or linear and the size can range from less than 4000 nucleotides in the case of the leviviruses to over 30 000 nucleotides in the case of the coronaviruses. In some cases, virion RNAs are translated to produce viral protein, while in others subgenomic RNAs need to be produced that serve as templates for the synthesis of viral products. Despite the differences in virion morphology and genome organization, all positive-strand RNA viruses that infect animals, with the exception of retroviruses, replicate in the cytoplasm of the infected host cell. This does not mean, however, that the nucleus (or its constituents) is an uninvolved bystander to the replication of these viruses. On the contrary, there is ample evidence that supports a role for nuclear factors in the replication of these viruses. In the first part of this chapter we summarize the data implicating the nucleus or nuclear components in the life cycle of positive-strand RNA viruses. We then focus our discussion on the recent finding that infection by members of the picornavirus family inhibits nuclear import, thus potentially causing the accumulation of nuclear proteins in the cytoplasm of infected cells. Finally, we discuss the implications that inhibition of nuclear import may have upon the ability of the host cell to mount an effective antiviral response.

Viruses and the Nucleus   Edited by Julian A. Hiscox
© 2006 John Wiley & Sons, Ltd

6.2   Replication of positive-strand RNA viruses in enucleated cells

For many years, researchers have questioned whether the nucleus has a role in the replication of positive-strand RNA viruses. Early studies demonstrated that picorna-viruses, such as foot and mouth disease virus and echovirus, and alphaviruses, such as Semliki Forest virus, were insensitive to the addition of actinomycin D or mitomycin C (Black and Brown, 1968; Reich *et al.*, 1961; Taylor, 1965), inhibitors of nuclear RNA or DNA synthesis, respectively. Initial attempts to determine whether nuclear macromolecular machines or nuclear factors played a role in poliovirus replication relied upon the ability to infect 'cytoplasmic fragments' prepared from whole cells by microdissection (Crocker *et al.*, 1964). This study revealed that poliovirus proteins and RNA could be synthesized in 'cytoplasmic fragments' lacking nuclei, but at a slower rate than in intact cells. These results suggested that nuclear factors contributed to efficient production of viral proteins and RNA. Unfortunately, the yield of virus from cytoplasmic fragments could not be assessed because of the presence of intact cells.

The development of a protocol for the efficient removal of cell nuclei by treatment with cytochalasin B (Prescott *et al.*, 1972) allowed for the generation of enucleated cell populations in which viral yield could be determined (Bossart and Bienz, 1979; Detjen *et al.*, 1978; Erwin and Brown, 1983; Follett *et al.*, 1975; Kos *et al.*, 1975; Pollack and Goldman, 1973). In two separate studies of enucleated African green monkey BSC-1 cells, the yield of infectious poliovirus was reduced to 20–40% of that from nucleated controls (Follett *et al.*, 1975; Pollack and Goldman, 1973). Replication of echovirus was more sensitive to enucleation, as yields dropped by greater than 90% (Follett *et al.*, 1975). When enucleated mouse L cells were transfected with infectious poliovirus RNA, the yield was reduced by 90% from that of control cells (Detjen *et al.*, 1978). In contrast to these studies, which demonstrated replication of poliovirus, albeit at reduced levels, Bossart and Bienz observed that enucleation of human Hep-2 cells completely prevented production of infectious poliovirus (Bossart and Bienz, 1979). Alphaviruses exhibited differential sensitivity to enucleation, with the yield of Sindbis virus being reduced up to 90% (Erwin and Brown, 1983; Kos *et al.*, 1975), while that of Semliki Forest virus was unaffected (Follett *et al.*, 1975). Japanese encephalitis virus, a flavivirus, failed to replicate to detectable levels in enucleated cells, indicating that replication of this virus required nuclear contributions (Kos *et al.*, 1975). Other flaviviruses may also require contributions from the nucleus, as immunofluorescence analysis demonstrated reduced viral protein synthesis for both dengue and West Nile virus in enucleated cells (Lad *et al.*, 1993). The ability of some viruses to replicate in enucleated cells has been interpreted as an indication that the nucleus is not required for replication. Those studies that observed a modest reduction in viral yield attributed it to a loss of cytoplasmic volume caused by the enucleation procedure, or to reduced metabolic activity of the enucleated cell population (Detjen *et al.*, 1978; Pollack and Goldman, 1973). Another possibility, however, is that the reduction of viral yield is directly attributable to the loss of nuclear components. In this scenario, the ability of enucleated cells to support reduced levels of replication would be due to the presence of residual nuclear factors left behind following the enucleation procedure. Nuclear factors could be present in the enucleated cell cytoplasm due to release from the nucleus during the enucleation procedure, to *de novo* synthesis directed by pre-existing cytoplasmic mRNAs or to trapping of the cytoplasmic fraction of nuclear shuttling proteins.

6.3 Localization of viral proteins to the nucleus

Indirect immunofluorescence and immuno-electron microscopy suggests that many positive-strand RNA viruses encode proteins that interact with the nucleus or nuclear components during infection (Table 6.1), e.g. the nucleoprotein of the avian infectious

Table 6.1   Positive-strand RNA virus proteins that localize to the nucleus

Family	Virus	Protein	Reference
Coronaviridae	Avian infectious bronchitis virus	Nucleoprotein	Hiscox *et al.*, 2001
	Transmissible gastroenteritis virus	Nucleoprotein	Wurm *et al.*, 2001
	Mouse hepatitis virus	Nucleoprotcin	Wurm *et al.*, 2001
Flaviviridae	Hepatitis C virus	NS3	Errington *et al.*, 1999
			Kim *et al.*, 1999
			Muramatsu *et al.*, 1997
			Wolk *et al.*, 2000
		Core	Yasui *et al.*, 1998
	Dengue virus	Core	Tadano *et al.*, 1989
		NS4B	Westaway *et al.*, 1997
		NS5	Kapoor *et al.*, 1995
	Yellow fever virus	NS5	Buckley *et al.*, 1992
Arteriviridae	Equine arteritis virus	NSP-1	Tijms *et al.*, 2002
		Nucleocapsid	Tijms *et al.*, 2002
	Porcine reproductive and respiratory syndrome virus	Nucleocapsid	Rowland *et al.*, 1999
Astroviridae	Human astrovirus	Orf 1a	Willcocks *et al.*, 1999
Togaviridae	Semliki Forest virus	nsP2	Peranen *et al.*, 1990
	Sindbis virus	nsP2	Liang *et al.*, 1995
Potyviridae	Tobacco etch virus	P3	Langenberg and Zhang, 1997
		Nla	Restrepo *et al.*, 1990
			Baunoch *et al.*, 1991
		Nlb	Restrepo *et al.*, 1990
			Baunoch *et al.*, 1991

bronchitis virus, a coronavirus, has been shown to localize, in part, to the nucleolus of infected cells (Hiscox *et al.*, 2001). The significance of this observation is strengthened by the finding that the nucleoproteins from other coronaviruses, such as transmissible gastroenteritis virus and mouse hepatitis virus, also localize to the nucleoli of infected cells (Wurm *et al.*, 2001). Similarly, structural and non-structural proteins from flaviviruses, such as the hepatitis C virus (HCV) and the dengue virus core protein (Tadano *et al.*, 1989; Yasui *et al.*, 1998), the dengue and yellow fever virus RNA-dependent RNA polymerase, NS5 (Buckley *et al.*, 1992; Kapoor *et al.*, 1995), the dengue NS4B (Westaway *et al.*, 1997) and HCV NS3 (Errington *et al.*, 1999; Kim *et al.*, 1999; Muramatsu *et al.*, 1997; Wolk *et al.*, 2000) proteins, have all been reported to localize under certain conditions to the nucleus or nucleoli of cells. In the case of the HCV NS3 protein, however, co-expression with NS4A restricts NS3 to the cytoplasm (Kim *et al.*, 1999; Muramatsu *et al.*, 1997; Wolk *et al.*, 2000). This finding illustrates that distinct viral proteins can have roles in the localization of other viral products, in this case probably by sequestration in membrane-bound replication complexes. While there are numerous examples of nuclear localization of viral proteins encoded by positive-strand RNA viruses (Table 6.1) (Baunoch *et al.*, 1991; Langenberg and Zhang, 1997; Liang *et al.*, 1995; Peranen *et al.*, 1990; Restrepo *et al.*, 1990; Rowland *et al.*, 1999; Tijms *et al.*, 2002; Willcocks *et al.*, 1999), it is not clear whether nuclear localization plays a role in the infectious cycle.

In some cases deletional and mutational analyses have identified nuclear localization signals (NLSs) responsible for nuclear import of viral proteins. NLSs have been identified in the NS5B and core protein from dengue virus (Forwood *et al.*, 1999; Wang *et al.*, 2002), the human astrovirus ORF1a gene (Willcocks *et al.*, 1999), the nsP2 protein of Semliki Forest virus (Fazakerley *et al.*, 2002) and the RNA-dependent RNA polymerase (NIb) and 3C-like protease (NIa) of tobacco etch virus (Carrington *et al.*, 1991; Li and Carrington, 1993; Restrepo *et al.*, 1990), to name a few. These results demonstrate that viruses exploit existing cellular trafficking pathways to deliver viral proteins to the nucleus. The presence, and in some cases conservation of these NLSs across serotypes (Carter, 1994; Fazakerley *et al.*, 2002; Wang *et al.*, 2002) suggests that there is some advantage to the virus in maintaining these NLSs and consequently nuclear localization of the viral protein. In most instances, the role of these proteins in the nucleus remains unknown; however, in the case of the Semliki Forest virus (SFV) nsP2 protein, the importance of the NLS has been demonstrated. The SFV nsP2 protein is involved in both polyprotein processing and RNA synthesis (Schlesinger and Schlesinger, 2001). SFV nsP2 contains an NLS and about half of nsP2 is found in the nucleus of infected cells (Peranen *et al.*, 1990). A single amino acid change that disrupts the NLS of nsP2 results in a virus that can replicate to wild-type levels in tissue culture (Rikkonen, 1996), but is severely attenuated in adult mice (Fazakerley *et al.*, 2002). When compared to wild-type virus, the mutant virus replicated to lower titres and did not cause the death of adult mice when injected intracerebrally (Fazakerley *et al.*, 2002). It remains to be seen whether this defect in pathogenesis is due to a lack of nuclear targeting of nsP2 or because of a disruption in other functions provided by nsP2.

6.4   Interaction of nuclear factors with viral proteins and nucleic acids

Perhaps a clue to the role of these viral proteins in the nucleus can be obtained from identifying cellular nuclear binding partners (Table 6.2). For example, the infectious bronchitis virus nucleoprotein not only localizes to the nucleolus, but interacts *in vitro* with nucleolin (Chen *et al.*, 2002), a major component of the nucleolus that has been implicated in a variety of cellular activities, including ribosomal RNA processing and transcription (reviewed in Ginisty *et al.*, 1999). Likewise, the core proteins of dengue and HCV, which are detected in the nucleus, interact with hnRNP K (Chang *et al.*, 2001; Hsieh *et al.*, 1998), a nuclear RNA binding protein implicated in transcriptional and translational regulation (Krecic and Swanson, 1999). The finding that the core protein from both these viruses not only localizes to the nucleus but also interacts with the same nuclear protein supports the notion that these events are important for viral replication.

Nuclear proteins have also been shown to interact with viral proteins and RNAs in the cytoplasm (Table 6.2). For example, the internal ribosomal entry site (IRES) of picornaviruses such as poliovirus, rhinovirus and encephalomyocarditis virus can be cross-linked to the nuclear proteins La and the polypyrimidine tract binding protein (PTB) (Borman *et al.*, 1993; Hellen *et al.*, 1993; Kaminski *et al.*, 1995; Kim and Jang, 1999; Meerovitch *et al.*, 1993; Rojas-Eisenring *et al.*, 1995). Furthermore, the addition of La or PTB to *in vitro* translation reactions stimulates translation from these IRES elements (Hellen *et al.*, 1993; Hunt *et al.*, 1999; Kim and Jang, 1999; Meerovitch *et al.*, 1993). La and PTB have also been shown to interact with the IRES and 3' NCR of HCV, suggesting that they may contribute to the translation of these viral genomes (Ali and Siddiqui, 1995, 1997; Ito and Lai, 1997; Spangberg *et al.*, 1999; Tsuchihara *et al.*, 1997). PTB also binds to the mouse hepatitis virus plus-strand leader RNA and appears to contribute to RNA synthesis (Li *et al.*, 1999). Similarly, La has also been shown to interact with the negative-strand RNA of Sindbis virus (Pardigon and Strauss, 1996) and the leader RNA of rhabdoviruses (Kurilla *et al.*, 1984; Kurilla and Keene, 1983; Wilusz *et al.*, 1983). Thus, many positive-strand RNA viruses may utilize functions provided by PTB and La for the translation and replication of their genomes.

The hnRNP A1 has been shown to bind to RNA sequences in the intergenic and 3'-untranslated regions (Huang and Lai, 2001; Li *et al.*, 1997) and to the nucleocapsid (Wang and Zhang, 1999) of mouse hepatitis virus, although this hnRNP does not appear to be absolutely required for viral replication (Shen and Masters, 2001). Cross-linking studies have revealed that the 5' end of Norwalk virus RNA can interact with La, PTB, PCBP-2 and hnRNP L (Gutierrez-Escolano *et al.*, 2000). Unfortunately, due to an inability to cultivate human noroviruses, such as Norwalk virus, it has not been possible to determine the significance of these interactions *in vivo*. If these same nuclear proteins interact with the 5' end of the recently identified murine norovirus 1 (Karst *et al.*, 2003) genome, it may now be possible to determine a role for these

Table 6.2 Nuclear proteins and their viral targets

Nuclear protein	Cellular functions[a]	Viral target	Function in viral replication	Reference
Nucleolin	rRNA transcription and processing, Pol II transcription, mRNA stability	Nucleoprotein of avian infectious bronchitis virus	?	Chen *et al.*, 2002
		3' NCR of poliovirus genome	RNA synthesis?	Waggoner and Sarnow, 1998
hnRNP K	Translational silencing, mRNA stability	Core protein of dengue virus	?	Chang *et al.*, 2001
		Core protein of HCV	?	Hsieh *et al.*, 1998
PTB	Splicing, polyadenylation	IRES of entero, rhino and cardiovirus genomes	Translation	Borman *et al.*, 1993 Hellen *et al.*, 1993 Kaminski *et al.*, 1995
		Plus-strand leader RNA of MHV	RNA synthesis?	Li *et al.*, 1999
		IRES, 3' NCR of HCV	Translation?	Ali and Siddiqui, 1995 Ito and Lai, 1997 Tsuchihara *et al.*, 1997
		5' End of Norwalk virus genome	?	Gutierrez-Escolano *et al.*, 2000
La	Pol III transcription, processing	IRES of entero-, rhino- and cardiovirus genomes	Translation	Kim and Jang, 1999 Meerovitch *et al.*, 1993 Rojas-Eisenring *et al.*, 1995
		IRES, 3' NCR of HCV	Translation	Ali and Siddiqui, 1997 Spangberg *et al.*, 1999

		5′ End of Norwalk virus genome	?	Gutierrez-Escolano et al., 2000
		3′ End of Sindbis virus (−) strand RNA	?	Pardigon and Strauss, 1996
		Leader RNA of rhabdoviruses	?	Kurilla et al., 1984 Kurilla and Keene, 1983 Wilusz et al., 1983
hnRNP A1	mRNA transport, splicing	Intergenic and 3′ untranslated region of MHV RNA	RNA synthesis?	Huang and Lai, 2001 Li et al., 1997
		Nucleoprotein of MHV	RNA synthesis?	Wang and Zhang, 1999
hnRNP C	Splicing, mRNA stability	5′ End of Norwalk virus genome	?	Gutierrez-Escolano et al., 2000
Sam68	Cell cycle control(?), mRNA transport	$3D^{pol}$ of poliovirus	?	McBride et al., 1996

[a] Function of nuclear protein in uninfected cells.

interactions in the replication or pathogenesis of noroviruses. The 3′ non-coding region of poliovirus can be cross-linked to nucleolin in cell extracts, and immuno-depletion of nucleolin from *in vitro* replication reactions resulted in a slight decrease in virus yield at early times (Waggoner and Sarnow, 1998). Interestingly, at later times yields from nucleolin-depleted extracts were equivalent to those from undepleted reactions, indicating that nucleolin may contribute to early events in RNA synthesis (Waggoner and Sarnow, 1998). More recent studies have suggested that nucleolin can also stimulate poliovirus translation through interaction with the IRES (Izumi *et al.*, 2001). Sam68, a KH domain-containing protein that has been implicated in RNA trafficking (Li *et al.*, 2002; Reddy *et al.*, 1999, 2000a, 2000b), has also been shown to interact with the poliovirus RNA-dependent RNA polymerase 3D, although a role for Sam68 in viral replication remains to be determined (McBride *et al.*, 1996).

While a definitive demonstration of a role for theses nuclear proteins in viral replication remains to be demonstrated, they are all RNA-binding proteins with diverse functions in RNA metabolism. These proteins have been implicated in the synthesis, processing, stability, transport and translation of cellular RNAs. Clearly, these are proteins whose functions an RNA virus might find useful for certain aspects of replication.

6.5   Cytoplasmic accumulation of nuclear proteins during infection

A few of these interactions, such as the HCV and dengue core proteins with hnRNP K (Chang *et al.*, 2001; Hsieh *et al.*, 1998; Tadano *et al.*, 1989; Yasui *et al.*, 1998), or the infectious bronchitis virus N protein with nucleolin (Chen *et al.*, 2002; Hiscox *et al.*, 2001), involve viral proteins that localize to the nucleus. Most, however, involve viral proteins or RNA that are not thought to enter the nucleus. This creates an apparent paradox: how can factors present in the nucleus interact with viral RNA or proteins located in the cytoplasm? One possible explanation is that most of the nuclear factors identified as interacting with viral targets are known to shuttle between the nuclear and cytoplasmic compartments. Proteins that have been shown to possess shuttling activity include La, nucleolin, PTB, hnRNP A1 and hnRNP K (Bachmann *et al.*, 1989; Borer *et al.*, 1989; Michael *et al.*, 1995; Pinol-Roma and Dreyfuss, 1992). Thus, even though predominantly nuclear at steady state, a small fraction of these proteins are present at any given time in the cytoplasm of cells and would be available for interaction with viral targets, perhaps in sufficient quantity to carry out some function in the viral life cycle.

Another explanation comes from the finding that, in some cases, viral infection causes the redistribution of nuclear factors to the cytoplasm. For example, infection with mouse hepatitis virus results in the relocalization of hnRNP A1 from predominantly nuclear to cytoplasmic (Li *et al.*, 1997). Similarly, in poliovirus-infected cells La, Sam68, PTB and nucleolin all relocalize to the cytoplasm following infection (Back *et al.*, 2002; McBride *et al.*, 1996; Meerovitch *et al.*, 1993; Waggoner and

Sarnow, 1998). Recently, infection by rhinovirus type 14 was shown to cause a similar redistribution of La, nucleolin and Sam68 (Gustin and Sarnow, 2002). Thus, these viruses appear to orchestrate the cytoplasmic accumulation of nuclear proteins, such that they are now available for binding to viral targets and can contribute to viral replication. Interestingly, relocalization is not limited to nuclear proteins known to interact with viral components as hnRNP A, C and K redistribute to the cytoplasm of cells infected with poliovirus or rhinovirus (Gustin and Sarnow, 2001, 2002). These findings suggest that the number of nuclear proteins available in the cytoplasm of poliovirus- and rhinovirus-infected cells may be much larger than currently thought. Despite the growing number of nuclear proteins that appear in the cytoplasm of cells infected with poliovirus and rhinovirus, infection does not result in the depletion of all proteins from the nucleus, as TBP, fibrillarin and SC35 remain in the nucleus following infection (Gustin and Sarnow, 2001, 2002; McBride *et al.*, 1996; Meerovitch *et al.*, 1993; Waggoner and Sarnow, 1998).

6.6   Disruption of nucleo-cytoplasmic trafficking by positive-strand RNA viruses

Recent experiments with picornavirus-infected cells have provided some clues to how viral infection might bring about the redistribution of nuclear proteins to the cytoplasm. These experiments were designed to determine whether infection resulted in a disruption to nucleo-cytoplasmic trafficking. Initially, the distribution of a fusion protein, consisting of the green fluorescent protein (GFP) fused to a classical NLS (GFP–NLS), was monitored in infected cells. The classical NLS is recognized by a heterodimeric import receptor consisting of importin-α and importin-β. Importin-α recognizes and binds to the NLS, while importin-β is responsible for delivering the complex to the nuclear pore (Nakielny and Dreyfuss, 1999). GFP–NLS was localized entirely to the nucleus of uninfected cells, but accumulated in the cytoplasm following infection with coxsackievirus, poliovirus or rhinovirus (Belov *et al.*, 2000; Gustin and Sarnow, 2001, 2002). Because the GFP–NLS fusion protein contained no other known targeting signals or interaction domains, these results suggested that the classical import pathway was disrupted in picornavirus-infected cells.

This same strategy was used to examine the status of the transportin import pathway. Transportin is the import receptor responsible for the nuclear targeting of cargos harbouring the M9 NLS. The M9 NLS was originally identified in hnRNP A1 as a glycine-rich, bidirectional shuttling signal (Pollard *et al.*, 1996). While transportin is responsible for the nuclear import of cargos containing the M9 NLS, the export receptor remains to be identified. As with the classical NLS, when the M9-NLS was fused to GFP the distribution of the resulting protein (GFP–M9NLS) was restricted to the nucleus. Upon infection with either rhinovirus or poliovirus, however, a significant amount of GFP–M9NLS accumulated in the cytoplasm, indicating that this pathway was also disrupted during infection (Gustin and Sarnow, 2001, 2002).

These results provide evidence that infection with entero- and rhinoviruses results in the relocalization of certain nuclear proteins to the cytoplasm. The reason for this relocalization could be a block in nuclear import or an enhancement of nuclear export. To test specifically whether nuclear import was inhibited, an *in vitro* import assay was employed. This assay relies on the ability of digitonin-permeabilized cells to support nuclear import when supplemented with soluble components such as Ran, import receptors and energy (Adam *et al.*, 1990). When mock-infected HeLa cells were permeabilized and incubated with an import cargo consisting of GFP fused to a classical NLS, an energy-regenerating system and rabbit reticulocyte lysate (RRL) as a source of soluble factors, significant accumulation of the GFP–NLS cargo in the nucleus was observed. When infected cells were permeabilized and incubated under identical conditions, however, very little of the GFP–NLS import cargo was observed in the nucleus. These results suggested that the classical import pathway was inhibited in poliovirus-infected cells. More recently, this same approach was used to demonstrate that rhinovirus infection also inhibits the classical nuclear import pathway (Gustin and Sarnow, 2002).

To deliver cargos to the appropriate cellular compartment, all transport pathways must traverse a large, macromolecular structure called the nuclear pore complex (NPC), found embedded in the nuclear envelope (Figure 6.1A).The vertebrate NPC is composed of approximately 30 different proteins and has a molecular weight of over 125 MDa (Cronshaw *et al.*, 2002; Stoffler *et al.*, 1999). Many of the proteins that make up the NPC contain a repeating FG amino acid motif and are thus defined as a family of proteins called 'nucleoporins' (Nups) (Ryan and Wente, 2000). For nuclear import to occur, receptor–cargo complexes must first dock at the cytoplasmic face of the NPC prior to translocation into the nucleus (Newmeyer and Forbes, 1988). When import reactions are carried out in the absence of energy, or at non-physiological temperatures, cargo–receptor complexes accumulate at the NPC, indicative of docking (Adam *et al.*, 1990). When poliovirus- or rhinovirus-infected cells were examined under these conditions, very little accumulation of import cargo at the NPC was observed (Gustin and Sarnow, 2001, 2002). These findings revealed that one of the earliest steps in the classical nuclear import pathway, viz. docking of heterotrimeric importin-α/β cargo complexes at the NPC, is disrupted in poliovirus- and rhinovirus-infected cells.

These results suggested that infection altered the NPC in some fashion. This was confirmed by the analysis of whole cell lysates from infected cells, which revealed that two components of the NPC, Nup153 and p62, were degraded during the course of infection (Figure 6.1B) (Gustin and Sarnow, 2001, 2002). In contrast Ran, importin-β and the cargo nucleolin were not degraded in infected cells (Figure 6.1B) (Gustin and Sarnow, 2001, 2002). Degradation of Nup153 and p62 was not due to the inhibition of translation or transcription that occurs in infected cells, as the addition of the translation elongation inhibitor cycloheximide or the transcription inhibitor actinomycin D did not change the relative abundance of Nup153 or p62 (Gustin and Sarnow, 2001, 2002). Importantly, the degradation of Nup153/p62 and the perturbation of NPCs in poliovirus-infected cells occurred at a time that coincided with the relocalization of nuclear proteins. Both Nup153 and p62 have been shown to interact with the classical import receptors (Nakielny *et al.*, 1999; Percipalle *et al.*, 1997; Shah *et al.*, 1998), suggesting that the degradation of Nup153, p62 or both could be responsible for the inhibition of the classical import pathway observed in poliovirus- and rhinovirus-infected cells.

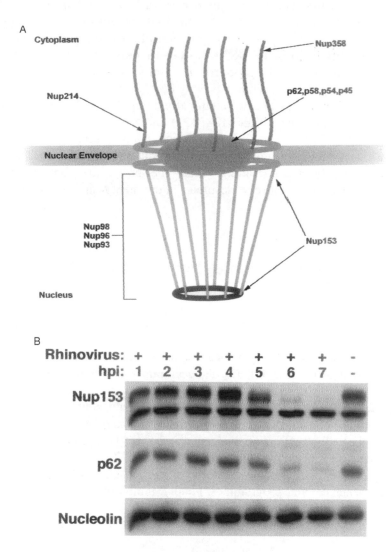

Figure 6.1 Degradation of NPC components in rhinovirus-infected HeLa cells. (A) Schematic representation of the vertebrate NPC. The approximate locations of various nucleoporins within the NPC are indicated. (B). 50 µg whole cell lysates prepared from mock-infected cells or cells that had been infected with rhinovirus type 14 for the indicated length of time were analysed by immunoblotting with monoclonal antibody 414 to detect Nup153 and p62, or MS3 to detect nucleolin. Adapted from the *Journal of Virology* (Gustin and Sarnow, 2002)

A variety of evidence suggests that other viruses may also disrupt nucleo-cytoplasmic trafficking. The observation that coxsackievirus causes the relocalization of a GFP–NLS fusion protein (Belov *et al.*, 2000) suggests that inhibition of nuclear import may be a common feature of enterovirus infections. The cytoplasmic accumulation of hnRNP A1 in cells infected with mouse hepatitis virus suggests that the transportin import pathway may be disrupted in corona-virus-infected cells (Li *et al.*, 1997). Recent findings with the hepatitis C virus non-structural protein 5A (NS5A) suggest that this flavivirus may also disrupt nucleo-cytoplasmic trafficking. NS5A has been reported to interact with karyo-pherin-β3, a member of the importin-β family of proteins that has been implicated in ribosomal protein import (Chung *et al.*, 2000; Jakel and Gorlich, 1998). Expression of karyopherin-β3 can functionally complement the loss of two yeast importin-β family members, PSE1 (KAP121) and KAP123 (Chung *et al.*, 2000). Interestingly, expression of NS5A in yeast blocks this complementation, suggesting that NS5A may inhibit the function of karyopherin-β3 (Chung *et al.*, 2000). Cumulatively, these findings raise the possibility that inhibition of nuclear import may be a more widespread phenomenon among positive-strand RNA viruses.

6.7 Advantages provided to the virus by inhibiting nuclear import

Two major consequences of inhibiting nuclear import that could be beneficial for viral infection can be envisaged. First, as discussed above, inhibition of import may contribute to the accumulation of nuclear factors in the cytoplasm of infected cells. Many of the cellular nuclear proteins that accumulate in the cytoplasm of infected cells are RNA-binding proteins with known functions in RNA-processing, stability, transport and translation. Clearly, these activities could contribute to aspects of viral replication such as translation, replication or packaging of the viral genome. In addition, the redistribution of nuclear proteins to the cytoplasm may prevent these proteins from carrying out their normal function. This could benefit the virus by preventing either operation of unwanted cellular biosynthetic activities or activation of detrimental cellular programmes. Finally, inhibition of nuclear import could prevent the ability of cells to successfully mount an antiviral response. Antiviral responses that could be inhibited in this manner include the interferon response and apoptosis.

Innate immunity mediated by interferon requires the activation of latent transcription factors in the cytoplasm, such as the signal transducers and activators of transcription (STATs) and interferon response factors (IRFs) (Stark *et al.*, 1998). Following activation, these transcription factors are imported into the nucleus, where they activate transcription of target genes encoding antiviral functions (Stark *et al.*, 1998).

Inhibition of nuclear import would be predicted to impair the ability of these transcription factors to enter the nuclear compartment, thus preventing the culminating step in this signal transduction cascade (Figure 6.2A). While intriguing, the time frame of inhibition of nuclear import is somewhat inconsistent with this scenario. In poliovirus-infected cells, relocalization is first apparent by 3 h post-infection and is complete by 4.5 h (Gustin and Sarnow, 2001; McBride *et al.*, 1996; Meerovitch *et al.*, 1993; Waggoner and Sarnow, 1998). In agreement with these findings, degradation of NPC proteins first becomes apparent by 3 h post-infection and is complete by about 5 h (Gustin and Sarnow, 2001). Rhinovirus, which has a slightly extended replicative cycle when compared to poliovirus (Stott and Killington, 1972), causes relocalization of cellular proteins and degradation of NPC proteins that is first observed 4–5 h following infection and is complete at 6–7 h (Gustin and Sarnow, 2002). Thus, the reported inhibition of nuclear import may occur too late in the infectious cycle to provide effective protection from the interferon response.

While this may be the case, it is worth noting that the characterization of trafficking pathways and NPC components in infected cells is far from complete. At least 15 distinct import pathways have been identified in vertebrates and this number is likely to grow (Gorlich and Kutay, 1999; Nakielny and Dreyfuss, 1999). While the classical pathway was analysed specifically in poliovirus- and rhinovirus-infected cells, and the relocalization of cargos specific to the transportin and hnRNP K pathways suggests that these routes are also inhibited, the status of other import pathways has not been examined. In addition, the NPC is composed of over 30 different proteins (Cronshaw *et al.*, 2002) and the status of only two of these factors, Nup153 and p62, has been examined (Figure 6.1A). Thus, it is possible that, very early during infection, as-yet unidentified NPC proteins or trafficking pathways are targeted that result in an attenuation of the interferon response.

A more likely viral defence mechanism to be impaired by a disruption of nucleo-cytoplasmic trafficking is the induction of programmed cell death, or apoptosis (Figure 6.2B reviewed in Knipe *et al.*, 2001). Recent studies have suggested that nuclear import is essential for carrying out certain apoptotic pathways. For example, nuclear import of protein kinase Cδ (PKCδ) is required for mitochondrial-dependent apoptosis (DeVries *et al.*, 2002). Mutations that disrupt the NLS in PKCδ prevent nuclear import and block the development of etoposide-induced apoptosis in rat parotid salivary acinar cells (DeVries *et al.*, 2002). Examination of apoptosis in ischaemic neurons revealed that active caspase-8 translocates to the nucleus, where it cleaves PARP-2. Caspase-independent programmed cell death also appears to rely on nuclear import of effector molecules. For example, induction of apoptosis by staurosporine in Rat-1 fibroblasts results in the release of apoptosis-inducing factor (AIF) from mitochondria and its translocation to the nucleus (Susin *et al.*, 1999). Microinjection of AIF into the cytoplasm of cells results in the development of typical apoptotic indicators, such as chromatin condensation and exposure of phosphatidylserine on the exterior of the cell (Susin *et al.*, 1999). A requirement for nuclear translocation was illustrated by the

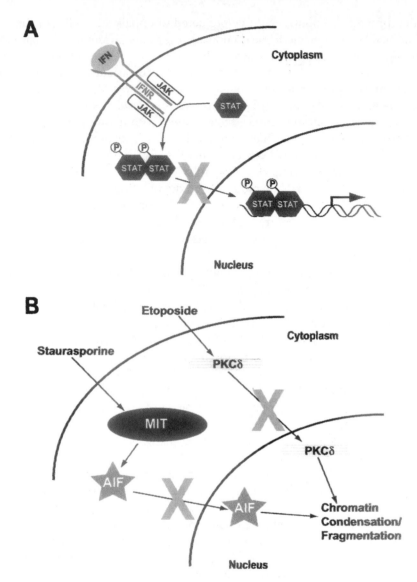

Figure 6.2   Disruption of antiviral responses by inhibition of nuclear import. (A) Interferon response. Binding of interferon to its receptor results in the phosphorylation and dimerization of STAT molecules in the cytoplasm. STAT dimers translocate into the nucleus and activate transcription of target genes. Inhibition of nuclear import may prevent nuclear import of activated STAT dimers. (B) Apoptotic response. Etoposide-induced apoptosis requires nuclear import of PKCδ. Not shown is the catalytic fragment of PKCδ that is generated by caspase cleavage and also imported into the nucleus. Staurosporine treatment induces mitochondria to release AIF into the cytoplasm, and it is then imported into the nucleus. Once in the nucleus, PKCδ and AIF cause condensation and fragmentation of cellular chromatin. Inhibition of import may prevent the accumulation of PKCδ and/or AIF in the nucleus and thus prevent completion of the apoptotic response

finding that co-injection of wheat germ agglutinin, an inhibitor of nuclear transport (Finlay *et al.*, 1987), prevented the development of apoptotic markers (Susin *et al.*, 1999). Similarly, cell-mediated killing by perforin requires nuclear uptake of granzyme proteases (Blink *et al.*, 1999). Overexpression of the anti-apoptotic factor, bcl-2, prevents nuclear accumulation of granzyme and the development of apoptosis (Jans *et al.*, 1999). Thus it is possible that, by inhibiting nuclear import, poliovirus might impair the apoptotic response and permit replication to high titres.

Evidence suggests that poliovirus does indeed encode anti-apoptotic functions. In most cell types, productive poliovirus infection results in the development of typical cytopathic effects with very little indication of apoptosis. However, if cells are infected in the presence of guanidine, a specific inhibitor of viral negative-strand RNA synthesis (Barton and Flanegan, 1997), typical apoptotic indicators are manifested, such as DNA fragmentation and cytoplasmic blebbing (Tolskaya *et al.*, 1995). Subsequent work has revealed that poliovirus-induced apoptosis results in the release of cytochrome *c* from mitochondria and the activation of caspases, including caspase-9 (Agol *et al.*, 1998; Belov *et al.*, 2003). During productive infection, cytochrome *c* is still released from mitochondria but active caspase-9 is not observed, suggesting a potential mechanism to explain how poliovirus prevents completion of the apoptotic programme (Belov *et al.*, 2003). In these studies, however, the status of other mitochondrial associated apoptotic factors, such as AIF or PKCδ, was not examined. Thus, it remains possible that inhibition of nuclear import may contribute to preventing induction of apoptosis in cells productively infected with poliovirus by blocking the nuclear accumulation of these or other proteins (Figure 6.2B).

6.8   Conclusions

Ample evidence implicates the nucleus and nuclear proteins in the replication and pathogenesis of positive-strand RNA viruses. Examples exist from most of the positive-strand RNA virus families, including plant viruses, of viral proteins accumulating in the nucleus. In some cases, the identification of nuclear binding partners suggests potential roles for these viral proteins in the nucleus. The discovery of a number of interactions between viral proteins or the RNA genome and nuclear factors is consistent with a role for nuclear components in the viral life cycle. In some cases, this possibility has been supported by the demonstration that nuclear proteins can enhance translation or replication of viral RNAs *in vitro* (Hellen *et al.*, 1993; Hunt *et al.*, 1999; Kim and Jang, 1999; Li *et al.*, 1999; Meerovitch *et al.*, 1993; Waggoner and Sarnow, 1998). The observation that viral infection can induce the redistribution of some of these nuclear proteins to the cytoplasm supports an *in vivo* role for these proteins in replication. The finding that rhinovirus and poliovirus inhibit nuclear import and cause the degradation of NPC proteins suggests a potential mechanism to account for the cytoplasmic accumulation of nuclear proteins during picornavirus infection.

Clearly, there are a number of questions that remain to be answered. Chief among these is whether nuclear components are required for viral replication. Answering this question will require the analysis of viruses harbouring mutations that specifically disrupt targeting to the nucleus or interactions with nuclear factors. Alternatively, the analysis of viral replication in cell lines that do not express the nuclear protein in question could also demonstrate a requirement for specific interactions. The development of efficient systems for the knockdown of gene expression in somatic cells using small interfering RNAs should facilitate this latter analysis.

The inhibition of nuclear import and degradation of NPC components described for poliovirus and rhinovirus raises many questions as well. While it is intriguing to propose that degradation of Nup153 and p62 is responsible for the inhibition of import, there is no direct evidence to support such a model. The identification of viral factors responsible for these events will no doubt help to clarify this issue. To fully understand the impact of infection on nucleo–cytoplasmic trafficking, it will be necessary to better characterize the NPC in infected cells. Recently, Cronshaw *et al.* reported the proteomic characterization of an enriched fraction of NPCs from rat liver (Cronshaw *et al.*, 2002). An analysis of the NPC proteome from infected cells would allow for a complete inventory of the changes caused by infection and allow for predictions regarding the status of specific trafficking pathways to be tested. Of equal importance is determining whether inhibition of nuclear import (or degradation of NPC components) contributes to viral replication directly, by facilitating relocalization of nuclear factors, or indirectly, by helping the virus to evade host defence mechanisms.

6.8.1   Method: *in vitro* nuclear import assay using digitonin-permeabilized HeLa cells

The development of an *in vitro* assay to monitor nuclear import in mammalian cells has provided a powerful system with which to dissect the biochemical underpinnings of nucleo-cytoplasmic transport (reviewed in Gorlich and Kutay, 1999). This assay relies on the ability of digitonin to selectively permeabilize the plasma membrane, due to its high cholesterol content, while leaving the nuclear envelope and the NPC intact (Adam *et al.*, 1990). Once permeabilized, the cells are washed to remove soluble components required for nuclear import, such as the small GTPase Ran, NTF2 and import receptors. Import reactions are then initiated by the addition of a cytosolic lysate that serves as a source of soluble factors, an import cargo and an energy source such as GTP. Typically, the import cargo is a protein containing an NLS that can be easily detected in the nuclei of cells by fluorescent microscopy. As the factors required for nuclear import have been identified, it has become possible to reconstitute nuclear import in permeabilized cells using purified proteins, thus adding a further refinement to this powerful assay (Adam and Gerace, 1991; Moore and Blobel, 1993).

Here, we describe the protocol used by our laboratory to monitor nuclear import in HeLa cell monolayers. As a convenient source of soluble components for this assay, we use rabbit reticulocyte lysate (RRL), although cytosolic fractions from other cell types can be prepared (Adam *et al.*, 1992). The import cargo (GST–NLS–EGFP) has been described and consists of a classical bipartite NLS fused to EGFP and glutathione S-transferase (GST) (Rosorius *et al.*, 1999).

Import cargo

The GST–NLS–EGFP import cargo was purified from *Escherichia coli* BL21 cells transformed with pGEX–NLS–GFP, as described by Rosorius *et al.* (1999). Following purification, GST–NLS–EGFP was extensively dialysed at 4°C against transport buffer (TB, 20 mM HEPES pH 7.3, 110 mM KOAc, 5 mM NaOAc, 1 mM MgOAc, 1 mM EGTA, 2 mM DTT and 1 µg/ml each of chymotrypsin, leupeptin, antipain and pepstatin). Dialysed GST–NLS–EGFP was stored at −80°C until needed.

Preparation of cytosol

Untreated or micrococcal nuclease-treated RRL can be purchased from Promega Corp. (Madison, WI) and both work well in our hands. Prior to use in the nuclear import assay, RRL was centrifuged at $100\,000 \times g$ for 30 min at 4°C and dialysed extensively against TB at 4°C. Dialysed RRL was dispensed into single-use aliquots of 100 µl, snap-frozen in liquid nitrogen and stored at −80°C.

Nuclear import assay

Two days prior to the import assay, HeLa cells were seeded onto 12 mm glass coverslips in 35 mm tissue culture dishes. The cells should be seeded at a density such that they are 50–70% confluent at the time they are processed for the import assay. The cells were washed once in PBS and once with ice-cold TB before permeabilizing for 5 min at 0°C in TB containing 40 µg/ml digitonin. Permeabilized cells were then washed once in ice-cold TB and inverted into a 20 µl import reaction consisting of TB supplemented with 50% RRL, 0.4 µM GST–NLS–EGFP, 1 mM ATP, 1 mM GTP, 5 mM creatine phosphate and 20 U/ml creatine kinase. Reactions were allowed to proceed for 30 min at 25°C and then were washed once in ice-cold TB. Following completion of the import reaction, the cells were fixed in 3% formaldehyde for 20 min at 25°C and then stained with Hoechst 33258 to visualize cellular chromatin. The fixed cells were then inverted into Vectashield mounting medium (Vector Labs) on glass slides and analysed by fluorescence microscopy.

Acknowledgements

Research in the authors' laboratories is supported by grants from the National Institutes of Health to P.S. (AI 25105) and the National Institutes of Health and the National Center for Research Resources Center of Biomedical Research Excellence to K.G. (P20 RR15587).

References

Adam, S. A. and Gerace, L. (1991). Cytosolic proteins that specifically bind nuclear location signals are receptors for nuclear import. *Cell* **66**, 837–847.

Adam, S. A., Marr, R. S. and Gerace, L. (1990). Nuclear protein import in permeabilized mammalian cells requires soluble cytoplasmic factors. *J Cell Biol* **111**, 807–816.

Adam, S. A., Sterne-Marr, R. and Gerace, L. (1992). Nuclear protein import using digitonin-permeabilized cells. *Methods Enzymol* **219**, 97–110.

Agol, V. I., Belov, G. A., Bienz, K., Egger, D., Kolesnikova, M. S., Raikhlin, N. T., Romanova, L. I., Smirnova, E. A. and Tolskaya, E. A. (1998). Two types of death of poliovirus-infected cells: caspase involvement in the apoptosis but not cytopathic effect. *Virology* **252**, 343–353.

Ali, N. and Siddiqui, A. (1995). Interaction of polypyrimidine tract-binding protein with the 5′ noncoding region of the hepatitis C virus RNA genome and its functional requirement in internal initiation of translation. *J Virol* **69**, 6367–6375.

Ali, N. and Siddiqui, A. (1997). The La antigen binds 5′ noncoding region of the hepatitis C virus RNA in the context of the initiator AUG codon and stimulates internal ribosome entry site-mediated translation. *Proc Natl Acad Sci USA* **94**, 2249–2254.

Bachmann, M., Falke, D., Schroder, H. C. and Muller, W. E. (1989). Intracellular distribution of the La antigen in CV-1 cells after herpes simplex virus type 1 infection compared with the localization of U small nuclear ribonucleoprotein particles. *J Gen Virol* **70**, 881–891.

Back, S. H., Kim, Y. K., Kim, W. J., Cho, S., Oh, H. R., Kim, J. E. and Jang, S. K. (2002). Translation of polioviral mRNA is inhibited by cleavage of polypyrimidine tract-binding proteins executed by polioviral 3C(pro). *J Virol* **76**, 2529–2542.

Barton, D. J. and Flanegan, J. B. (1997). Synchronous replication of poliovirus RNA: initiation of negative-strand RNA synthesis requires the guanidine-inhibited activity of protein 2C. *J Virol* **71**, 8482–8489.

Baunoch, D. A., Das, P., Browning, M. E. and Hari, V. (1991). A temporal study of the expression of the capsid, cytoplasmic inclusion and nuclear inclusion proteins of tobacco etch polyvirus in infected plants. *J Gen Virol* **72**(3), 487–492.

Belov, G. A., Evstafieva, A. G., Rubtsov, Y. P., Mikitas, O. V., Vartapetian, A. B. and Agol, V. I. (2000). Early alteration of nucleocytoplasmic traffic induced by some RNA viruses. *Virology* **275**, 244–248.

Belov, G. A., Romanova, L. I., Tolskaya, E. A., Kolesnikova, M. S., Lazebnik, Y. A. and Agol, V. I. (2003). The major apoptotic pathway activated and suppressed by poliovirus. *J Virol* **77**, 45–56.

Black, D. N. and Brown, F. (1968). The influence of mitomycin C, actinomycin D and ultraviolet light on the replication of the viruses of foot-and-mouth disease and vesicular stomatitis. *J Gen Virol* **3**, 453–457.

Blink, E. J., Trapani, J. A. and Jans, D. A. (1999). Perforin-dependent nuclear targeting of granzymes: a central role in the nuclear events of granule-exocytosis-mediated apoptosis? *Immunol Cell Biol* **77**, 206–215.

Borer, R. A., Lehner, C. F., Eppenberger, H. M. and Nigg, E. A. (1989). Major nucleolar proteins shuttle between nucleus and cytoplasm. *Cell* **56**, 379–390.

Borman, A., Howell, M. T., Patton, J. G. and Jackson, R. J. (1993). The involvement of a spliceosome component in internal initiation of human rhinovirus RNA translation. *J Gen Virol* **74**(9), 1775–1788.

Bossart, W. and Bienz, K. (1979). Virus replication, cytopathology, and lysosomal enzyme response in enucleated HEp-2 cells infected with poliovirus. *Virology* **92**, 331–339.

Buckley, A., Gaidamovich, S., Turchinskaya, A. and Gould, E. A. (1992). Monoclonal antibodies identify the NS5 yellow fever virus non-structural protein in the nuclei of infected cells. *J Gen Virol* **73**(5), 1125–1130.

Carrington, J. C., Freed, D. D. and Leinicke, A. J. (1991). Bipartite signal sequence mediates nuclear translocation of the plant polyviral NIa protein. *Plant Cell* **3**, 953–962.

Carter, M. J. (1994). Genomic organization and expression of astroviruses and caliciviruses. *Arch Virol Suppl* **9**, 429–439.

Chang, C. J., Luh, H. W., Wang, S. H., Lin, H. J., Lee, S. C. and Hu, S. T. (2001). The heterogeneous nuclear ribonucleoprotein K (hnRNP K) interacts with dengue virus core protein. *DNA Cell Biol* **20**, 569–577.

Chen, H., Wurm, T., Britton, P., Brooks, G. and Hiscox, J. A. (2002). Interaction of the coronavirus nucleoprotein with nucleolar antigens and the host cell. *J Virol* **76**, 5233–5250.

Chung, K. M., Lee, J., Kim, J. E., Song, O. K., Cho, S., Lim, J., Seedorf, M., Hahm, B. and Jang, S. K. (2000). Nonstructural protein 5A of hepatitis C virus inhibits the function of karyopherin beta3. *J Virol* **74**, 5233–5241.

Crocker, T. T., Pfendt, E. and Spendlove, R. (1964). Poliovirus: growth in non-nucleate cytoplasm. *Science* **145**, 401–403.

Cronshaw, J. M., Krutchinsky, A. N., Zhang, W., Chait, B. T. and Matunis, M. J. (2002). Proteomic analysis of the mammalian nuclear pore complex. *J Cell Biol* **158**, 915–927.

Detjen, B. M., Lucas, J. and Wimmer, E. (1978). Poliovirus single-stranded RNA and double-stranded RNA: differential infectivity in enucleate cells. *J Virol* **27**, 582–586.

DeVries, T. A., Neville, M. C. and Reyland, M. E. (2002). Nuclear import of PKCdelta is required for apoptosis: identification of a novel nuclear import sequence. *EMBO J* **21**, 6050–6060.

Errington, W., Wardell, A. D., McDonald, S., Goldin, R. D. and McGarvey, M. J. (1999). Subcellular localisation of NS3 in HCV-infected hepatocytes. *J Med Virol* **59**, 456–462.

Erwin, C. and Brown, D. T. (1983). Requirement of cell nucleus for Sindbis virus replication in cultured *Aedes albopictus* cells. *J Virol* **45**, 792–799.

Fazakerley, J. K., Boyd, A., Mikkola, M. L. and Kaariainen, L. (2002). A single amino acid change in the nuclear localization sequence of the nsP2 protein affects the neurovirulence of Semliki Forest virus. *J Virol* **76**, 392–396.

Finlay, D. R., Newmeyer, D. D., Price, T. M. and Forbes, D. J. (1987). Inhibition of *in vitro* nuclear transport by a lectin that binds to nuclear pores. *J Cell Biol* **104**, 189–200.

Follett, E. A., Pringle, C. R. and Pennington, T. H. (1975). Virus development in enucleate cells: echovirus, poliovirus, pseudorabies virus, reovirus, respiratory syncytial virus and Semliki Forest virus. *J Gen Virol* **26**, 183–196.

Forwood, J. K., Brooks, A., Briggs, L. J., Xiao, C. Y., Jans, D. A. and Vasudevan, S. G. (1999). The 37-amino-acid interdomain of dengue virus NS5 protein contains a functional NLS and inhibitory CK2 site. *Biochem Biophys Res Commun* **257**, 731–737.

Ginisty, H., Sicard, H., Roger, B. and Bouvet, P. (1999). Structure and functions of nucleolin. *J Cell Sci* **112**(6), 761–772.

Gorlich, D. and Kutay, U. (1999). Transport between the cell nucleus and the cytoplasm. *Annu Rev Cell Dev Biol* **15**, 607–660.

Gustin, K. E. and Sarnow, P. (2001). Effects of poliovirus infection on nucleocytoplasmic trafficking and nuclear pore complex composition. *EMBO J* **20**, 240–249.

Gustin, K. E. and Sarnow, P. (2002). Inhibition of nuclear import and alteration of nuclear pore complex composition by rhinovirus. *J Virol* **76**, 8787–8796.

Gutierrez-Escolano, A. L., Brito, Z. U., del Angel, R. M. and Jiang, X. (2000). Interaction of cellular proteins with the 5′ end of Norwalk virus genomic RNA. *J Virol* **74**, 8558–8562.

Hellen, C. U., Witherell, G. W., Schmid, M., Shin, S. H., Pestova, T. V., Gil, A. and Wimmer, E. (1993). A cytoplasmic 57-kDa protein that is required for translation of picornavirus RNA by internal ribosomal entry is identical to the nuclear pyrimidine tract-binding protein. *Proc Natl Acad Sci USA* **90**, 7642–7646.

Hiscox, J. A., Wurm, T., Wilson, L., Britton, P., Cavanagh, D. and Brooks, G. (2001). The coronavirus infectious bronchitis virus nucleoprotein localizes to the nucleolus. *J Virol* **75**, 506–512.

Hsieh, T. Y., Matsumoto, M., Chou, H. C., Schneider, R., Hwang, S. B., Lee, A. S. and Lai, M. M. (1998). Hepatitis C virus core protein interacts with heterogeneous nuclear ribonucleoprotein K. *J Biol Chem* **273**, 17651–17659.

Huang, P. and Lai, M. M. (2001). Heterogeneous nuclear ribonucleoprotein a1 binds to the 3′-untranslated region and mediates potential 5′–3′-end cross talks of mouse hepatitis virus RNA. *J Virol* **75**, 5009–5017.

Hunt, S. L., Skern, T., Liebig, H. D., Kuechler, E. and Jackson, R. J. (1999). Rhinovirus 2A proteinase mediated stimulation of rhinovirus RNA translation is additive to the stimulation effected by cellular RNA binding proteins. *Virus Res* **62**, 119–128.

Ito, T. and Lai, M. M. (1997). Determination of the secondary structure of and cellular protein binding to the 3′-untranslated region of the hepatitis C virus RNA genome. *J Virol* **71**, 8698–8706.

Izumi, R. E., Valdez, B., Banerjee, R., Srivastava, M. and Dasgupta, A. (2001). Nucleolin stimulates viral internal ribosome entry site-mediated translation. *Virus Res* **76**, 17–29.

Jakel, S. and Gorlich, D. (1998). Importin beta, transportin, RanBP5 and RanBP7 mediate nuclear import of ribosomal proteins in mammalian cells. *EMBO J* **17**, 4491–4502.

Jans, D. A., Sutton, V. R., Jans, P., Froelich, C. J. and Trapani, J. A. (1999). BCL-2 blocks perforin-induced nuclear translocation of granzymes concomitant with protection against the nuclear events of apoptosis. *J Biol Chem* **274**, 3953–3961.

Kaminski, A., Hunt, S. L., Patton, J. G. and Jackson, R. J. (1995). Direct evidence that polypyrimidine tract binding protein (PTB) is essential for internal initiation of translation of encephalomyocarditis virus RNA. *RNA* **1**, 924–938.

Kapoor, M., Zhang, L., Ramachandra, M., Kusukawa, J., Ebner, K. E. and Padmanabhan, R. (1995). Association between NS3 and NS5 proteins of dengue virus type 2 in the putative RNA replicase is linked to differential phosphorylation of NS5. *J Biol Chem* **270**, 19100–19106.

Karst, S. M., Wobus, C. E., Lay, M., Davidson, J. and Virgin, H. W. T. (2003). STAT1-dependent innate immunity to a Norwalk-like virus. *Science* **299**, 1575–1578.

Kim, J. E., Song, W. K., Chung, K. M., Back, S. H. and Jang, S. K. (1999). Subcellular localization of hepatitis C viral proteins in mammalian cells. *Arch Virol* **144**, 329–343.

Kim, Y. K. and Jang, S. K. (1999). La protein is required for efficient translation driven by encephalomyocarditis virus internal ribosomal entry site. *J Gen Virol* **80**(12), 3159–3166.

Knipe, D. M., Samuel, C. E. and Palese, P. (2001). Virus–host cell interactions. In *Field's Virology*, 4th edn, pp. 133–70, D. M. Knipe, P. M. Howley *et al.* (eds). Philadelphia, PA: Lippincott Williams and Wilkins.

Kos, K. A., Osborne, B. A. and Goldsby, R. A. (1975). Inhibition of group B arbovirus antigen production and replication in cells enucleated with cytochalasin B. *J Virol* **15**, 913–917.

Krecic, A. M. and Swanson, M. S. (1999). hnRNP complexes: composition, structure, and function. *Curr Opin Cell Biol* **11**, 363–371.

Kurilla, M. G., Cabradilla, C. D., Holloway, B. P. and Keene, J. D. (1984). Nucleotide sequence and host La protein interactions of rabies virus leader RNA. *J Virol* **50**, 773–778.

Kurilla, M. G. and Keene, J. D. (1983). The leader RNA of vesicular stomatitis virus is bound by a cellular protein reactive with anti-La lupus antibodies. *Cell* **34**, 837–845.

Lad, V. J., Gupta, A. K., Ghosh, S. N. and Banerjee, K. (1993). Immunofluorescence studies on the replication of some arboviruses in nucleated and enucleated cells. *Acta Virol* **37**, 79–83.

Langenberg, W. G. and Zhang, L. (1997). Immunocytology shows the presence of tobacco etch virus P3 protein in nuclear inclusions. *J Struct Biol* **118**, 243–247.

Li, H. P., Huang, P., Park, S. and Lai, M. M. (1999). Polypyrimidine tract-binding protein binds to the leader RNA of mouse hepatitis virus and serves as a regulator of viral transcription. *J Virol* **73**, 772–777.

Li, H. P., Zhang, X., Duncan, R., Comai, L. and Lai, M. M. (1997). Heterogeneous nuclear ribonucleoprotein A1 binds to the transcription-regulatory region of mouse hepatitis virus RNA. *Proc Natl Acad Sci USA* **94**, 9544–9549.

Li, J., Liu, Y., Kim, B. O. and He, J. J. (2002). Direct participation of Sam68, the 68-kDa Src-associated protein in mitosis, in the CRM1-mediated Rev nuclear export pathway. *J Virol* **76**, 8374–8382.

Li, X. H. and Carrington, J. C. (1993). Nuclear transport of tobacco etch potyviral RNA-dependent RNA polymerase is highly sensitive to sequence alterations. *Virology* **193**, 951–958.

Liang, F., Qu, J., Zhang, X., Chen, J., Ding, M. and Zhai, Z. (1995). [The distribution of SbV nonstructural protein 2 (nsP2) in host cell]. *Wei Sheng Wu Xue Bao* **35**, 260–263.

McBride, A. E., Schlegel, A. and Kirkegaard, K. (1996). Human protein Sam68 relocalization and interaction with poliovirus RNA polymerase in infected cells. *Proc Natl Acad Sci USA* **93**, 2296–2301.

Meerovitch, K., Svitkin, Y. V., Lee, H. S., Lejbkowicz, F., Kenan, D. J., Chan, E. K., Agol, V. I., Keene, J. D. and Sonenberg, N. (1993). La autoantigen enhances and corrects aberrant translation of poliovirus RNA in reticulocyte lysate. *J Virol* **67**, 3798–3807.

Michael, W. M., Siomi, H., Choi, M., Pinol-Roma, S., Nakielny, S., Liu, Q. and Dreyfuss, G. (1995). Signal sequences that target nuclear import and nuclear export of pre-mRNA-binding proteins. *Cold Spring Harb Symp Quant Biol* **60**, 663–668.

Moore, M. S. and Blobel, G. (1993). The GTP-binding protein Ran/TC4 is required for protein import into the nucleus. *Nature* **365**, 661–663.

Muramatsu, S., Ishido, S., Fujita, T., Itoh, M. and Hotta, H. (1997). Nuclear localization of the NS3 protein of hepatitis C virus and factors affecting the localization. *J Virol* **71**, 4954–4961.

Nakielny, S. and Dreyfuss, G. (1999). Transport of proteins and RNAs in and out of the nucleus. *Cell* **99**, 677–690.

Nakielny, S., Shaikh, S., Burke, B. and Dreyfuss, G. (1999). Nup153 is an M9-containing mobile nucleoporin with a novel Ran-binding domain. *EMBO J* **18**, 1982–1995.

Newmeyer, D. D. and Forbes, D. J. (1988). Nuclear import can be separated into distinct steps *in vitro*: nuclear pore binding and translocation. *Cell* **52**, 641–653.

Pardigon, N. and Strauss, J. H. (1996). Mosquito homolog of the La autoantigen binds to Sindbis virus RNA. *J Virol* **70**, 1173–1181.

Peranen, J., Rikkonen, M., Liljestrom, P. and Kaariainen, L. (1990). Nuclear localization of Semliki Forest virus-specific nonstructural protein nsP2. *J Virol* **64**, 1888–1896.

Percipalle, P., Clarkson, W. D., Kent, H. M., Rhodes, D. and Stewart, M. (1997). Molecular interactions between the importin alpha/beta heterodimer and proteins involved in vertebrate nuclear protein import. *J Mol Biol* **266**, 722–732.

Pinol-Roma, S. and Dreyfuss, G. (1992). Shuttling of pre-mRNA binding proteins between nucleus and cytoplasm. *Nature* **355**, 730–732.

Pollack, R. and Goldman, R. (1973). Synthesis of infective poliovirus in BSC-1 monkey cells enucleated with cytochalasin B. *Science* **179**, 915–916.

Pollard, V. W., Michael, W. M., Nakielny, S., Siomi, M. C., Wang, F. and Dreyfuss, G. (1996). A novel receptor-mediated nuclear protein import pathway. *Cell* **86**, 985–994.

Prescott, D. M., Myerson, D. and Wallace, J. (1972). Enucleation of mammalian cells with cytochalasin B. *Exp Cell Res* **71**, 480–485.

Reddy, T. R., Tang, H., Xu, W. and Wong-Staal, F. (2000a). Sam68, RNA helicase A and Tap cooperate in the post-transcriptional regulation of human immunodeficiency virus and type D retroviral mRNA [in process citation]. *Oncogene* **19**, 3570–3575.

Reddy, T. R., Xu, W., Mau, J. K., Goodwin, C. D., Suhasini, M., Tang, H., Frimpong, K., Rose, D. W. and Wong-Staal, F. (1999). Inhibition of HIV replication by dominant negative mutants of Sam68, a functional homolog of HIV-1 Rev. *Nat Med* **5**, 635–642.

Reddy, T. R., Xu, W. D. and Wong-Staal, F. (2000b). General effect of Sam68 on Rev/Rex regulated expression of complex retroviruses. *Oncogene* **19**, 4071–4074.

Reich, E., Franklin, R. M., Shatkin, A. J. and Tatum, E. L. (1961). Effect of actinomycin D on cellular nucleic acid synthesis and vrus production. *Science* **134**, 556–557.

Restrepo, M. A., Freed, D. D. and Carrington, J. C. (1990). Nuclear transport of plant polyviral proteins. *Plant Cell* **2**, 987–998.

Rikkonen, M. (1996). Functional significance of the nuclear-targeting and NTP-binding motifs of Semliki Forest virus nonstructural protein nsP2. *Virology* **218**, 352–361.

Rojas-Eisenring, I. A., Cajero-Juarez, M. and del Angel, R. M. (1995). Cell proteins bind to a linear polypyrimidine-rich sequence within the 5′-untranslated region of rhinovirus 14 RNA. *J Virol* **69**, 6819–6824.

Rosorius, O., Heger, P., Stelz, G., Hirschmann, N., Hauber, J. and Stauber, R. H. (1999). Direct observation of nucleocytoplasmic transport by microinjection of GFP-tagged proteins in living cells. *Biotechniques* **27**, 350–355.

Rowland, R. R., Kervin, R., Kuckleburg, C., Sperlich, A. and Benfield, D. A. (1999). The localization of porcine reproductive and respiratory syndrome virus nucleocapsid protein to the nucleolus of infected cells and identification of a potential nucleolar localization signal sequence. *Virus Res* **64**, 1–12.

Ryan, K. J. and Wente, S. R. (2000). The nuclear pore complex: a protein machine bridging the nucleus and cytoplasm. *Curr Opin Cell Biol* **12**, 361–371.

Schlesinger, S. and Schlesinger, M. J. (2001). Togaviridae: the viruses and their replication. In *Field's Virology*, 4th edn, pp. 895–916, D. M. Knipe and P. M. Howley (eds). Philadelphia, PA: Lippincott/Raven.

Shah, S., Tugendreich, S. and Forbes, D. (1998). Major binding sites for the nuclear import receptor are the internal nucleoporin Nup153 and the adjacent nuclear filament protein Tpr. *J Cell Biol* **141**, 31–49.

Shen, X. and Masters, P. S. (2001). Evaluation of the role of heterogeneous nuclear ribonucleo-protein A1 as a host factor in murine coronavirus discontinuous transcription and genome replication. *Proc Natl Acad Sci USA* **98**, 2717–2722.

Spangberg, K., Goobar-Larsson, L., Wahren-Herlenius, M. and Schwartz, S. (1999). The La protein from human liver cells interacts specifically with the U-rich region in the hepatitis C virus 3′ untranslated region. *J Hum Virol* **2**, 296–307.

Stark, G. R., Kerr, I. M., Williams, B. R., Silverman, R. H. and Schreiber, R. D. (1998). How cells respond to interferons. *Annu Rev Biochem* **67**, 227–264.

Stoffler, D., Fahrenkrog, B. and Aebi, U. (1999). The nuclear pore complex: from molecular architecture to functional dynamics. *Curr Opin Cell Biol* **11**, 391–401.

Stott, E. J. and Killington, R. A. (1972). Rhinoviruses. *Annu Rev Microbiol* **26**, 503–524.

Susin, S. A., Lorenzo, H. K., Zamzami, N., Marzo, I., Snow, B. E., Brothers, G. M., Mangion, J., Jacotot, E., Costantini, P., Loeffler, M., Larochette, N., Goodlett, D. R., Aebersold, R., Siderovski, D. P., Penninger, J. M. and Kroemer, G. (1999). Molecular characterization of mitochondrial apoptosis-inducing factor. *Nature* **397**, 441–446.

Tadano, M., Makino, Y., Fukunaga, T., Okuno, Y. and Fukai, K. (1989). Detection of dengue 4 virus core protein in the nucleus. I. A monoclonal antibody to dengue 4 virus reacts with the antigen in the nucleus and cytoplasm. *J Gen Virol* **70**(6), 1409–1415.

Taylor, J. (1965). Studies on the mechanism of action of interferon I. Interferon action and RNA synthesis in chick embryo fibroblasts infected with Semliki Forest virus. *Virology* **25**, 340–349.

Tijms, M. A., van der Meer, Y. and Snijder, E. J. (2002). Nuclear localization of non-structural protein 1 and nucleocapsid protein of equine arteritis virus. *J Gen Virol* **83**, 795–800.

Tolskaya, E. A., Romanova, L. I., Kolesnikova, M. S., Ivannikova, T. A., Smirnova, E. A., Raikhlin, N. T. and Agol, V. I. (1995). Apoptosis-inducing and apoptosis-preventing functions of poliovirus. *J Virol* **69**, 1181–1189.

Tsuchihara, K., Tanaka, T., Hijikata, M., Kuge, S., Toyoda, H., Nomoto, A., Yamamoto, N. and Shimotohno, K. (1997). Specific interaction of polypyrimidine tract-binding protein with the extreme 3′-terminal structure of the hepatitis C virus genome, the 3′X. *J Virol* **71**, 6720–6726.

van Regenmortel, M. H. V., Fauquet, C. M., Bishop, D. H. L., Carstens, E. B., Estes, M. K., Lemon, S. M., Maniloff, J., Mayo, M. A., McGeoch, D. J., Pringle, C. R. and Wickner, R. B. (2000). Virus taxonomy: classification and nomenclature of viruses. In *Seventh report of the international committee on taxonomy of viruses*. Virology Division International Union of Microbiological Societies (eds). San Diego, CA: Academic Press.

Waggoner, S. and Sarnow, P. (1998). Viral ribonucleoprotein complex formation and nucleolar–cytoplasmic relocalization of nucleolin in poliovirus-infected cells. *J Virol* **72**, 6699–6709.

Wang, S. H., Syu, W. J., Huang, K. J., Lei, H. Y., Yao, C. W., King, C. C. and Hu, S. T. (2002). Intracellular localization and determination of a nuclear localization signal of the core protein of dengue virus. *J Gen Virol* **83**, 3093–3102.

Wang, Y. and Zhang, X. (1999). The nucleocapsid protein of coronavirus mouse hepatitis virus interacts with the cellular heterogeneous nuclear ribonucleoprotein A1 *in vitro* and *in vivo*. *Virology* **265**, 96–109.

Westaway, E. G., Khromykh, A. A., Kenney, M. T., Mackenzie, J. M. and Jones, M. K. (1997). Proteins C and NS4B of the flavivirus Kunjin translocate independently into the nucleus. *Virology* **234**, 31–41.

Willcocks, M. M., Boxall, A. S. and Carter, M. J. (1999). Processing and intracellular location of human astrovirus non-structural proteins. *J Gen Virol* **80**(10), 2607–2611.

Wilusz, J., Kurilla, M. G. and Keene, J. D. (1983). A host protein (La) binds to a unique species of minus-sense leader RNA during replication of vesicular stomatitis virus. *Proc Natl Acad Sci USA* **80**, 5827–5831.

Wolk, B., Sansonno, D., Krausslich, H. G., Dammacco, F., Rice, C. M., Blum, H. E. and Moradpour, D. (2000). Subcellular localization, stability, and trans-cleavage competence of the hepatitis C virus NS3-NS4A complex expressed in tetracycline-regulated cell lines. *J Virol* **74**, 2293–2304.

Wurm, T., Chen, H., Hodgson, T., Britton, P., Brooks, G. and Hiscox, J. A. (2001). Localization to the nucleolus is a common feature of coronavirus nucleoproteins, and the protein may disrupt host cell division. *J Virol* **75**, 9345–9356.

Yasui, K., Wakita, T., Tsukiyama-Kohara, K., Funahashi, S. I., Ichikawa, M., Kajita, T., Moradpour, D., Wands, J. R. and Kohara, M. (1998). The native form and maturation process of hepatitis C virus core protein. *J Virol* **72**, 6048–6055.

7 Viruses and the Nucleolus

David A. Matthews[1] and Julian A. Hiscox[2]

[1]Division of Virology, Department of Pathology and Microbiology, School of Medical Sciences, University of Bristol, Bristol, UK
[2]School of Biochemistry and Microbiology, University of Leeds, Leeds, UK

7.1   Introduction

The eukaryotic nucleus contains a number of domains or subcompartments, which include nucleoli, Cajal bodies, nuclear speckles, gems, transcription and replication foci, and chromosome territories (Lamond and Earnshaw, 1998). For many years the exclusive function of the nucleolus was thought to be ribosomal rRNA synthesis and ribosome biogenesis. In the past few years, however, the nucleolus has been implicated in many aspects of cell biology, including functions such as gene silencing, senescence, telomerase activity and cell cycle regulation (Carmo-Fonseca *et al.*, 2000; Olson *et al.*, 2000; Pederson, 1998; Zhang *et al.*, 2004). More recently the nucleolus has been proposed to act as stress sensor for the cell (Rubbi and Milner, 2003). As a consequence of infection or a deliberate process, a number of viruses interact with the nucleolus and its components and a wide variety of examples can be found from DNA, retro- and RNA viruses. As we shall see, the HIV–nucleolus relationship is the best understood at the present time. However, with the advent of proteomics coupled with confocal microscopy and live cell imagery, the role of the nucleolus in virus infection and how viruses subvert nucleolar functions is being rapidly opened to investigation. Given the pivotal role of the nucleolus in host cell function, it is not surprising that viruses interact with this structure and its components. This chapter seeks to review the nucleolus and how viruses interact with this most fascinating of subnuclear structures.

Viruses and the Nucleus   Edited by Julian A. Hiscox

7.1.1   Structure of the nucleolus

With the advent of light microscopy the nucleolus was one of the first subcellular structures described, due to its high refractive index and protein content. Electron microscopy revealed that the nucleolus consisted of at least three different regions; the outer granular component (GC), the inner dense fibrillar component (DFC) and the innermost fibrillar centre (FC) (Gerbi and Borovjagin, 1997). Each of these regions has been implicated in different functions and found to contain different proteins. Synthesis of the primary 45S rRNA molecule from rDNA may occur in the FC, and nascent rRNA molecules are distributed in a Christmas tree or globular pattern. Processing of 45S rRNA (into 5.8S, 18S and 28S) occurs in the DFC, whilst rRNA assembly into ribosomal subunits occurs in the GC, prior to their export into the cytoplasm. An additional recognized feature is the peri-nucleolar compartment, which has been implicated in RNA metabolism (Huang *et al.*, 1998).

7.1.2   Composition of the nucleolus

Investigation of the protein content of the nucleolus and how this is dynamic in response to a variety of cellular conditions has rapidly advanced with the application of robust purification strategies and proteomic analysis. Such studies initially revealed that nucleoli purified from HeLa cells contained at least 271 proteins and that the protein content was dynamic and varied with metabolic conditions, such as transcription activity and cell cycle (Andersen *et al.*, 2002). That the nucleolar proteome would change in a virus-infected cell therefore seems inevitable. The distribution of different proteins in the nucleolus is shown in Figure 7.1 (adapted from Andersen *et al.*, 2002). Since the Anderson *et al.* paper,

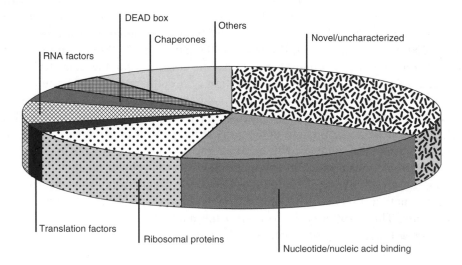

Figure 7.1   Proteomic analysis of the nucleolus has revealed that this dynamic subnuclear structure contains over 400 proteins, which can be grouped together into different functional classes. Data taken from Andersen *et al.* (2002), Leung *et al.* (2003), Leung and Lamond (2003)

several other groups have conducted proteomic analysis of the nucleolus, e.g. Scherl *et al.* (2002) identified some 213 different proteins and, together with the previously mentioned study, this added up to over 350 unique proteins. More recent data and revised estimates suggest that the nucleolus could contain over 400 proteins (Leung *et al.*, 2003; Leung and Lamond, 2003) and although new proteins have been discovered, there has been little change in the overall distribution of proteins shown in Figure 7.1. However, the problem now arises as to what constitutes a nucleolar protein, as many proteins traffic in and out of the nucleolus to other subnuclear structures and also to the cytoplasm. The fact that the nucleolus contains many nucleotide/nucleic acid binding proteins and potential helicases illustrates its fundamental role in RNA processing. Surprisingly, perhaps, the nucleolus has recently been shown to contain translation factors, and it may be possible that these play a role in nuclear translation (Iborra *et al.*, 2001).

7.1.3 Major nucleolar proteins

Three of the most abundant and well-understood proteins in the nucleolus are nucleolin, fibrillarin and B23. These three are major rRNA-associated proteins and ribosome assembly factors. They each appear to have multiple roles and their functions can be modified in a number of ways. Electron microscopy and immunofluorescence analysis showed that B23 is predominantly located in the GC region of the nucleolus, whereas nuclcolin and fibrillarin are mainly present in the DFC. Using fluorescent protein fusions, Chen and Huang (2001) have shown that these proteins exchange rapidly between the nucleoplasm and the nucleolus. Thus, their localization within the nucleolus is dynamic rather than static.

7.1.3.1 *Nucleolin (C23)*

Nucleolin (first called C23) represents approximately 10% of the total nucleolar protein content and is highly phosphorylated and methylated and also can be ADP-ribosylated (Ginisty *et al.*, 1999). While mammalian nucleolin has a predicted molecular mass of approximately 77 kDa (depending on the species), the apparent molecular mass is 100–110 kDa, and has been attributed to the amino acid composition of the N-terminal domain, which is highly phosphorylated (Ginisty *et al.*, 1999). Nucleolin has the potential to bind to multiple RNA targets and proteins, such as the tumour suppressor p53 and telomerase and this may reflect its variety of functions (Daniely *et al.*, 2002; Ginisty *et al.*, 2001; Ginisty *et al.*, 1999; Khurts *et al.*, 2004). This multifunctional nature of nucleolin and the myriad of protein interactions it can undertake, illustrates an overriding principal of nucleolar architecture, that it is an amalgamation of very large protein:protein and protein:nucleic acids interactions, and disruption to any part of these can cause disease. One of the main functions of nucleolin is facilitating the first cleavage step of rRNA in the presence of U3 snoRNP. Nucleolin may also function as a chaperone for correct folding in pre-rRNA processing, and has been implicated as a repressor of transcription

(Yang *et al.*, 1994). Nucleolin is reported to shuttle between the nucleolus and the cyto-plasm (Borer *et al.*, 1989) and is thought to enter the nucleus and nucleolus via an interaction with B23 (Li *et al.*, 1996). Considering the pleiotrophic nature of nucleolin, it is not surprising that it is regulated by a variety of methods, such as proteolysis, methylation and phosphorylation by a variety of kinases, such as PKC-α, CKII and P34cdc2.

7.1.3.2   B23 (nucleophosmin/numatrin)

B23 (also called numatrin, nucleophosmin or NO38) is widely distributed amongst different species with approximately the same molecular mass of 35–40 kDa (Maridor and Nigg, 1990; Shaw and Jordan, 1995). Like nucleolin and fibrillarin, B23 has been implicated in ribosome assembly (Dumbar *et al.*, 1989), binding to other nucleolar proteins, nucleo-cytoplasmic shuttling (Li *et al.*, 1996) and possibly regulating transcription of rDNA by mediating structural changes in chromatin (Okuwaki *et al.*, 2001).

Two isoforms of the protein are expressed. The major form, B23.1, is predominately located in the nucleolus, and the minor form, B23.2, is located in the nucleoplasm (Okuwaki *et al.*, 2002). The two molecules are almost identical, except that B23.1 has an additional 35 amino acids at the C-terminus. These C-terminal 35 amino acids are required for the nucleic acid binding activity (i.e. B23.2 cannot bind nucleic acids). Recent evidence suggests that B23.2 can hetero-oligomerize with B23.1 to inactivate the RNA-binding property of B23.1 (Okuwaki *et al.*, 2002). In addition, Okuwaki *et al.* also proposed that phosphorylation by cell cycle kinases also ablate the RNA-binding activity of B23.1. Such inactivation of RNA binding would prevent B23.1 from playing a role in rRNA biogenesis. Similar to nucleolin, B23 can also form multiple protein:protein interactions. For example, B23 can interact with ARF and recruit this protein to the nucleolus to inhibit its function (Korgaonkar *et al.*, 2005; Zhang, 2004) and regulate GADD45a nuclear translocation, which is in turn involved in the G2/M phase progression (Gao *et al.*, 2005).

7.1.3.3   Fibrillarin

Fibrillarin is most abundant within the DFC (Kill, 1996). It is a highly conserved three-domain structure, with a central RNA-binding domain flanked by a N-terminal Gly/DMA-rich domain and a C-terminal α-helical domain (Aris and Blobel, 1991). Fibrillarin was originally identified in serum from patients suffering from the autoimmune disease scleroderma pigmentosum (O'Connor and Brian, 2000). It is has been established to be crucial for nucleolar assembly at the end of telophase (Dousset *et al.*, 2000; Savino *et al.*, 2001), onset of rDNA transcription (Sirri *et al.*, 2002), processing of rRNA and splicing of snoRNA (Tollervey *et al.*, 1993), the latter taking place in the Cajal bodies. Immunoblot analysis revealed a size of 34 kDa and a basic pI of 8.5, although size varies slightly between fibrillarin from different species.

7.1.4 The nucleolus is dispersed and reformed during the cell cycle

During interphase in higher eukaryotic cells, the number of nucleoli vary, depending on the stage of the cell cycle, but at the start of mitosis the rRNA synthesis is inhibited and the nucleolus is dispersed. The nucleolus reforms at the end of mitosis as synthesis and processing of rRNA are re-started (Hernandez-Verdun *et al.*, 2002). Interestingly, correct nucleolar reformation requires both processing and re-initiation of rRNA synthesis (Sirri *et al.*, 2002).

Synthesis of rRNA is mediated by RNA pol I and initiation requires a number of factors, including upstream binding factor (UBF) and the selectivity factor SL1, a multimeric complex. During mitosis, rRNA genes are silenced by phosphorylation of the transcription factor complex, SL 1. This silencing is accompanied by disruption of the nuclelolus and dispersal of nucleolar antigens. The initiation complex plus pol I remains associated with the rDNA throughout mitosis, forming what are known as the nucleolar organising regions (NORs).

The proteins involved in rRNA processing and ribosome assembly are distributed differently, usually at the periphery of mitotic chromosomes during prophase and metaphase. However, as mitosis proceeds to telophase the processing proteins are concentrated into discrete areas termed pre-nucleolar bodies (PNBs). During the transition from telophase to the M/G_1 boundary, the re-initiation of rRNA synthesis correlates with the PNBs and the NORs recombining to form the interphase nucleolus.

7.1.5 The nucleolus and the cell cycle

The nucleolus and associated proteins are implicated in (and regulated by) the cell cycle (Carmo-Fonseca *et al.*, 2000). The concentration of nucleolin and B23 (Sirri *et al.*, 1997) and the distribution of fibrillarin are dependent on the cell cycle (Azum-Gelade *et al.*, 1994; Fomproix *et al.*, 1998). Nucleolin is stable in prolifcrative cells, but undergoes sclf-cleavage in quiescent cells (Chen *et al.*, 1991), is regulated at the level of transcription and post-translation (Bicknell *et al.*, 2005) and has been suggested to be involved in the regulation of cell growth and proliferation (Srivastava and Pollard, 1999).

That the nucleolus is affected by the cell cycle is perhaps not surprising. However, it has recently become clear that, in addition to being regulated by the cell cycle, the nucleolus also plays a role in regulating the cell cycle. This idea has been pioneered in yeast studies, which have shown that exit from mitosis is controlled by sequestering factors within the nucleolus, in particular the localization of cdc14 within the nucleolus by the .RENT complex (regulator of nucleolar silencing and telophase exit; Kasulke *et al.*, 2002; Shou *et al.*, 1999). This is analogous to sequestering transcription factors in the cytoplasm to prevent their function in the nucleus. More recently, Jin *et al.* (2003) have shown that MDM2, a protein that localizes to the nucleolus, can promote the turnover of the CDK inhibitor p21. Cdk9 is a member of the cyclin-dependent kinase family and is involved in transcription elongation; two isoforms exist, the 55 kDa isoform, which has localizes to the nucleolus, and a 42 kDa isoform, which is present in the nucleoplasm. Liu and Herrmann (2004) suggest that the distribution of these molecules may affect their access to substrates and hence their respective functions.

Under a variety of different cellular stresses p53 can guard against aberrant cell division by inducing cell cycle arrest or apoptosis, and nucleolar disruption is a common factor in all cellular stresses that induce p53 (Rubbi and Milner, 2003). The latter authors therefore suggested that the impairment of nucleolar function might stabilize p53 by preventing its degradation. Interestingly, they proposed that the nucleolus may act as a stress sensor and may regulate low levels of p53, whose abundance increases immediately in response to stress. They also point out that an important implication of their model is that when nucleolar function is disrupted by virus infection (i.e. a stress), then this will cause the stabilization of p53. They suggest that this is why a large number of viruses inactivate p53. In a somewhat similar vein, Jin *et al.* (2004) proposed that changes in the amount of the ribosomal protein L23 (which is present in the nucleolus) in response to changing growth conditions could interact with the HDM2–p53 pathway by altering nucleolar architecture.

Cajal bodies associate with nucleoli, contain many similar components such as fibrillarin (Platani *et al.*, 2000; Snaar *et al.*, 2000) and they sequester CDK2 and cyclin E in a cell cycle-dependent manner (Liu *et al.*, 2000). Moreover, Liu *et al.* (2000) showed that CDK2 and p80 relocated from the Cajal bodies to the nucleolar periphery upon treatment with CDK2 inhibitors. In this case, however, it is unclear why these proteins are localized to the Cajal bodies and or the perinucleolar regions. More recently, however, it has been reported that CDK2-cyclin E-mediated phosphorylation modulates an interaction between B23 and the centrosomes, controlling duplication of the centrosome (Tokuyama *et al.*, 2001).

7.1.5.1   The nucleolus and cancer

Given the dynamic interrelationship between the nucleolus and cell cycle, it is unsurprising that the nucleolus has been associated with cancer, either involved in the disorder directly or as a diagnostic indicator for certain forms of the disease (Zimber *et al.*, 2004). Werner syndrome (WS) is a rare disease whose symptoms become apparent in after the first 10 years of life. Affected individuals show many signs of premature ageing and the first real sign of the disease is the absence of a growth spurt during adolescence, followed by cancer, osteoporosis, accelerated atherosclerosis and diabetes mellitus. The defective gene thought to be responsible for this disease encodes a mutant protein called WRN. The wild-type protein contains the seven signature motifs characteristic of DNA and RNA helicases and it should therefore come as no surprise that this protein is predominately located in the nucleolus (Marciniak *et al.*, 1998) and may function in longevity. Studies using fluorescent-tagged fusion proteins demonstrated that human WRN contained both a nuclear localization signal and mapped a nucleolar localization signal to a 144 amino acid region which is independent of the nuclear localization signal (von Kobbe and Bohr, 2002).

The nucleolus itself is used as a diagnostic indicator for certain cancers, particularly in tumours associated with clear-cell carcinoma of the kidney, certain ovarian tumours and malignant melanoma. Careful analysis of the morphology and number of nucleoli found within a cell, and also analysing nucleolar proteins themselves, can be useful in differentiating between malignant and benign cells (Derenzini *et al.*, 2004; Ghazizadeh *et al.*, 1997; Sirri *et al.*, 2000; Trabucco *et al.*, 1994; Zergeroglu *et al.*, 2001).

7.2   Techniques used to examine the nucleolus

Nucleoli can be visualized using immunofluoresence by mounting preparations in propidium iodide to visualize nuclear DNA and regions of rRNA synthesis (Hiscox *et al.*, 2001), or using antibodies to nucleolar proteins (Matthews, 2001; Wurm *et al.*, 2001) or transfecting cells with plasmids that express nucleolar antigens tagged with fusion proteins, such as green fluorescent protein (GFP) (Costes *et al.*, 2004; Platani *et al.*, 2000; Snaar *et al.*, 2000).

As a word of caution, some groups have reported that charged proteins can translocate across cell membranes and accumulate in the nucleus after fixation, although data has been presented to the contrary (Lundberg and Lundberg, 2001; Lundberg and Johansson, 2002; Lundberg *et al.*, 2003; O'Hare and Elliot, 2001). This raises the possibility that the observed nucleolar localization of proteins could be a post-fixation artefact. To discount this possibility, we routinely fix cells under a variety of conditions, including either 4% formaldehyde followed by non-ionic detergent or 100% methanol at −20°C. Incontrovertible evidence can be obtained using live cell imaging with appropriately tagged fusion proteins, e.g. EGFP, ECFP and YFP, although the presence of the fusion tag could interfere with nuclear import if the protein is not actively imported through the nuclear pore complex.

In addition to microscopy, subcellular fractions corresponding to cytoplasmic and nuclear and/or nucleolar extracts can be obtained. These can be purified using a variety of protocols including published procedures, e.g. nucleolar preparation from cell culture (Andersen *et al.*, 2002) or virus infected cells (Emmett *et al.*, 2005) and in combination with nuclear/cytoplasmic fractionation kits (e.g. BIORAD, Novagen). Together with appropriate markers to cytoplasmic, nuclear and nucleolar proteins, the integrity of preparations can be verified by western blot. The subcellular localization of the protein of interest can then be determined by western blot.

7.3   Nucleolar localization signals (NoLS)

A number of factors can determine whether a protein localizes to the nucleolus. Soluble proteins of less than 40–60 kDa can diffuse passively into the nucleoplasm through the nuclear pore complex, and could in principle diffuse in and out of the nucleolar compartment (Peters, 1986; Richardson *et al.*, 1988). Non-specific RNA binding proteins that diffuse into the nucleus may therefore be expected to become concentrated in the nucleolus because a large amount of rRNA is present. The transport

RPRRRATTRRRTTTGTRRRRRRR   RRRRRRR   NO MOTIF

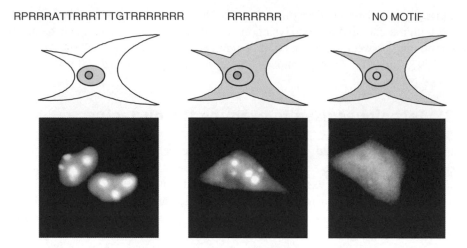

Figure 7.2   EGFP fusion proteins can detect localization signals. Three sequences (derived from adenovirus protein V) are fused to EGFP and the localization of the fusion is observed as either 'nuclear and nucleolar', 'nucleolar accumulation' only or diffuse throughout the cell. In this case the nuclear and nucleolar localization is clearly separated

of larger proteins through the nuclear pore is an active process requiring ATP/GTP and nuclear localization signals (NLSs), which also make up (in part) nucleolar localization signals (NoLS). GTP may be part of a process controlling nucleolar localization (Tsai and McKay, 2005). In many cases NoLSs are also called nucleolar retention signals (NoRSs) and the terms have been used interchangeable, although they may have subtle differences. NLSs include the 'pat4' motif, which consists of a continuous stretch of four basic amino acids (arginine and lysine) and the 'pat7' motif, which starts with a proline and is followed within three residues by a segment containing three basic residues out of four (Garcia-Bustos *et al.*, 1991), or a bipartite signal (Robbins *et al.*, 1991). Localization of a protein to the nucleolus is probably a result of targeting to the nucleus via NLSs followed by an interaction between the newly imported protein (via the NoLS) and components that make up the nucleolus (Carmo-Fonseca *et al.*, 2000; Shaw and Jordan, 1995) (Figure 7.2). An example of a protein that localizes to the nucleolus in this manner is nucleolin, which contains a bipartite NLS and associates with rRNA in the nucleolus via RNA-binding domains (Schmidt-Zachmann and Nigg, 1993).

7.4   Viral interactions with the nucleolus

One of the first documented interactions of viruses with the nucleolus is work by Nigel Dimmock, where he showed that influenza virus A proteins were found in the nucleolus and that the architecture of this structure was altered as virus infection progressed (Compans and Dimmock, 1969; Dimmock, 1969). Since this time a number of viral proteins have been shown to localize to the nucleolus, with examples from animal retroviruses and DNA viruses, and animal and plant RNA viruses (Table 7.1) (Figure 7.3). Why viral proteins localize to the nucleolus has not been

Table 7.1 Examples of viral proteins that localize to the nucleolus and sub-nuclear compartments (modified and updated from Hiscox, 2002)

Virus	Protein	Nucleolus	Interaction with nucleolar antigens	Reference(s)
RNA Viruses				
BDV	Replication complex	+		(Pyper *et al.*, 1998)
CMV	3A	+		(Mackenzie and Tremaine, 1988)
	Capsid	+		(Lin *et al.*, 1996)
Coronavirus	Nucleoprotein	+	+	(Hiscox *et al.*, 2001; Ning *et al.*, 2003; Wurm *et al.*, 2001)
HDV	Delta antigen Viral RNA	+	+	(Lee *et al.*, 1998)
HCV	Core, NS5B	+	+	(Falcon *et al.*, 2003; Hirano *et al.*, 2003)
Influenza A virus	NP	+		(Davey *et al.*, 1985; Dimmock, 1969)
NDV	Matrix protein	+		(Peeples *et al.*, 1992)
Poliovirus	(3′ NCR)		+	(Waggoner and Sarnow, 1998)
Arteriviruses	Nucleoprotein	+	+	(Rowland *et al.*, 1999, 2003; Tijms *et al.*, 2002, Yoo *et al.*, 2003)
SFV	Capsid protein	+		(Favre *et al.*, 1994)
Retrovirus				
HIV-1	Rev	+	+	(Dundr *et al.*, 1995; Fankhauser *et al.*, 1991)
	Tat	+		(Siomi *et al.*, 1990)
HTLV-1	Rex	+	+	(Adachi *et al.*, 1993; Siomi *et al.*, 1988)
DNA viruses				
Adenovirus	IVa2	+		(Lutz *et al.*, 1996)
	V	+	+	(Matthews, 2001; Matthews and Russell, 1998)

Table 7.1   (*continued*)

Virus	Protein	Nucleolus	Interaction with nucleolar antigens	Reference(s)
	Mu	+		(Lee *et al.*, 2004b)
	VII	+		(Lee *et al.*, 2003b)
EBV	EBNA5	+		(Szekely *et al.*, 1995)
HSV-1	Us11	+		(MacLean *et al.*, 1987)
	ORF57	+		(Goodwin *et al.*, 1999; Goodwin and Whitehouse, 2001)
	ICP27	+		(Mears *et al.*, 1995)
MDV	MEQ	+		(Liu *et al.*, 1997)

BDV, Borna disease virus; BVDV, bovine viral diarrhoea virus; CMV, cucumber mosaic virus; EBV, Epstein–Barr virus; HDV, hepatitis delta virus; HIV, human immunodeficiency virus; HSV, herpes simplex virus; HTLV, human T-cell leukaemia virus; MDV, Marek's disease virus; NDV, Newcastle disease virus; PRRSV, porcine reproductive and respiratory syndrome virus; SFV, Semliki Forest virus.

precisely determined although, given the multifunctional role of the nucleolus, several activities could be targeted, including cellular transcription (Liu *et al.*, 1997; Puvion-Dutilleul and Christensen, 1993), virus transcription (Matthews, 2001), virus translation (Aminev *et al.*, 2003a 2003b; Hiscox *et al.*, 2001) or cell division (Wurm *et al.*, 2001).

7.4.1   Viral NoLS

Several viral NoLS have been identified (Table 7.2), either by sequence comparison to known NoLSs or experiments where candidate NoLSs have been used to target fusion proteins to the nucleolus (e.g. arteriviruses; Rowland *et al.*, 1999). Viral NoLSs can also possess either pat 4 or pat7 NLS motifs, e.g. the coronaviruses (Hiscox *et al.*, 2001; Wurm *et al.*, 2001), or stretches of basic residues, for example adenovirus protein V (Matthews, 2001) or MDV MEQ protein (Liu *et al.*, 1997). Interestingly, Matthews (2001) identified a possible bipartite NoLS in adenovirus protein V (Table 7.2, amino acids 157–184). Some viral NoLS have been shown to control interaction with nucleolin, e.g. HDV large antigen, which contain two nucleolin-binding domains (Table 7.2) and determine targeting of the protein to the nucleolus (Lee *et al.*, 1998).

Certainly studies with viral proteins that localize to the nucleolus are providing novel insights into the delineation between nuclear localization and nucleolar retention. For example, microinjection studies with herpes simplex virus type 1

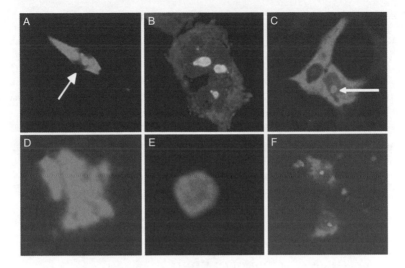

Figure 7.3 Examples of different virus effects on the nucleolus. (A) HVS ORF57 causes the redistribution of importin-α to the nucleolus (arrowed). (B) Redistribution of nucleolar protein B23 to the cytoplasm in cells expressing adenovirus protein V. (C) Coronavirus nucleoprotein localizes to the nucleolus (arrowed). Fibrillarin is redistributed from a globular structure (D) to the perinucleolar region in coronavirus-infected cells (E, F)

US11 protein, a protein that undergoes nuclear and nucleolar localization, revealed that nucleolar localization was regulated by the phosphorylation state of the protein and a previously uncharacterized proline-rich motif (Catez *et al.*, 2002). The nuclear and nucleolar localization signals in adenovirus core protein VII (preVII) have also yielded useful information (Lee *et al.*, 2003). The main NoLS in preVII was shown to be an arginine-rich region in the centre of the protein. Site-directed mutation in a GFP-tagged version of this NoLS showed that the charge: mass ratio of a protein, rather than any sequence specific interaction, was responsible for nucleolar targeting. A recent bioinformatics analysis of cellular nucleolar proteins indicates that they have a higher frequency of basic amino acids (6.3% Arg, 8.2% Lys) compared to nuclear (5.8% Arg, 6.8% Lys) and cytoplasmic (5.8% Arg, 6.0% Lys) proteins (Leung *et al.*, 2003).

Certainly several NoLSs have been identified in cellular proteins. Signals can vary, for example the short arginine rich motif RKKRKKK in the NF-kappaB inducing kinase NIK (Birbach *et al.*, 2004) or the short IMRRRGL motif of human angiogenin protein (Lixin *et al.*, 2001). Conversely, NoRSs can be composed of some twenty or so amino acids such as the MQRKPTIRRKNLRLRRK motif identified in survivin-deltaEx3 protein and the RSRKYTSWYVALKR motif of the 18-kDa fibroblast growth factor-2 (Sheng *et al.*, 2004). In many cases proteins localizing to the both the cytoplasm/ nucleus/nucleolus contain a variety of these signals to determine sub-cellular localization.

Table 7.2 Amino acid sequence analysis of examples of virus proteins that contain nucleolar localization signals (NoLS) or signals that mediate an interaction with nucleolin

Protein	(Amino acid position)	NoLS	Reference
Adenovirus protein V	23–42 315–337 159–182	KKEEQDYKPRKLKRV<u>KKKKK</u> RPRRRATTRRRTTTGT<u>RRRRRRR</u> *<u>KR</u>GLKRESGDLAPTVQLMVP<u>KRQR</u>L*	(Matthews, 2001)
Adenovirus protein preVII	93–112 127–141	RRYAKMKRRRRRVARRHRRR RARRTGRRAAMRARR	(Lee *et al.*, 2003)
Adenovirus protein preMu	32–50	MRRAHHRRRRASHRRMRGG	(Lee *et al.*, 2004)
BVDV capsid protein	71–91	HNKNK<u>PPESRKKL</u>EKALLAW	(Ross *et al.*, 2001)
HDV antigen	35–50 51–65	**RKLKKKIKKL**EEDNPWC LGNIKGIIG**KKDK**DGC	(Lee *et al.*, 1998)
IBV nucleoprotein	350–369	GNSPAPRQQR<u>PKKEKKLKK</u>Q	(Hiscox *et al.*, 2001)
MDV MEQ protein	62–78	<u>RRRKR</u>NRDAA<u>RRRRR</u>KQ	(Liu *et al.*, 1997)
SFV capsid protein	73–90 92–105	KP<u>KKKK</u>TTKPKPKTQPKK KKKDKQAD<u>KKKKK</u>P	(Favre *et al.*, 1994)
TEGV nucleoprotein	331–350	RPSEVAKEQ<u>RKRK</u>SRSKSAE	(Wurm *et al.*, 2001)

Abbreviations as in Table 7.1, together with: IBV, infectious bronchitis virus; TGEV, transmissible gastroenteritis virus. Potential NuLS are underlined. The nucleolin-binding sites in HDV antigen are shown in bold.

7.4.2 Viruses can redistribute and interact with nucleolar proteins

Virus infection can also result in the redistribution, or viral proteins become associated with, nucleolar proteins (Table 7.1). We will review these phenomena by first examining viruses that replicate primarily in the nucleus. Virus replication within the nucleus might be expected to result in widespread disruption of nuclear structure. Thus, demonstrating a direct link between phenomenon (i.e. nucleolar disruption) and utility (i.e. enhancing virus proliferation) is problematic. On the other hand, for viruses that replicate in the cytoplasm, disruption of the nulcleolus during replication might reflect a more deliberate attempt to sequester cellular proteins for the purpose of viral replication.

Although not directly linked to the nucleolus on first inspection, several viruses interact with nucleolin (or nucleolin-like molecules) found on the cell surface. The function of cell surface nucleolin is uncertain, although it has been implicated in a variety of processes, including endocytosis of lactoferrin (Legrand *et al.*, 2004). Coxsackie B viruses may use such a molecule as a receptor (or co-receptor) (Raab de Verdugo *et al.*, 1995) and cell surface nucleolin has been shown to be involved in the aggregation of HIV particles on the cell surface (Nisole *et al.*, 2002a), perhaps facilitating entry of the virus into the host cell. Certainly two independent studies have shown that the C-terminal portion of nucleolin may be involved in the anchorage of retrovirus particles to the cell surface (Nisole *et al.*, 2002b), most likely via the nucleocapsid portion of Gag, at least as found in murine leukaemia virus (MLV) (Bacharach *et al.*, 2000).

7.4.3   Viruses that replicate in the nucleus interact with and disrupt the nucleolus

Adenovirus infection results in disruption of the nucleolus (Puvion-Dutilleul and Christensen, 1993). Synthesis of rRNA is disrupted at late times in adenovirus-infected cells, as is transport and processing of rRNA (Castiglia and Flint, 1983). How this is achieved is unknown, but adenovirus disrupts fibrillarin distribution and treating uninfected cells with antibody to fibrillarin prevents its translocation to nucleoli and results in reduction or inhibition of pol I transcription (Fomproix *et al.*, 1998). In addition, adenovirus infection redistributes upstream binding factor (UBF) from the nucleolus and localizes UBF adjacent to DBP-rich centres of viral DNA replication (Matthews, unpublished observations).

Adenovirus infection also results in the redistribution of nucleolin and B23, a phenomenon that has been attributed, in part, to a viral DNA-binding protein called protein V (Matthews, 2001). Protein V is a histone-like protein that is thought to play a role in condensing the virus DNA inside the capsid and contains at least three independent nucleolar targeting regions (Matthews, 2001). There are two other viral proteins involved in condensing the viral DNA, called proteins Mu and VII. Interestingly, these proteins also contain nucleolar localization signals (Lee *et al.*, 2003, 2004). Moreover, adenovirus protein IVa2 contains a nucleolar localization signal (Lutz *et al.*, 1996). This protein is thought to play a role in transcription control and viral DNA packaging. Finally the adenovirus genome is covalently linked to a virus-coded protein known as 'pre-terminal protein'. This protein is intimately involved in origin recognition and initiation of DNA replication and appears to harbour a nucleolar targeting sequence. Thus, adenovirus makes five different proteins that contain nucleolar localization signals. The roles these proteins play in nucleolar dysfunction are yet to be elucidated.

More recently, Okuwaki *et al.* (2001) have shown that B23 stimulates adenovirus replication. In addition, B23 has been reported to interact and co-localize with the viral ssDNA binding protein (DBP) in infected cells (Okuwaki *et al.*, 2001; Matthews, unpublished observations). This viral protein is intimately associated with the viral replication machinery (Hay *et al.*, 1995), modulates transcription and can determine host range of the virus. Although there is clearly a relationship between adenovirus and nucleolar antigens the precise functional nature of this relationship remains to be understood.

Hepatitis delta virus (HDV) is a negative-strand RNA virus and a subviral satellite of hepatitis B virus (HBV), which replicates in the nucleus of infected cells. Only two proteins are thought to be coded by HDV, known as small and large hepatitis delta virus antigens (HDAg). The two proteins are identical, except that large HDAg contains an additional 19 amino acids at the C-terminus. These two proteins regulate HDV RNA replication. Nucleolar targeting of HDAg has been shown, and this protein has been implicated in binding both nucleolin and B23 (Lee *et al.*, 1998; Huang *et al.*, 2001). It is proposed that HDAg interacts with both these proteins in a complex that promotes viral RNA replication, presumably as a result of nucleolin's RNA helicase activity. In addition, the upregulation of B23 expression by HDAg expression strengthens the proposal that B23 also plays a role in replication of the virus's RNA genome.

In addition there have been a number of observations that imply a relationship between the nucleolus and viruses that replicate in the nucleus. Adeno-associated virus (AAV) major capsid protein binds to and co-localizes with nucleolin in infected cells. Marek's disease virus (MDV) is a highly oncogenic herpesvirus that codes for an oncogenic protein called MEQ. This protein is homologous to the bZIP family of transcription factors and is localized to the nucleolus of infected cells (Liu *et al.*, 1997).

In latent Epstein–Barr virus (EBV) infection, the viral genome is maintained in latently infected cells, in part through a protein called Epstein–Barr nuclear antigen 1 (EBNA 1). This protein helps to ensure that the latent viral episome is efficiently segregated upon cell division. This is achieved though an interaction with EBNA 1 binding protein (EBP2), a nucleolar protein. The yeast homologue of EBP2 is also a nucleolar protein required for rRNA processing (Huber *et al.*, 2000). EBV also sythesizes a small RNA molecule called EBER-1 that associates with ribosomal protein L22 and relocalizes it from the nucleolus to the nucleus. The functional significance of this is unknown (Toczyski *et al.*, 1994).

Herpes simplex virus type 1 (HSV-1) has been shown to trigger retention of rRNA in infected cells (Besse and Puvion-Dutilleul, 1996) and specific interactions between HSV-1 infection and nucleolar antigens are known. For example, L22 is a cellular ribosomal protein localized in the nucleolus early during infection, but late in infection it translocates form the nucleolus to the nucleus, where it interacts with ICP4, a HSV-1 protein involved in transcriptional control (Leopardi and Roizman, 1996a; Leopardi and Roizman, 1996b). HSV-1 makes an RNA-binding protein, Us11, that interacts with ribosomes and, as a consequence, is targeted to the nucleolus (Roller *et al.*, 1996). The Us11 protein is incorporated into the virus particles and it is thought that shortly after infection this protein is released from the virus particle and associates with ribosomes. HSV-1 also produces RNA species that can associate with ribosomes in the cytoplasm, where, it is speculated, they might affect ribosome function. Finally, HSV-1 modulates splicing and mRNA transport via the virus coded ICP27 protein; this protein has also been shown to contain nucleolar-targeting motifs (Mears *et al.*, 1995). In HSV-1-infected cells the viral $\gamma_1 34.5$ protein is responsible for preventing protein synthesis shutdown, as mediated by double-stranded RNA-dependent protein kinase PKR. Recently this protein has been shown to shuttle between the cytoplasm, nucleus and nucleolus (Cheng *et al.*, 2002).

There are relatively few examples of RNA viruses that replicate in the nucleus, the most notable being influenza virus. There is, however, a suggestion that this virus, too, may have an interaction with the nucleolus via a protein called NS1. This protein is an RNA-binding protein that antagonizes interferon α/β. NS1 associated with the nucleolus when purified and injected into *Xenopus* oocytes (Davey *et al.*, 1985). Ward *et al.* (1994) showed that NS1 expressed in yeast associated with nucleoli. Unfortunately, whilst there are many examples of interactions between nuclear replicating viruses and the nucleolus, the true functional significance of these interactions remains elusive. With that said, interesting results have been reported with the replication of RNAs from the potato spindle tuber viroid (PSTVd). In this case, positive-sense RNAs are sequestered into the nucleolus, while negative-sense RNAs accumulate in the nucleoplasm. This phenomenon potentially plays a vital role in the replication of the virus genome and provides a mechanism for gene expression (Qi and Ding, 2003). The presence of the positive-strand RNA also causes a redistribution of small nucleolar RNA. With this particular viroid, the positive-sense RNA would appear to be spread throughout the nucleolus, rather than confined to any particular region (Harders *et al.*, 1989). Certain proteins of maize fine streak virus, a plant rhabdovirus which replicates in the nucleus, also localize to the nucleolus, although the function of this is unknown (Tsai *et al.*, 2005).

7.4.4   HIV rev uses the nucleolus to transport RNA

Tat, Rev and Rex proteins localize to the nucleolus and contain NoLS (Cochrane *et al.*, 1990; Kubota *et al.*, 1989). Rev protein co-localizes with B23 in the granular and dense fibrillar regions of the nucleolus (Dundr *et al.*, 1995), and rRNA synthesis may be critical for the nucleolar localization of both proteins (Dundr *et al.*, 1995). The HIV Rev protein is responsible for the export of partially or un-spliced HIV mRNA. This is mediated in part by the exportin-1 (CRM) nuclear export pathway. However, Michienzi *et al.* (2000) used anti-HIV ribozymes fused to small nucleolar RNA to markedly reduce replication of HIV. These fusion ribozymes were apparently exclusively localized to the nucleolus and clearly blocked HIV replication. The implication is that Rev is trafficking HIV RNA through the nucleolus, where the nucleolar-localized ribozymes were able to cleave the target RNA. Rev protein also undergoes multimerization in the nucleolus and this step is crucial to nuclear–cytoplasmic transport (Daelemans *et al.*, 2004). A potential parallel is seen in *Herpesvirus Saimiri*, which codes for a nucleo–cytoplasmic shuttle protein, ORF-57, that is responsible for the export of some of the virus's transcripts. This protein has been shown to localize to the nucleolus and to trigger the aberrant accumulation of importin-α in the nucleolus when overexpressed in transient transfections (Goodwin *et al.*, 1999; Goodwin and Whitehouse, 2001).

In addition, HIV Tat protein has been shown to be nucleolar. This protein binds to a motif in HIV RNA known as TAR. The Tat–TAR interaction raises the processivity of RNA pol II and thus the transcription efficiency of HIV RNA is increased. Michienzi *et al.* (2002) used TAR decoys fused to small nucleolar RNAs to target the TAR to the nucleolus or fused the TAR decoy to a nucleoplasmic RNA. In this instance, the TAR decoys worked better when they were localized to the nucleolus. This data suggests that the decoy works best if it is targeted to the Tat present in the nucleolus. The implication from both these studies is that the nucleolus is an important functional site for HIV replication.

7.4.5   Cytoplasmic viruses also target proteins to the nucleolus

Many viruses, especially RNA viruses, appear to replicate mainly or almost exclusively in the cytoplasm of the host cell. This raises the question of why these viruses might want to interact with a nuclear structure. However, many of these cytoplasmic viruses do appear to interact on some level with the nucleolus (Hiscox, 2002, 2003). Nuclear localization of positive-strand RNA virus proteins was first recognized in the alphavirus, Semliki Forest virus (SFV). In this case both the SFV capsid (C) protein (Jakob, 1994, 1995; Michel *et al.*, 1990) and nsP2 (Peranen *et al.*, 1990) were observed to localize to the nucleus and/or nucleolus and found to contain NLSs that resembled cellular motifs. Indeed, C protein was shown to contain two nucleolar targeting signals in the N-terminal region (Favre *et al.*, 1994). The functional relevance of why these proteins would localize to the nucleus or nucleolus, and how this relates to their function in virus replication, are both unknown. nsP2 is involved in the regulation of negative-strand RNA synthesis (Sawicki and Sawicki, 1993; Suopanki *et al.*, 1998) and C protein is involved in nucleocapsid assembly and viral RNA binding (Owen and Kuhn, 1996; Weiss *et al.*, 1989). However, C protein also associates with ribosomes to promote disassembly and assembly of the virus particle (Ulmanen *et al.*, 1976; Wengler and Wengler, 1984). A conserved ribosome binding site (RBS) was identified in the C protein of alphaviruses (Wengler *et al.*, 1992). The C protein has a M_r of 33 000 and may be expected to diffuse through the NPC before associating with the nucleolus. However, C protein accumulation in the nucleus is energy-dependent, thus suggesting that transport across the NPC is active (Michel *et al.*, 1990). The C protein may localize preferentially to nucleoli via an interaction between the RBS and newly synthesized rRNA or ribosomal subunits. Although a recombinant SFV whose nsP2 contained altered NLS was reported to have identical properties to wild-type virus (Rikkonen *et al.*, 1994), it was reported recently that this change affects the neurovirulence of SFV, implying that this could be due to changes in processes involved in RNA replication and/or the nuclear transport of nsP2 (Fazakerley *et al.*, 2002). Nuclear/nucleolar localization of Japanese encephalitis virus core protein has been shown to be important for virus replication (Mori *et al.*, 2005).

Nucleolin is prevented from entering the nucleus in poliovirus-infected cells and nucleolin has been shown to interact with the poliovirus 3′ non-coding region (NCR) (Waggoner and Sarnow, 1998). This interaction was shown to stimulate viral genome replication in *in vitro* assays (Waggoner and Sarnow, 1998). In addition, nucleolin was shown to stimulate IRES-mediated translation of the poliovirus genome (Izumi *et al.*, 2001), which may explain why picornavirus directs the accumulation of nucleolin in the cytoplasm. Intriguing and exciting is the recent work, demonstrating the targeting of EMCV proteins to the nucleolus and the accumulation of these proteins in pre-ribosomal subunits, by Aminev *et al.* (2003a, 2003b), who speculate that this might be part of a virus strategy to recruit ribosomes directly onto the virus IRES to promote translation of the viral polyprotein.

Biochemical analysis using pull-down assays demonstrated that several positive-strand RNA virus proteins interact with nucleolar proteins, including nucleolin and B23, e.g. expression and purification of recombinant infectious bronchitis virus (IBV) nucleoprotein, a protein which binds viral RNA with high affinity (Chen *et al.*, 2005), could be used to pull down nucleolin from preparations of purified nuclei (Chen *et al.*, 2002). Nucleolin forms a direct interaction with hepatitis C virus NS5B protein and the C-terminal region of nucleolin was shown to inhibit the RNA-dependent RNA polymerase activity of this protein in a dose-dependent manner (Hirano *et al.*, 2003).

Plant viruses also interact with the nucleolus (Ryabov *et al.*, 1998; Taliansky and Robinson, 2003) and the structure may be involved in systemic virus infection, e.g. nucleolar localization of umbravirus ORF3 protein has been correlated with its ability to form cytoplasmic ribonucleoprotein particles and transport viral RNA long distances via the phloem (Kim *et al.*, 2004). This protein has a defined NoLS which corresponds to an arginine-rich region, and is therefore similar to mammalian viral NoLSs (Ryabov *et al.*, 2004).

7.5   The nucleolus, viruses and autoimmunity

A number of nucleolar proteins have been associated with autoimmunity. Fibrillarin has been associated with the disease scleroderma (Falkner *et al.*, 2000; Takeuchi *et al.*, 1995) and autoimmunity to nucleolin with systematic lupus erythematosus (Jarjour *et al.*, 1992). Viral antigens that localize to the nucleolus of infected cells and displace (or mimic) nucleolar antigens might therefore stimulate the host immune response into producing antibodies to viral proteins that could have the potential to bind with the equivalent host proteins through structural mimicry. For example, poliovirus infection, which prevents nucleolin from entering the nucleus (Waggoner and Sarnow, 1998), may cause an increase in nucleolin in the cytoplasm and concomitant increase in cell surface expression. Thus, understanding the basis of viral protein interactions with the nucleolar milieu, and its consequences, may extend our understanding of how such autoimmune conditions develop.

7.6 Summary

Clearly, interaction of viruses with the nucleolus and its proteins is a pan-virus phenomenon, with examples being found from different DNA, RNA and retroviruses. Whilst in most cases the functional relevance of these interactions is still being eluci-dated, several patterns are emerging and, unsurprisingly, are related directly to either facilitating virus replication or to nucleolar function. Four areas rapidly lend themselves to comment; subversion of transcription, processing of viral RNA, stimulation of translation of virus mRNAs and cell cycle control. Diverse viruses, such as adenoviruses and coronaviruses, which reorganize fibrillarin to the perinucleolar region may do so to disrupt cellular transcription. Those viruses whose proteins are translated from mRNAs containing an IRES may recruit nucleolar factors, such as nucleolin, to facilitate this and, as we have seen, perhaps directly recruit ribosomal subunits from the nucleolus and load these onto viral mRNAs. Perhaps the best characterized functional interaction of viruses with the nucleolus is with HIV, where transit and processing of HIV RNA occurs in the nucleolus, together with the necessary steps to allow export of this RNA through the nucleoplasm and nuclear pore complex and ultimately into the cytoplasm. Lastly, correlations can be drawn between viral proteins that localize to the nucleolus and their involvement in cell cycle control. Whether there is a functional relatedness between the two is currently unknown. Certainly in many cases efficient virus replication depends on the host cell being in the correct stage of the cell cycle. Therefore, a number of viral proteins could exert cell cycle perturbations by targeting cell cycle factors in the nucleolus and/or by causing disruptions to nucleolar architecture. Over the next few years the myriad of cell biological and biophysical techniques that are now available will play an absolutely crucial role in elucidating the relationship between viruses and the nucleolus.

References

Adachi, Y., Copeland, T. D., Hatanaka, M. and Oroszlan, S. (1993). Nucleolar targeting signal of Rex protein of human T-cell leukemia virus type I specifically binds to nucleolar shuttle protein B-23. *J. Biol. Chem.* **268**, 13930–13934.

Aminev, A. G., Amineva, S. P. and Palmenberg, A. C. (2003a). Encephalomyocarditis viral protein 2A localizes to nucleoli and inhibits cap-dependent mRNA translation. *Virus Res.* **95**, 45–57.

Aminev, A. G., Amineva, S. P. and Palmenberg, A. C. (2003b). Encephalomyocarditis virus (EMCV) proteins 2A and 3BCD localize to nuclei and inhibit cellular mRNA transcription but not rRNA transcription. *Virus Res.* **95**, 59–73.

Andersen, J. S., Lyon, C. E., Fox, A. H., Leung, A. K. L., Lam, Y. W., Steen, H., Mann, M. and Lamond, A. I. (2002). Directed proteomic analysis of the human nucleolus. *Current Biol.* **12**, 1–11.

Aris, J. P. and Blobel, G. (1991). cDNA cloning and sequencing of human fibrillarin, a conserved nucleolar protein recognized by autoimmune antisera. *Proc. Natl. Acad. Sci. (USA)* **88**, 931–935.

Azum-Gelade, M.-C., Noaillac-Depeyre, J., Caizergues-Ferrer, M. and Gas, N. (1994). Cell cycle redistribution of U3 snRNA and fibrillarin. *J. Cell Sci.* **107**, 463–475.

Bacharach, E., Gonsky, J., Alin, K., Orlova, M. and Goff, S. P. (2000). The carboxy-terminal fragment of nucleolin interacts with the nucleocapsid domain of retroviral Gag proteins and inhibits virion assembly. *J. Virol.* **74**, 11027–11039.

Besse, S. and Puvion-Dutilleul, F. (1996). Intranuclear retention of ribosomal RNAs in response to herpes simplex virus type 1 infection. *J. Cell Sci.* **109**, 119–129.

Bicknell, K. A., Brooks, G., Kaiser, P., Chen, H., Dove, B. K. and Hiscox, J. A. (2005). Nucleolin is regulated both at the level of transcription and translation. *Biochem. Biophys. Res. Comm.* **332**, 817–822.

Birbach, A., Bailey, S. T., Ghosh, S. and Schmid, J. A. (2004). Cytosolic, nuclear and nucleolar localization signals determine subcellular distribution and activity of the NF-kappaB inducing kinase NIK. *J. Cell Sci.* **117**, 3615–3624.

Borer, R. A., Lehner, C. F., Eppenberger, H. M. and Nigg, E. A. (1989). Major nucleolar proteins shuttle between nucleus and cytoplasm. *Cell* **56**, 379–390.

Carmo-Fonseca, M., Mendes-Soares, L. and Campos, I. (2000). To be or not to be in the nucleolus. *Nature Cell Biology* **2**, E107-E112.

Castiglia, C. L. and Flint, S. J. (1983). Effects of adenovirus infeciton on rRNA synthesis and maturation in HeLa cells. *Mol. Cell. Biol.* **3**, 662–671.

Catez, F., Erard, M., Schaerer-Uthurralt, N., Kindbeiter, K., Madjar, J. J. and Diaz, J. J. (2002). Unique motif for nucleolar retention and nuclear export regulated by phosphorylation. *Mol. Cell. Biol.* **22**, 1126–1139.

Chen, C.-M., Chiang, S.-Y. and Yeh, N.-H. (1991). Increased stability of nucleolin in proliferating cells by inhibition of its self-cleaving activity. *J. Biol. Chem.* **266**, 7754–7758.

Chen, D. and Huang, S. (2001). Nucleolar components involved in ribosome biogenesis cycle between the nucleolus and nucleoplasm in interphase cells. *J. Cell Biol.* **153**, 169–176.

Chen, H., Gill, A., Dove, B. K., Emmett, S. R., Kemp, F. C., Ritchie, M. A., Dee, M. and Hiscox, J. A. (2005). Mass spectroscopic characterisation of the coronavirus infectious bronchitis virus nucleoprotein and elucidation of the role of phosphorylation in RNA binding using surface plasmon resonance. *J. Virol.* **79**, 1164–1179.

Chen, H., Wurm, T., Britton, P., Brooks, G. and Hiscox, J. A. (2002). Interaction of the coronavirus nucleoprotein with nucleolar antigens and the host cell. *J. Virol.* **76**, 5233–5250.

Cheng, G., Brett, M. E. and He, B. (2002). Signals that dictate nuclear, nucleolar, and cytoplasmic shuttling of the gamma(1)34.5 protein of herpes simplex virus type 1. *J. Virol.* **76**, 9434–9445.

Cochrane, A. W., Perkins, A. and Rosen, C. A. (1990). Identification of sequences important in the nucleolar localization of human immunodeficiency virus Rev: relevance of nucleolar localization to function. *J. Virol.* **64**, 881–885.

Compans, R. W. and Dimmock, N. J. (1969). An electron microscopic study of single-cycle infection of chick embryo fibroblasts by influenza virus. *Virology* **39**, 499–515.

Costes, S. V., Daelemans, D., Cho, E. H., Dobbin, Z., Pavlakis, G. and Lockett, S. (2004). Automatic and quantitative measurement of protein-protein colocalization in live cells. *Biophys. J.* **86**, 3993–4003.

Daelemans, D., Costes, S. V., Cho, E. H., Erwin-Cohen, R. A., Lockett, S. and Pavlakis, G. N. (2004). In vivo HIV-1 Rev multimerization in the nucleolus and cytoplasm by fluorescence resonance energy transfer. *J. Biol. Chem.*, M407713200.

Daniely, Y., Dimitrova, D. D. and Borowiec, J. A. (2002). Stress-dependent nucleolin mobilization mediated by p53-nucleolin complex formation. *Mol Cell Biol* **22**, 6014–6022.

Davey, J., Colman, A. and Dimmock, N. J. (1985). Location of influenza virus M, NP, and NS1 proteins in microinjected cells. *J. Gen. Virol.* **66**, 2319–2334.

Derenzini, M., Ceccarelli, C., Santini, D., Taffurelli, M. and Trere, D. (2004). The prognostic value of the AgNOR parameter in human breast cancer depends on the pRb and p53 status. *J. Clin. Pathol.* **57**, 755–761.

Dimmock, N. J. (1969). New virus-specific antigens in cells infected with influenza virus. *Virology* **39**, 224–234.

Dousset, T., Wang, C., Verheggen, C., Chen, D., Hernandez-Verdun, D. and Huang, S. (2000). Initiation of nucleolar assembly is independent of RNA polymerase I transcription. *Mol. Biol. Cell.* **11**, 2705–2717.

Dumbar, T. S., Gentry, G. A. and Olson, M. O. (1989). Interaction of nucleolar phosphoprotein B23 with nucleic acids. *Biochemistry* **28**, 9495–9501.

Dundr, M., Lena, G. H., Hammarskjold, M. L., Rekosh, D., Helga-Maria, C. and Olson, M. O. (1995). The roles of nucleolar structure and function in the subcellular localisation of the HIV-1 rev protein. *J. Cell Sci.* **108**, 2811–2823.

Emmett, S. R., Dove, B., Mahoney, L., Wurm, T. and Hiscox, J. A. (2005). The cell cycle and virus infection. *Methods Mol. Biol.* **296**, 197–218.

Falcon, V., Acosta-Rivero, N., Chinea, G., de la Rosa, M. C., Menendez, I., Duenas-Carrera, S., Gra, B., Rodriguez, A., Tsutsumi, V., Shibayama, M., Luna-Munoz, J., Miranda-Sanchez, M. M., Morales-Grillo, J. and Kouri, J. (2003). Nuclear localization of nucleocapsid-like particles and HCV core protein in hepatocytes of a chronically HCV-infected patient. *Biochem. Biophys. Res. Commun.* **310**, 54–58.

Falkner, D., Wilson, J., Fertig, N., Clawson, K., Medsger, T. A. and Morel, P. A. (2000). Studies of HLA-DR and DQ alleles in systemic sclerosis patients with autoantibodies to RNA polymerases and U3-RNP (fibrillarin). *J. Rheum.* **27**, 1196–1202.

Fankhauser, C., Izaurralde, E., Adachi, Y., Wingfield, P. and Laemmli, U. K. (1991). Specific complex of human immunodeficiency virus type 1 Rev and nucleolar B23 proteins: Dissociation by the Rev response element. *Mol. Cell. Biol.* **11**, 2567–2575.

Favre, D., Studer, E. and Michel, M. R. (1994). Two nucleolar targeting signals present in the N-temrinal part of Semliki Forest virus capsid protein. *Arch. Virol.* **137**, 149–155.

Fazakerley, J. K., Boyd, A., Mikkola, M. L. and Kaariainen, L. (2002). A single amino acid change in the nuclear localization sequence of the nsP2 protein affects the neurovirulence of semliki forest virus. *J. Virol.* **76**, 392–396.

Fomproix, N., Gebrane-Younes, J. and Hernandez-Verdun, D. (1998). Effects of anti-fibrillarin antibodies on building of functional nucleoli at the end of mitosis. *J. Cell Sci.* **111**, 359–372.

Gao, H., Jin, S., Song, Y., Fu, M., Wang, M., Liu, Z., Wu, M. and Zhan, Q. (2005). B23 regulates GADD45a nuclear translocation and contributes to GADD45a-induced cell cycle G2-M arrest. *J. Biol. Chem.* **280**, 10988–10996.

Garcia-Bustos, J., Heitman, J. and Hall, M. N. (1991). Nuclear protein localization. *Biochem. Biochim. Biophys. Acta.* **1071**, 83–101.

Gerbi, S. A. and Borovjagin, A. (1997). U3 snoRNA may recycle through different compartments of the nucleolus. *Chromosoma* **105**, 401–406.

Ghazizadeh, M., Sasaki, Y., Araki, T., Konishi, H. and Aihara, K. (1997). Prognostic value of proliferative activity of ovarian carcinoma as revealed by PCNA and AgNOR analyses. *Am. J. Clin. Pathol.* **107**, 451–458.

Ginisty, H., Amalric, F. and Bouvet, P. (2001). Two different combinations of RNA-binding domains determine the RNA binding specificity of nucleolin. *J. Biol. Chem.* **276**, 14338–14343.

Ginisty, H., Sicard, H., Roger, B. and Bouvet, P. (1999). Structure and functions of nucleolin. *J. Cell Sci.* **112**, 761–772.

Goodwin, D. J., Hall, K. T., Stevenson, A. J., Markham, A. F. and Whitehouse, A. (1999). The open reading frame 57 gene product of herpesvirus saimiri shuttles between the nucleus and cytoplasm and is involved in viral RNA nuclear export. *J. Virol.* **73**, 10519–10524.

Goodwin, D. J. and Whitehouse, A. (2001). A gamma-2 herpesvirus nucleocytoplasmic shuttle protein interacts with importin alpha 1 and alpha 5. *J. Biol. Chem.* **276**, 19905–19912.

Harders, J., Lukacs, N., Robert-Nicoud, M., Jovin, T. M. and Riesner, D. (1989). Imaging of viroids in nuclei from tomato leaf tissue by in situ hybridization and confocal laser scanning microscopy. *Embo J.* **8**, 3941–3949.

Hay, R. T., Freeman, A., Leith, I., Monaghan, A. and Webster, A. (1995). Molecular interactions during adenovirus DNA replication. *Curr. Top. Microbiol. Immunol.* **199**, 31–48.

Hernandez-Verdun, D., Roussel, P. and Gebrane-Younes, J. (2002). Emerging concepts of nucleolar assembly. *J. Cell. Sci.* **115**, 2265–2270.

Hirano, M., Kaneko, S., Yamashita, T., Luo, H., Qin, W., Shirota, Y., Nomura, T., Kobayashi, K. and Murakami, S. (2003). Direct interaction between nucleolin and hepatitis C virus NS5B. *J. Biol. Chem.* **278**, 5109–5115.

Hiscox, J. A. (2002). Brief review: The nucleolus – a gateway to viral infection? *Arch. Virol.* **147**, 1077–1089.

Hiscox, J. A. (2003). The interaction of animal cytoplasmic RNA viruses with the nucleus to facilitate replication. *Virus Res.* **95**, 13–22.

Hiscox, J. A., Wurm, T., Wilson, L., Cavanagh, D., Britton, P. and Brooks, G. (2001). The coronavirus infectious bronchitis virus nucleoprotein localizes to the nucleolus. *J. Virol.* **75**, 506–512.

Huang, S., Deerinck, T. J., Ellisman, M. H. and Spector, D. L. (1998). The perinucleolar compartment and transcription. *J. Cell Biol.* **143**, 35–47.

Huang, W. H., Yung, B. Y., Syu, W. J. and Lee, Y. H. (2001). The nucleolar phosphoprotein B23 interacts with hepatitis delta antigens and modulates the hepatitis delta virus RNA replication. *J. Biol. Chem.* **276**, 25166–25175.

Huber, M. D., Dworet, J. H., Shire, K., Frappier, L. and McAlear, M. A. (2000). The budding yeast homolog of the human EBNA1-binding protein 2 (Ebp2p) is an essential nucleolar protein required for pre-rRNA processing. *J. Biol. Chem.* **275**, 28764–28773.

Iborra, F. J., Jackson, D. A. and Cook, P. R. (2001). Coupled transcription and translation within nuclei of mammalian cells. *Science* **293**, 1139–1142.

Izumi, R. E., Valdez, B., Banerjee, R., Srivastava, M. and Dasgupta, A. (2001). Nucleolin stimulates viral internal ribosome entry site-mediated translation. *Virus Res.* **76**, 17–29.

Jakob, R. (1994). Nucleolar accumulation of Semliki Forest virus nucleocapsid C protein: influence of metabolic status, cytoskeleton and receptors. *J. Med. Micro.* **40**, 389–392.

Jakob, R. (1995). Electroporation-mediated delivery of nucleolar targeting sequences from Semliki Forest virus nucleocapsid protein. *Prep. Biochem.* **25**, 99–117.

Jarjour, W. N., Minota, S., Roubey, R. A., Mimura, T. and Winfield, J. B. (1992). Autoantibodies to nucleolin cross-react with histone H1 in systematic lupus erythematosus. *Molecular Biology* **16**, 263–266.

Jin, A., Itahana, K., O'Keefe, K. and Zhang, Y. (2004). Inhibition of HDM2 and Activation of p53 by Ribosomal Protein L23. *Mol. Cell. Biol.* **24**, 7669–7680.

Jin, Y., Lee, H., Zeng, S. X., Dai, M. S. and Lu, H. (2003). MDM2 promotes p21waf1/cip1 proteasomal turnover independently of ubiquitylation. *Embo J.* **22**, 6365–6377.

Kasulke, D., Seitz, S. and Ehrenhofer-Murray, A. E. (2002). A role for the Saccharomyces cerevisiae RENT complex protein Net1 in HMR silencing. *Genetics* **161**, 1411–1423.

Khurts, S., Masutomi, K., Delgermaa, L., Arai, K., Oishi, N., Mizuno, H., Hayashi, N., Hahn, W. C. and Murakami, S. (2004). Nucleolin interacts with telomerase. *J. Biol. Chem.* **279**, 51508–51515.

Kill, I. R. (1996). Localisation of the Ki-67 antigen within the nucleolus. Evidence for a fibrillarin-deficient region of the dense fibrillar component. *J. Cell. Sci.* **109**, 1253–1263.

Kim, S. H., Ryabov, E. V., Brown, J. W. and Taliansky, M. (2004). Involvement of the nucleolus in plant virus systemic infection. *Biochem. Soc. Trans.* **32**, 557–560.

Korgaonkar, C., Hagen, J., Tompkins, V., Frazier, A. A., Allamargot, C., Quelle, F. W. and Quelle, D. E. (2005). Nucleophosmin (B23) targets ARF to nucleoli and inhibits its function. *Mol. Cell. Biol.* **25**, 1258–1271.

Kubota, S., Siomi, H., Satoh, T., Endo, S.-I., Maki, M. and Hatanaka, M. (1989). Effects of chimeric mutants of human immunodeficiency virus type 1 Rev and human T-cell leukemia virus type I Rex on nucleolar targeting signal. *J. Virol.* **65**, 2452–2456.

Lam, Y. W., Trinkle-Mulcahy, L. and Lamond, A. I. (2005). The nucleolus. *J. Cell. Sci.* **118**, 1335–1337.

Lamond, A. I. and Earnshaw, W. C. (1998). Structure and function in the nucleus. *Science* **280**, 547–553.

Lee, C. H., Chang, S. C., Chen, C. J. and Chang, M. F. (1998). The nucleolin binding activity of hepatitis delta antigen is associated with nucleolus targeting. *J. Biol. Chem.* **273**, 7650–7656.

Lee, T. W., Blair, G. E. and Matthews, D. A. (2003). Adenovirus core protein VII contains distinct sequences that mediate targeting to the nucleus and nucleolus, and colocalization with human chromosomes. *J. Gen. Virol.* **84**, 3423–3428.

Lee, T. W., Lawrence, F. J., Dauksaite, V., Akusjarvi, G., Blair, G. E. and Matthews, D. A. (2004). Precursor of human adenovirus core polypeptide Mu targets the nucleolus and modulates the expression of E2 proteins. *J. Gen. Virol.* **85**, 185–196.

Legrand, D., Vigie, K., Said, E. A., Elass, E., Masson, M., Slomianny, M. C., Carpentier, M., Briand, J. P., Mazurier, J. and Hovanessian, A. G. (2004). Surface nucleolin participates in both the binding and endocytosis of lactoferrin in target cells. *Eur. J. Biochem.* **271**, 303–317.

Leopardi, R. and Roizman, B. (1996a). Functional interaction and colocalization of the herpes simplex virus 1 major regulatory protein ICP4 with EAP, a nucleolar-ribosomal protein. *Proc. Natl. Acad. Sci. USA* **93**, 4572–4576.

Leopardi, R. and Roizman, B. (1996b). The herpes simplex virus major regulatory protein ICP4 blocks apoptosis induced by the virus or by hyperthermia. *Proc. Natl. Acad. Sci. USA* **93**, 9583–9587.

Leung, A. K., Andersen, J. S., Mann, M. and Lamond, A. I. (2003). Bioinformatic analysis of the nucleolus. *Biochem. J.* **376**, 553–569.

Leung, A. K. and Lamond, A. I. (2003). The dynamics of the nucleolus. *Crit. Rev. Eukaryot. Gene. Expr.* **13**, 39–54.

Li, Y. P., Busch, R. K., Valdez, B. C. and Busch, H. (1996). C23 interacts with B23, a putative nucleolar-localization-signal-binding protein. *Euro. J. Biochem.* **237**, 153–158.

Lin, N.-S., Hsieh, C.-E. and Hsu, Y.-H. (1996). Capsid protein of cucumber mosaic virus accumulates in the nuclei and at the periphery of the nucleoli in infected cells. *Arch. Virol.* **141**, 727–732.

Liu, H. and Herrmann, C. H. (2004). Differential localization and expression of the Cdk9 42k and 55 k isoforms. *J. Cell Physiol.* **203**, 251–260.

Liu, J.-L., Hebert, M. B., Ye, Y., Templeton, D. J., King, H.-J. and Matera, A. G. (2000). Cell cycle-dependent localization of the CDK2-cyclin E complex in Cajal (coiled) bodies. *J. Cell Sci.* **113**, 1543–1552.

Liu, J.-L., Lee, L. F., Ye, Y., Qian, Z. and Kung, H.-J. (1997). Nucleolar and nuclear localization properties of a herpesvirus bZIP oncoprotein, MEQ. *J. Virol.* **71**, 3188–3196.

Lixin, R., Efthymiadis, A., Henderson, B. and Jans, D. A. (2001). Novel properties of the nucleolar targeting signal of human angiogenin. *Biochem. Biophys. Res. Commun.* **284**, 185–193.

Lundberg, M. and M, J. (2001). Is VP22 nuclear homing an artifact? *Nature Biotechnology* **19**, 713.

Lundberg, M. and Johansson, M. (2002). Positively charged DNA-binding proteins cause apparent cell membrane translocation. *Biochem. Biophys. Res. Comm.* **291**, 367–371.

Lundberg, M., Wikstrom, S. and Johansson, M. (2003). Cell surface adherence and endocytosis of protein transduction domains. *Molecular Therapy* **8**, 143–150.

Lutz, P., Puvion-Dutilleul, F., Lutz, Y. and Kedinger, C. (1996). Nucleoplasmic and nucleolar distribution of the adenovirus IVa2 gene product. *J. Virol.* **70**, 3449–3460.

Mackenzie, D. J. and Tremaine, J. H. (1988). Ultrastructural location of non-structural protein 3A of cucumber mosaic virus in infected tissue using monoclonal antibodies to a cloned chimeric fusion protein. *J. Gen. Virol.* **69**, 2387–2395.

MacLean, C. A., Rixon, F. J. and Marsden, H. S. (1987). The products of gene Us11 of herpes simplex virus type 1 are DNA-binding and localize to the nucleoli of infected cells. *J. Gen. Virol.* **68**, 1921–1937.

Marciniak, R. A., Lombard, D. B., Johnson, F. B. and Guarente, L. (1998). Nucleolar localisation of the Werner syndrome protein in human cells. *Proc. Natl. Acad. Sci. (USA)* **95**, 6887–6892.

Maridor, G. and Nigg, E. A. (1990). cDNA sequences of chicken nucleolin/C23 and no38/B23, two major nucleolar proteins. *NAR* **18**, 1286.

Matthews, D. A. (2001). Adenovirus protein V induces redistribution of nucleolin and B23 from nucleolus to cytoplasm. *J. Virol.* **75**, 1031–1038.

Matthews, D. A. and Russell, W. C. (1998). Adenovirus core protein V is delivered by the invading virus to the nucleolus of the infected cell and later in infection is associated with nucleoli. *J. Gen. Virol.* **79**, 1671–1675.

Mears, W. E., Lam, V. and Rice, S. A. (1995). Identification of nuclear and nucleolar localization signals in the herpes simplex virus regulatory protein ICP27. *J. Virol.* **69**, 935–947.

Michel, M. R., Elgizoli, M., Dai, Y., Jakob, R., Koblet, H. and Arrigo, A. P. (1990). Karyophilic properties of Semliki Forest virus nucleocapsid protein. *J. Virol.* **64**, 5123–5131.

Michienzi, A., Cagnon, L., Bahner, I. and Rossi, J. J. (2000). Ribozyme-mediated inhibition of HIV 1 suggests nucleolar trafficking of HIV-1 RNA. *Proc. Natl. Acad. Sci. USA* **97**, 8955–8960.

Michienzi, A., Li, S., Zaia, J. A. and Rossi, J. J. (2002). A nucleolar TAR decoy inhibitor of HIV-1 replication. *Proc. Natl. Acad. Sci. (USA)* **99**, 14047–14052.

Mori, Y., Okabayashi, T., Yamashita, T., Zhao, Z., Wakita, T., Yasui, K., Hasebe, F., Tadano, M., Konishi, E., Moriishi, K. and Matsuura, Y. (2005). Nuclear localization of Japanese encephalitis virus core protein enhances viral replication. *J. Virol.* **79**, 3448–3458.

Ning, Q., Lakatoo, S., Liu, M. F., Yang, W. M., Wang, Z. M., Phillips, M. J. and Levy, G. A. (2003). Induction of prothrombinase fgl2 by the nucleocapsid protein of virulent mouse hepatitis virus is dependent on host hepatic nuclear factor-4 alpha. *J. Biol. Chem.* **278**, 15541–15549.

Nisole, S., Krust, B. and Hovanessian, A. G. (2002a). Anchorage of HIV on permissive cells leads to coaggregation of viral particles with surface nucleolin at membrane raft microdomains. *Exp. Cell Res.* **276**, 155–173.

Nisole, S., Said, E. A., Mische, C., Prevost, M.-C., Krust, B., Bouvet, P., Bianco, A., Briand, J.-P. and Hovanessian, A. G. (2002b). The anti-HIV pentameric pseudopeptide HB-19 binds the C-terminal end of nucleolin and prevents anchorage of virus particles in the plasma membrane of target cells. *J. Biol. Chem.* **277**, 20877–20886.

O'Connor, J. B. and Brian, D. A. (2000). Downstream ribosomal entry for translation of coronavirus TGEV gene 3b. *Virology* **269**, 172–182.

O'Hare, P. and Elliot, G. (2001). Reply to 'Is VP22 nuclear homing an artifact?' *Nature Biotech.* **19**, 714.

Okuwaki, M., Iwamatsu, A., Tsujimoto, M. and Nagata, K. (2001). Identification of nucleo-phosmin/B23, an acidic nucleolar protein, as a stimulatory factor for in vitro replication of adenovirus DNA complexed with viral basic core proteins. *J. Mol. Biol.* **311**, 41–55.

Okuwaki, M., Tsujimoto, M. and Nagata, K. (2002). The RNA binding activity of a ribosome biogenesis factor, nucleophosmin/B23, is modulated by phosphorylation with a cell cycle-dependent kinase and by association with its subtype. *Mol. Biol. Cell.* **13**, 2016–2030.

Olson, M. O., Dundr, M. and Szebeni, A. (2000). The nucleolus: an old factory with unexpected capabilities. *Trends in Cell Biology* **10**, 189–196.

Owen, K. E. and Kuhn, R. J. (1996). Identification of a region in the Sindbis virus nucleocapsid protein that is involved in specificity of RNA encapsidation. *J. Virol.* **70**, 2757–2763.

Pederson, T. (1998). The plurifunctional nucleolus. *NAR* **26**, 3871–3876.

Peeples, M. E., Wang, C., Gupta, K. C. and Coleman, N. (1992). Nuclear entry and nucleolar localization of the Newcastle disease virus (NDV) matrix protein occur early in infection and do not require other NDV proteins. *J. Virol.* **66**, 3263–3269.

Peranen, J., Rikkonen, M., Liljestrom, P. and Kaariainen, L. (1990). Nuclear localization of Semliki Forest virus-specific nonstructural protein nsP2. *J. Virol.* **64**, 1888–1896.

Peters, R. (1986). Flourescence microphotolysis to measure nucleocytoplasmic transport and intracellular mobility. *Biochim. Biophys. Acta* **864**, 305–359.

Platani, M., Goldberg, I., Swedlow, J. R. and Lamond, A. I. (2000). In vivo analysis of Cajal body movement, separation, and joining in live human cells. *J. Cell Biol.* **151**, 1561–1574.

Puvion-Dutilleul, F. and Christensen, M. E. (1993). Alterations of fibrillarin distribution and nucleolar ultrastructure induced by adenovirus infection. *European J. Cell Biol.* **61**, 168–176.

Pyper, J. M., Clements, J. E. and Zink, M. C. (1998). The nucleolus is the site of Borna disease virus RNA transcription and replication. *J. Virol.* **72**, 7697–7702.

Qi, Y. and Ding, B. (2003). Differential subnuclear localization of RNA strands of opposite polarity derived from an autonomously replicating viroid. *Plant Cell* **15**, 2566–2577.

Raab de Verdugo, U., Selinka, H.-C., Huber, M., Kramer, B., Kellermann, J., Hofschneider, P. H. and Kandolf, R. (1995). Characterization of a 100-kilodalton binding protein for the six serotypes of coxsackie B viruses. *J. Virol.* **69**, 6751–6757.

Richardson, W. D., Mills, A. D., Dilworth, S. M., Laskey, R. A. and Dingwall, C. (1988). Nuclear protein migration involves two steps: rapid binding at the nuclear envelope followed by slower translocation through nuclear pores. *Cell* **52**, 655–664.

Rikkonen, M., Peranen, J. and Kaariainen, L. (1994). Nuclear targeting of Semliki Forest virus nsP2. *Arch. Virol. Suppl* **9**, 369–77.

Robbins, J., Dilworth, S. M., Laskey, R. A. and Dingwall, C. (1991). Two interdependent basic domains in nucleoplasmin nuclear targetting sequence: identification of a class of bipartite nuclear targeting sequence. *Cell* **64**, 615–623.

Roller, R. J., Monk, L. L., Stuart, D. and Roizman, B. (1996). Structure and function in the herpes simplex virus 1 RNA-binding protein U(s)11: mapping of the domain required for ribosomal and nucleolar association and RNA binding in vitro. *J. Virol.* **70**, 2842–2851.

Ross, C., Hiscox, J. A. and McCauley, J. W. (2001). Unpublished data.

Rowland, R. R., Kerwin, R., Kuckleburg, C., Sperlich, A. and Benfield, D. A. (1999). The localisation of porcine reproductive and respiratory syndrome virus nucleocapsid protein to the nucleolus of infected cells and identification of a potential nucleolar localization signal sequence. *Virus Res.* **64**, 1–12.

Rowland, R. R. R., Schneider, P., Fang, Y., Wootton, S., Yoo, D. and Benfield, D. A. (2003). Peptide domains involved in the localization of the porcine reproductive and respiratory syndrome virus nucleocapsid protein to the nucleolus. *Virology* **316**, 135–145.

Rubbi, C. P. and Milner, J. (2003). Disruption of the nucleolus mediates stabilization of p53 in response to DNA damage and other stresses. *Embo J.* **22**, 6068–6077.

Ryabov, E. V., Kim, S. H. and Taliansky, M. (2004). Identification of a nuclear localization signal and nuclear export signal of the umbraviral long-distance RNA movement protein. *J. Gen. Virol.* **85**, 1329–1333.

Ryabov, E. V., Oparka, K. J., Santa Cruz, S., Robinson, D. J. and Taliansky, M. E. (1998). Intracellular location of two groundnut rosette umbravirus proteins delivered by PVX and TMV vectors. *Virology* **242**, 303–313.

Savino, T. M., Gebrane-Younes, J., De Mey, J., Sibarita, J. B. and Hernandez-Verdun, D. (2001). Nucleolar assembly of the rRNA processing machinery in living cells. *J. Cell. Biol.* **153**, 1097–1110.

Sawicki, D. L. and Sawicki, S. G. (1993). A second nonstructural protein functions in the regulation of alphavirus negative-strand RNA synthesis. *J. Virol.* **67**, 3605–3610.

Scherl, A., Coute, Y., Deon, C., Calle, A., Kindbeiter, K., Sanchez, J.-C., Greco, A., Hochstrasser, D. and Diaz, J.-J. (2002). Functional proteomic analysis of human nucleolus. *Mol. Biol. Cell* **13**, 4100–4109.

Schmidt-Zachmann, M. S. and Nigg, E. A. (1993). Protein localization to the nucleolus: a search for targetting domains in nucleolin. *J. Cell Sci.* **105**, 799–806.

Shaw, P. J. and Jordan, E. G. (1995). The nucleolus. *Annu. Rev. Cell Dev. Biol.* **11**, 93–121.

Sheng, Z., Lewis, J. A. and Chirico, W. J. (2004). Nuclear and nucleolar localization of 18-kDa fibroblast growth factor-2 is controlled by C-terminal signals. *J. Biol. Chem.* **279**, 40153–40160.

Shou, W., Seol, J. H., Shevchenko, A., Baskerville, C., Moazed, D., Chen, Z. W., Jang, J., Charbonneau, H. and Deshaies, R. J. (1999). Exit from mitosis is triggered by Tem1-dependent release of the protein phosphatase Cdc14 from nucleolar RENT complex. *Cell* **97**, 233–244.

Siomi, H., Shida, H., Maki, H. M. and Hatanaka, M. (1990). Effects of a highly basic region of human immunodeficiency virus Tat protein on nucleolar localization. *J. Virol.* **64**, 1803–1807.

Siomi, H., Shida, H., Nam, S. H., Nosaka, T., Maki, M. and Hatanaka, M. (1988). Sequence requirements for nucleolar localization of human T-cell leukemia virus type I pX protein, which regulates viral RNA processing. *Cell* **55**, 197–209.

Sirri, V., Hernandez-Verdun, D. and Roussel, P. (2002). Cyclin-dependent kinases govern formation and maintenance of the nucleolus. *J. Cell. Biol.* **156**, 969–981.

Sirri, V., Roussel, P., Gendron, M. C. and Hernandez-Verdun, D. (1997). Amount of the two major Ag-NOR proteins, nucleolin and protein B23 is cell-cycle dependent. *Cytometry* **28**, 147–156.

Sirri, V., Roussel, P. and Hernandez-Verdun, D. (2000). The AgNOR proteins: qualitative and quantitative changes during the cell cycle. *Micron* **31**, 121–126.

Snaar, S., Wiesmeijer, K., Jochemsen, A. G., Tanke, H. J. and Dirks, R. W. (2000). Mutational analysis of fibrillarin and its mobility in living human cells. *J. Cell. Biol.* **151**, 653–662.

Srivastava, M. and Pollard, H. B. (1999). Molecular dissection of nucleolin's role in growth and cell proliferation: new insights. *FASEB J.* **13**, 1911–1922.

Suopanki, J., Sawicki, D. L., Sawicki, S. G. and Kaariainen, L. (1998). Regulation of alphavirus 26S mRNA transcription by replicase component nsP2. *J. Gen. Virol.* **79**, 309–319.

Szekely, L., Jiang, W., Porkrovskaja, K., Wiman, K., Kelin, G. and Ringertz, N. (1995). Reversible nucleolar translocation of Epstein-Barr virus-encoded EBNA-5 and hsp70 proteins after exposure to heat shock or cell density congestion. *J. Gen. Virol.* **76**, 2423–2432.

Takeuchi, K., Turley, S. J., Tan, E. M. and Pollard, K. M. (1995). Analysis of the autoantibody response to fibrillarin in human-disease and murine models of autoimmunity. *J. Immunol.* **154**, 961–971.

Taliansky, M. E. and Robinson, D. J. (2003). Molecular biology of umbraviruses: phantom warriors. *J. Gen. Virol* **84**, 1951–1960.

Tijms, M. A., van der Meer, Y. and Snijder, E. J. (2002). Nuclear localization of non-structural protein 1 and nucleocapsid protein of equine arteritis virus. *J. Gen. Virol.* **83**, 795–800.

Toczyski, D. P., Matera, A. G., Ward, D. C. and Steitz, J. A. (1994). The Epstein-Barr virus (EBV) small RNA EBER1 binds and relocalizes ribosomal protein L22 in EBV-infected human B lymphocytes. *Proc. Natl. Acad. Sci USA* **91**, 3463–3467.

Tokuyama, Y., Horn, H. F., Kawamura, K., Tarapore, P. and Fukasawa, K. (2001). Specific phosphorylation of nucleophosmin on Thr(199) by cyclin-dependent kinase 2-cyclin E and its role in centrosome duplication. *J. Biol. Chem.* **276**, 21529–21537.

Tollervey, D., Lehtonen, H., Jansen, R., Kern, H. and Hurt, E. C. (1993). Temperature-sensitive mutations demonstrate roles for yeast fibrillarin in pre-rRNA procesing, pre-rRNA methylation, and ribosome assembly. *Cell* **72**, 443–457.

Trabucco, S., Varcaccio-Garofalo, G., Botticella, M. A., De Stefano, R., Capursi, T. and Resta, L. (1994). Expression of AgNORs in serous ovarian tumors. *Eur. J. Gynaecol. Oncol.* **15**, 222–229.

Tsai, C. W., Redinbaugh, M. G., Willie, K. J., Reed, S., Goodin, M. and Hogenhout, S. A. (2005). Complete genome sequence and in planta subcellular localization of maize fine streak virus proteins. *J. Virol.* **79**, 5304–5314.

Tsai, R. Y. L. and McKay, R. D. G. (2005). A multistep, GTP-driven mechanism controlling the dynamic cycling of nucleostemin. *J. Cell Biol.* **168**, 179–184.

Ulmanen, I., Soderlund, H. and Kaarianen, L. (1976). Semliki forest virus capsid protein associates with 60S ribosomal subunit in infected cells. *J. Virol.* **20**, 203–210.

von Kobbe, C. and Bohr, V. A. (2002). A nucleolar targetting sequence in the Werner syndrome protein resides within residues 949–1092. *J. Cell Sci.* **115**, 3901–3907.

Waggoner, S. and Sarnow, P. (1998). Viral ribonucleoprotein complex formation and nucleolar-cytoplasmic relocalization of nucleolin in poliovirus-infected cells. *J. Virol.* **72**, 6699–6709.

Ward, A. C., Azad, A. A. and Macreadie, I. G. (1994). Expression and characterisation of the influenza A virus non-structural protein NS1 in yeast. *Arch. Virol.* **138**, 299–314.

Weiss, B., Nitschko, H., Ghattas, I., Wright, R. and Schlesinger, S. (1989). Evidence for specificity in the encapsidation of Sindbis virus RNAs. *J. Virol.* **63**, 5310–5318.

Wengler, G. and Wengler, G. (1984). Identification of a transfer of viral core protein to cellular ribosomes during the early stages of alphavirus infection. *Virology* **134**, 435–442.

Wengler, G., Wurkner, D. and Wengler, G. (1992). Identification of a sequence element in the alphavirus core protein which mediates interaction of cores with ribosomes and the disassembly of cores. *Virology* **191**, 880–888.

Wurm, T., Chen, H., Britton, P., Brooks, G. and Hiscox, J. A. (2001). Localisation to the nucleolus is a common feature of coronavirus nucleoproteins and the protein may disrupt host cell division. *J. Virol.* **75**, 9345–9356.

Yang, T. H., Tsai, W. H., Lee, Y. M., Lei, H. Y., Lai, M. Y., Chen, D. S., Yeh, N. H. and Lee, S. C. (1994). Purification and characterization of nucleolin as a transcriptional repressor. *Mol. Cell. Biol.* **14**, 6068–6074.

Yoo, D., Wootton, S. K., Li, G., Song, C. and Rowland, R. R. (2003). Colocalization and interaction of the porcine arterivirus nucleocapsid protein with the small nucleolar RNA-associated protein fibrillarin. *J. Virol.* **77**, 12173–12183.

Zergeroglu, S., Aksakal, O., Demirturk, F. and Gokmen, O. (2001). Prognostic importance of the nucleolar organizer region score in ovarian epithelial tumors. *Gynecol. Obstet. Invest.* **51**, 60–63.

Zhang, S., Hemmerich, P. and Grosse, F. (2004). Nucleolar localization of the human telomeric repeat binding factor 2 (TRF2). *J. Cell. Sci.* **117**, 3935–3945.

Zhang, Y. (2004). The ARF-B23 connection: implications for growth control and cancer treatment. *Cell Cycle* **3**, 259–262.

Zimber, A., Nguyen, Q. D. and Gespach, C. (2004). Nuclear bodies and compartments: functional roles and cellular signalling in health and disease. *Cell Signal.* **16**, 1085–1104.

8 Virus Interactions with PML Nuclear Bodies

Keith N. Leppard and John Dimmock

Department of Biological Sciences, University of Warwick,
Coventry, UK

8.1 Introduction

Viruses that invade a cell have both to contend with cell defence mechanisms and to modify and recruit essential cell components to aid their replication. The nucleus is a key functional entity within the eukaryotic cell with which viruses may have to interact, irrespective of the cell compartment in which they replicate. To fulfil its roles in the cell, the nucleus must be a highly organized structure. Although our understanding of this organization is incomplete, a variety of substructures have been identified, among which are nucleoli, microspeckles, coiled bodies (see Chapters 3 and 7), gems (Cajal bodies) and ND10. Many of these substructures are observed by indirect immunofluorescence, and so are effectively defined as regions of localization of specific antigens. ND10, also known as PML oncogenic domains (PODs), PML nuclear bodies or Kr bodies, are implicated in several cellular functions and are also targeted during infection by a variety of viruses. This chapter considers the range of viral interactions with ND10 and the possible reasons for these in the light of what is currently known about the structure and function of ND10 in uninfected cells.

8.2 ND10 in uninfected cells

8.2.1 Discovery and composition of ND10

ND10 were originally described as dense granular bodies by electron microscopic analysis and were subsequently recognized as autoantigenic targets in patients with

Viruses and the Nucleus   Edited by Julian A. Hiscox
© 2006 John Wiley & Sons, Ltd

primary biliary cirrhosis (PBC). They are components of most cell types, where they are visualized by immunofluorescence as discrete dots numbering 10–20/nucleus (hence nuclear domain 10) (Dyck *et al.*, 1994; Koken *et al.*, 1994; Weis *et al.*, 1994). These dots are heterogeneous in number and size, dependent on cell type and the stage of the cell cycle (Koken *et al.*, 1995; Terris *et al.*, 1995) and sometimes appear hollow (Jiang and Ringertz, 1997). Under high magnification light microscopy and electron microscopy, larger ND10 appear to have a 'doughnut-like' shape, consisting of a dense fibrillar ring around a less dense central core, with many pores and channels throughout (Boisvert *et al.*, 2000); this appearance represents a cross-sectional view through a spherical structure.

A large number of proteins have been localized to ND10 (see Negorev and Maul, 2001, for a recent review). Two proteins, the promyelocytic leukaemia protein (PML) (Dyck *et al.*, 1994; Koken *et al.*, 1994; Weis *et al.*, 1994) and Sp100 (Szostecki *et al.*, 1990), are consistent and therefore diagnostic components of ND10. Indeed, ND10 as such do not exist in the absence of PML as other ND10 proteins, such as CREB binding protein (CBP) (Doucas *et al.*, 1999; LaMorte *et al.*, 1998) and Daxx (Ishov *et al.*, 1999; Li *et al.*, 2000), as well as Sp100 itself, depend on PML for their localization (Ishov *et al.*, 1999; Zhong *et al.*, 2000a). Some ND10 components only localize to these structures upon overexpression (e.g. BRCA1; Maul *et al.*, 1998) or under specific circumstances (e.g. retinoblastoma protein in its inactive, hyperphosphorylated form; Alcalay *et al.*, 1998) so that the composition of ND10 varies both within and between cells. It follows, therefore, that the ND10 within a cell may be functionally heterogeneous.

8.2.2   The promyelocytic leukaemia (PML) protein

PML protein is produced as a series of isoforms by translation from a family of mRNAs that derives from the *PML* gene primary transcript by differential splicing (Fagioli *et al.*, 1992; reviewed in Jensen *et al.*, 2001). These isoforms differ by the presence or absence of distinct C-terminal domains of up to 300 amino acids, as well as the potential omission of internal exons 4 and 5. All PML isoforms share an N-terminal domain containing three cysteine-rich zinc-binding domains. The first of these makes up a RING finger, a motif that is shared by a large family of proteins of diverse function (Borden, 2000; Freemont, 1993). The other two cysteine-rich regions are known as the B1 and B2 boxes. These are followed by an α-helical coiled-coil domain that is involved in dimerization of the protein. This RING-B-box-coiled-coil (RBCC) motif is common to a subgroup of RING finger proteins (Reymond *et al.*, 2001). In many earlier papers, general conclusions about PML function were drawn from experiments involving only a single specific PML cDNA. However, as might have been expected, evidence is increasing that the structural diversity in PML protein is matched by functional diversity.

PML protein diversity is further increased by post-translational coupling to SUMO-1, a ubiquitin-homology family member. The linkage of SUMO-1 to proteins is analogous to that of ubiquitin (reviewed by Wilson and Rangasamy, 2001). However, in SUMO-1, the lysine residue found in ubiquitin to be the target for polyubiquitination is not present, suggesting that poly-sumoylation does not occur. SUMO-1 has a calculated mass of 11.5 kDa and has been shown to be covalently linked to PML at one or more of three lysine residues (Duprez *et al.*, 1999; Kamitani *et al.*, 1998; Müller *et al.*, 1998; Sternsdorf *et al.*, 1997). Sumoylation of PML is regulated by phosphorylation, as inhibition of protein phosphatases causes loss of SUMO modification (Everett *et al.*, 1999b; Müller *et al.*, 1998). Thus, detecting PML by western blot analysis reveals a complex family of isoforms derived from both alternatively spliced transcripts and the effects of secondary modifications (Sternsdorf *et al.*, 1997).

Endogenous PML is localized primarily to ND10, although there is also some diffuse nuclear staining. Looking at individual isoforms by transfection, PML I–VI all associate with ND10 (Bischof *et al.*, 2002), although their numbers and sizes vary (Beech *et al.*, 2005); PML VII isoforms that lack the nuclear localization signal are cytoplasmic (Flenghi *et al.*, 1995). The PML RBCC motif is crucial for assembly of the protein into ND10, as mutations in this element result in a diffuse nuclear PML distribution (Borden *et al.*, 1996; Fagioli *et al.*, 1998).

Assembly of the normal multiprotein complex at ND10, and possibly the assembly of PML itself into ND10, also depend on SUMO-1 modification of PML. Circumstances that either increased (arsenic trioxide treatment) or decreased (cells in mitosis) the organization of PML into ND10 caused a corresponding increase or decrease in the amount of sumoylated PML and, in the latter case, of co-localizing Sp100 (Everett *et al.*, 1999b; Müller *et al.*, 1998): this is correlative evidence. Direct evidence comes from transfecting cDNA encoding non-sumoylatable, so-called 3K mutant forms of PML. PML-III(3K) formed nuclear aggregates rather than normal ND10 when expressed in PML-null fibroblasts and these did not recruit Sp100 or Daxx (Lehembre *et al.*, 2001; Zhong *et al.*, 2000a). However, in a similar experiment with HEp2 cells, PML-II(3K) formed what were judged to be ND10, although Daxx recruitment to them again failed (Ishov *et al.*, 1999). Similarly, PML-III(3K) also gave apparently normal ND10 in PML-null fibroblasts for a third group (Lallemand-Breitenbach *et al.*, 2001), although the structure of these ND10 did not alter in response to arsenic trioxide in the manner of normal ND10 and so were functionally, if not structurally, abnormal. Finally, overexpression of a SUMO-specific protease, which was sufficient to deplete the nucleus of SUMO-1 staining and to substantially destroy SUMO-modified forms of PML, had no effect on the integrity of ND10 detected by PML staining (Bailey and O'Hare, 2002). Thus, non-sumoylatable PML forms nuclear bodies that may or may not resemble normal ND10, but which certainly lack many of the other proteins normally localized to ND10.

8.2.3   Dynamics of ND10 organization

Recent data show that ND10 are not static entities, but are exchanging components with their surroundings constantly under normal growth. These studies rely on expressing fluorescent protein fusions with ND10 components to follow their movement in real time. In 293 cells at 22°C, tagged CBP equilibrated rapidly between ND10 and the adjacent nucleoplasm while tagged Sp100 and PML (possibly isoform VI) were essentially static (Boisvert *et al.*, 2001). A second study employed CBP, Sp100 and an unidentified PML isoform with distinct fluorescent protein tags in human fibroblasts and U2OS cells at 37°C (Wiesmeijer *et al.*, 2002). Sp100 or PML staining within an individual bleached ND10 recovered within 5–10 min, while CBP recovered within 10 s. The discrepancy between these studies regarding PML and Sp100 movement may be attributed either to different incubation temperatures, cell types or possibly the isoforms of PML and Sp100 expressed. With this proviso, it is clear that peripheral ND10 components, such as CBP, are in rapid dynamic equilibrium with the diffuse nuclear compartment, while the presumed core ND10 components exchange more slowly.

It has also been suggested that ND10 themselves can move within the nucleus. Tagged Sp100 expressed in BHK cells and observed at 37 °C over several minutes defined three movement classes of ND10: stationary, localized movement and rapid, directional movement (Muratani *et al.*, 2002). This last category typically comprised some of the smaller ND10 in the cell; they could be seen to both fuse with and split from larger bodies and their movement depended on metabolic energy. However, when tagged Sp100 and PML were used together, the larger foci of Sp100 co-localized with PML, while smaller ones did not, and only these Sp100-only foci moved during the 2 h time period (Wiesmeijer *et al.*, 2002). Given that the capacity of ND10 for Sp100 is saturable (Negorev *et al.*, 2001), the mobile Sp100 bodies may be artefacts of overexpressing the protein beyond the capacity of ND10 for Sp100 binding.

8.2.4 Changes to ND10 during the cell cycle

The average size and number of ND10 varies during the cell cycle (Koken *et al.*, 1995; Terris *et al.*, 1995; see Chapters 2 and 9 for information on the cell cycle). During S phase, PML and Sp100 are co-localized to ND10, although both proteins also have a diffuse nuclear component. Later, in G_2, many cells have an increased number of ND10, while in cells entering prophase the diffuse portion of PML is excluded from condensed chromosomes and Sp100 no longer co-localizes precisely with PML. At the beginning of mitosis, ND10 (defined by PML staining) become disrupted. With the breakdown of the nuclear envelope, PML forms irregular accumulations where no Sp100 is present. ND10 only gradually reform during early G_1, when a proportion of the PML can still be seen in the cytoplasm in punctate bodies (Everett *et al.*, 1999b).

Changes to PML post-translational modification probably regulate PML localization, and hence ND10 assembly/disassembly, during the cell cycle. PML and SUMO-1 co-localize at ND10 during interphase, but as ND10 break down, separation occurs. Any remaining PML foci during mitosis have been shown by immunofluorescence not to co-localize with SUMO-1 (Everett *et al.*, 1999b). Following synchronized cells through mitosis, the three prominent SUMO-1 modified forms of PML transiently disappear and are replaced by a novel form similar to that produced by inhibiting the action of phosphatases on asynchronous cells (Everett *et al.*, 1999b); this treatment also inhibited conjugation of SUMO-1 to overexpressed PML in an engineered cell line (Müller *et al.*, 1998). Moreover, the mitosis-specific isoform of PML is highly labile but is stabilized by phosphatase inhibitors, supporting the hypothesis that it is a phosphorylated form of PML.

8.2.5 Changes in ND10 in response to other factors

The organization of ND10 is sensitive to numerous external factors. Heat shock or cadmium treatment results in redistribution of the ND10 into a microspeckled appearance, consisting of hundreds of small dots amongst the chromatin (Maul *et al.*, 1995). This change is rapid and independent of protein synthesis. During recovery, the ND10 components regroup into discrete structures, passing through an intermediate track-like stage. These observations might result from a process of active transport, with ND10 components travelling along nuclear filaments. ND10 are similarly disrupted after exposure to ionizing radiation, with components localizing to double-strand DNA breaks (Carbone *et al.*, 2002). Amino acid starvation also affects ND10, in this case reducing their number but increasing their size, resulting in structures termed 'large bodies' (Kamei, 1996). Finally, interferon treatment causes an increase in the number and size of ND10. Both type I(α,β) and type II(γ) interferons (IFNs) strongly enhance transcription of the *PML* and *Sp100* genes (Guldner *et al.*, 1992; Lavau *et al.*, 1995) and may alter the isoform pattern of expressed PML (Grötzinger *et al.*, 1996). IFN treatment also causes an increased localization of Daxx to ND10 (Lehembre *et al.*, 2001). Thus, IFN may affect PML and ND10 function, both qualitatively and quantitatively (see Section 8.3).

ND10 are also perturbed in specific disease states. First, in acute promyelocytic leukaemia, a chromosomal translocation produces a fusion of PML with the retinoic acid receptor (RARα), expression of which disrupts ND10, resulting in a microspeckled appearance of PML (Dyck *et al.*, 1994; Koken *et al.*, 1994; Weis *et al.*, 1994). The significance of this disruption to the disease process is suggested by the finding that retinoic acid treatment can both revert the organization of ND10 in leukaemic cells to normal and achieve clinical remission. Second, in hepatocellular carcinoma, PML is overexpressed and its localization is altered to become partly cytoplasmic (Terris *et al.*, 1995). Infection by a number of viruses also modifies the organization of ND10, as discussed in detail in Section 8.4.

8.3 Functions of ND10

A detailed survey of all the data concerning possible functions of ND10 is beyond the scope of this chapter. Ideas about these functions are summarized below. Readers are referred to several recent reviews on aspects of this topic for more detailed information (Borden, 2002; Gottifredi and Prives, 2001; Negorev and Maul, 2001; Pearson and Pelicci, 2001; Regad and Chelbi-Alix, 2001; Zhong *et al.*, 2000b).

One broad function that has been ascribed to ND10 is in the linked areas of cell cycle regulation and senescence. As already discussed, ND10 are systematically altered during the cell cycle and they have been noted as sites of localization of key cell cycle regulators. They have also been implicated in regulating cyclin D1 expression (Lai and Borden, 2000). Finally, overexpression of PML reduces the cell growth rate and the proportion of cells in S phase (Koken *et al.*, 1995; Mu *et al.*, 1997), while PML-null cells show the opposite effect (Wang *et al.*, 1998a). Thus, it is conceivable that key cell cycle regulatory steps take place at ND10. An extreme form of cell cycle regulation is senescence, whereby cells permanently cease division and drop out of the cycle altogether. Overexpression of the right PML isoform (PML-IV) induced normal fibroblasts to senesce, although localization of the exogenous PML to ND10 was not necessary for this effect (Bischof *et al.*, 2002).

ND10 have also been implicated in apoptosis and the DNA damage response. Cells from PML knockout mice were unable to apoptose in response to Fas activation, tumour necrosis factor or interferons α/β or γ and were defective in caspase activation (Wang *et al.*, 1998b). Also, both p53 activation (part of the DNA damage response) in fibroblasts and DNA damage-induced apoptosis in thymocytes were impaired in the absence of PML (Guo *et al.*, 2000). Several proteins involved in responding to and repairing DNA damage localize to ND10 (Lombard and Guarente, 2000), as do actual sites of single- or double-strand DNA breaks (Carbone *et al.*, 2002). A DNA damage checkpoint kinase phosphorylates PML-IV and interacts with it *in vivo* (Yang *et al.*, 2002). Finally, a subset of ND10 participates in telomere maintenance in certain cell lines where telomerase has not been reactivated (Grobelny *et al.*, 2000), a reaction that may have characteristics of DNA repair.

Possibly underlying all these putative functional roles of ND10 is a role in regulating gene expression. PML has been shown to exert both repression and activation effects on transcription in different contexts. Specific PML isoforms have been shown to interact with transcription factors such as Sp1 (Vallian *et al.*, 1998), to recruit co-activators such as CBP into ND10 (LaMorte *et al.*, 1998), but also to interact with histone deacetylases (HDAC) that repress transcription (Wu *et al.*, 2001b). PML-IV has been implicated in regulating activation of p53 (Fogal *et al.*, 2000; Guo *et al.*, 2000) but localization to ND10 was not required for this (Bischof *et al.*, 2002). Binding of an HDAC to PML-IV both antagonized the activation of p53, by deacetylating it, and rescued cells from senescence induction (Langley *et al.*, 2002). However, despite all these data, there is conflicting evidence as to whether or not ND10 are themselves sites of active transcription (Boisvert *et al.*, 2000; LaMorte *et al.*, 1998). Lastly, it has been suggested that PML plays a role in regulating the export of cyclin D1 mRNA from the nucleus (Lai and Borden, 2000).

The final putative role of ND10 to be mentioned here is in the interferon response and antigen processing for adaptive immunity. As already discussed, various ND10 proteins are induced by interferons (see Section 8.2.5) and PML is required for apoptosis induction by interferon (see above). Moreover, the interferon-induced antiviral protein Mx1 binds to ND10 components (Engelhardt *et al.*, 2001) and interferon affects the activity and localization of Daxx (Lehembre *et al.*, 2001). There is also evidence to implicate ND10 in the interferon pathway from studies of specific viruses (see Section 8.4).

It has been reported that interferon treatment of the human epidermoid carcinoma cell line, HEp2, increased the proportion of ND10 associated with sites of active transcription from 30% to 80% (Kiesslich *et al.*, 2002), which is particularly interesting, given the observation that the major histocompatibility gene cluster, transcription of which is known to be interferon-responsive, showed a highly non-random close proximity to ND10 at all stages of the cell cycle (Shiels *et al.*, 2001). Developing the theme that ND10 might be involved in the upregulation of adaptive immune responses by interferon, a peptide representing a known MHC class I epitope accumulated at ND10 in the form of insoluble polyubiquitinated material when proteasome activity was inhibited, as did proteasomes themselves (Anton *et al.*, 1999). Proteasomes accumulated at ND10 when induced by interferon (Fabunmi *et al.*, 2001; Rockel and von Mikecz, 2002), and the proteasome activator, PA28, also showed interferon γ-enhanced accumulation at ND10 (Fabunmi *et al.*, 2001). Taken together, these data suggest that ND10 are intimately involved in the adaptive response to nuclear antigens.

There has been an ongoing debate as to whether the role of ND10 is active or passive. Thus, while there is evidence for important functions of ND10 or their components, an alternative model has ND10 as inert structures, so-called nuclear depots, with the function of storing excess nuclear components until such time as they are required. One of the principal arguments for this idea is the diversity of proteins that can be found to localize at ND10, particularly or only when overexpressed (Negorev and Maul, 2001). Indeed, an extreme view would be that ND10 represent dead-end deposits of excess aggregated protein that are either being chaperoned back to a functional structure or being held for destruction. Although there is a wealth of data that argues for a functional role of ND10 or their components in important cell processes, whether these functions have to occur at ND10, or whether the relevant components can function equally well at other locations in the nucleus, is the key question and very difficult to answer.

8.4   ND10 and virus infection

Infection by certain viruses can bring about changes in the appearance, distribution or composition of ND10. As discussed in detail below, these viruses include examples from various families that are evolutionarily unrelated, that have either RNA or DNA genomes, and that employ very different strategies for exploitation of the host cell to produce progeny virus particles. It is this diversity that is most suggestive of a fundamental significance to the interaction between viruses and ND10, although it should be recognized that most of the work so far has focused on nuclear-replicating DNA viruses. The nature of the interactions with ND10 that are made by a range of viruses is discussed below; general hypotheses to explain these findings are considered in Section 8.5.

8.4.1   Herpes simplex virus type 1 (HSV-1)

HSV-1, a nuclear-replicating DNA virus from the *alphaherpesvirus* subfamily, infects epithelial cells and establishes latency in neurons, causing cold sores on reactivation. It was the first virus to be shown to have effects upon ND10 and remains the most well-characterized example. The HSV-1 immediate-early protein ICP0 is an activator of gene expression and has also been linked to the reactivation of virus from latency (reviewed in Everett, 2000b). ICP0 precisely co-localizes with PML in ND10, which subsequently are disrupted with PML staining becoming diffuse in the nucleus, an effect for which ICP0 is necessary and sufficient (Everett and Maul, 1994; Maul and Everett, 1994; Maul *et al.*, 1993). ICP0 also causes a loss of SUMO-modified PML and Sp100 isoforms that is dependent upon active proteasomes (Chelbi-Alix and de Thé, 1999; Everett *et al.*, 1998; Müller and Dejean, 1999; Parkinson and Everett, 2000). In addition to PML, other selected proteins are also degraded (Everett *et al.*, 1999a). Ultimately, amounts of PML in the cell are significantly reduced through the action of ICP0, dependent upon ICP0 having an intact RING motif. These effects of ICP0 on PML and ND10 are now known to reflect its activity as an E3 ubiquitin ligase, via which it can target selected proteins for degradation by proteasomes. First, ICP0 induced and co-localized with polyubiquitin in infected cells, dependent on an intact RING motif (Everett, 2000a). Second, both full-length ICP0 and its isolated RING motif had E3 ubiquitin ligase activity *in vitro* and co-localized with E1 and E2 components of the ubiquitination system on transfection (Boutell *et al.*, 2002).

It is still not certain what is the relationship of de-sumoylation of PML and other ICP0 targets to their eventual destruction via the ubiquitin pathway. In principle, sumoylation of PML might protect it from ubiquitination, such that de-sumoylation would be the trigger for degradation. However, only a minor proportion of PML is sumoylated *in vivo* and PML de-sumoylation in response to other triggers (e.g. adenovirus infection; Section 8.4.5) does not lead to its degradation. Alternatively, ubiquitination of PML might proceed irrespective of its sumoylation status in the presence of an E3 ligase of the relevant specificity. SUMO-modified PML disappears rapidly in the presence of ICP0, in comparison with unmodified forms (Everett *et al.*, 1998). If de-sumoylation does precede degradation, it may be relevant that ICP0 exactly co-localizes with SENP-1, a SUMO-specific protease, when co-transfected with it (Bailey and O'Hare, 2002). It is also significant that ICP0 both co-immunoprecipitates a ubiquitin-specific protease, HAUSP, in infected cells, and causes a transient increase in its association with ND10 (Everett *et al.*, 1997). HAUSP specificity does not extend to sumoylated proteins (Everett *et al.*, 1998), discounting a role in de-sumoylation, and HAUSP binding is not required for the effects of ICP0 on ND10. However, evidence from ICP0 mutants unable to make this interaction indicates that it is important to the outcome of infection (Everett *et al.*, 1999c). Possibly, HAUSP normally acts to clear any conjugated ubiquitin from either PML or SUMO-PML, so stabilizing it.

Association between HSV-1 and ND10 is a very early event in infection. At 3 h post-infection, a time prior to ND10 disruption by ICP0, viral DNA was found by *in situ* hybridization to be closely juxtaposed with ND10 (Maul *et al.*, 1996); this association did not require any new RNA or protein synthesis to occur after infection. Using an ICP0 mutant virus unable to disrupt ND10, this association could still be visualized into the late phase of infection, when replication had begun (Ishov and Maul, 1996). In full agreement with these findings, two of the viral tegument proteins (that enter the cell as part of the infecting virion), when expressed as fluorescent fusion proteins, were found to accumulate together very early after infection in nuclear dots that were distinct from, but closely adjacent to, ND10 (Hutchinson *et al.*, 2002). Similarly, using the same technology, foci of the early transcription factor, ICP4, were seen to establish adjacent to the foci of ICP0 that marked ND10, and some of these developed into replication centres (Everett *et al.*, 2003). Such foci were not seen in ICP4 expression vector-transfected cells, indicating that they require other factors found only in infected cells for their formation; these might plausibly include the tegument proteins that form similar foci (Hutchinson *et al.*, 2002). Thus, it is clear that sites immediately adjacent to ND10 represent the initial location of incoming HSV-1 genomes and associated proteins, and that these recruit newly expressed viral proteins as they develop into virus replication centres.

Progression of incoming genomes into active replication centres is favoured by their localization close to ND10. Using a fluorescent protein fusion with a transcription factor that bound specifically to the DNA of an HSV-1 artificial infectious replicon, those replicons that had associated with ND10 were found to be more likely to progress than those that had not (Sourvinos and Everett, 2002). In this same study, replicons that carried an active, immediate-early transcription factor-dependent gene were more likely to make an association with ND10 in the first place. However, this finding is in apparent conflict with the earlier evidence that association of HSV genomes with ND10 did not require new gene expression from the virus. Either way, the central finding is that juxtaposition of infecting genomes with ND10 appears beneficial to the outcome of infection.

How do these various implications fit with studies of ICP0 mutants *in vivo*? Such mutants grew very poorly in comparison with wild-type in a mouse eye model, but growth was substantially (although not completely) restored in mice lacking type I interferon receptors (Leib *et al.*, 1999), indicating that one function of ICP0 is to overcome a limitation to infection due to an interferon response. Similarly, ICP0 mutants were hypersensitive to interferon when grown on a cell line that was restrictive for such mutants (Vero) but not in U2OS cells, where the ICP0-null phenotype is minimal (Mossman *et al.*, 2000). Also, ICP0 was necessary and sufficient to block the induction of an interferon response by HSV infection in cell culture (Eidson *et al.*, 2002) and was shown to block IRF3 and IRF7-mediated activation of interferon-responsive genes via its RING domain (Lin *et al.*, 2004). Hence, it is clear that ICP0 does counteract the interferon response of its host.

Is the role of ICP0 in countering interferon linked to its effects on ND10? First, there was a correlation between loss of polyubiquitin induction by ICP0 RING mutations and loss of ability to support virus growth in restrictive cell lines (Everett, 2000a), the same situation that creates interferon hypersensitivity (Mossman *et al.*, 2000). Second, both arsenic trioxide (which causes hypersumoylation of PML and enlarged ND10) and the proteasome inhibitor MG132 (which blocks PML degradation) impaired the establishment of replication centres and the growth of virus, especially at low multiplicity of infection, where it has been noted that ICP0 function is most important (Burkham *et al.*, 2001). Third, the inhibitory effect of interferon on viral gene expression and virus yield was impaired in PML-null as compared with PML-positive mouse fibroblasts (Chee *et al.*, 2003). These results suggest that it is the PML degradation activity of ICP0 that is required to facilitate virus growth and, by extension, to overcome the interferon response. However, one conflicting piece of data is that, although overexpression of PML-VI blocked the dispersal of ND10 in response to infection, it did not affect virus yield (Lopez *et al.*, 2002). Possibly, PML isoforms differ in their importance in the interferon response to infection and thus PML-VI overexpression is not able to mimic that response.

8.4.2   Human cytomegalovirus (HCMV)

The second herpesvirus to be shown to interact with ND10 was HCMV, also a nuclear replicating DNA virus but from the *betaherpesvirus* subfamily; it causes mainly inapparent infections. Infection of normal fibroblasts by HCMV dispersed ND10, as viewed by PML fluorescence, to give a diffuse nuclear stain (Kelly *et al.*, 1995); this effect depended on immediate-early or early viral protein synthesis. This result was subsequently repeated and the effect attributed to the immediate early protein IE1 (Ahn and Hayward, 1997; Korioth *et al.*, 1996). After HCMV-induced ND10 dispersal, PML has been reported to associate with mitotic chromosomes (Wilkinson *et al.*, 1998). This latter effect may be mediated via IE1, which shows a similar affinity (LaFemina *et al.*, 1989). Both IE1 and IE2 localized transiently to ND10, prior to their disruption (Ahn and Hayward, 1997; Ishov *et al.*, 1997; Korioth *et al.*, 1996); this emerging structure in the infected cell is a site of viral transcription (Ishov *et al.*, 1997). Viral replication domains subsequently develop from these sites, adjacent to former ND10 (Ahn *et al.*, 1999). The tegument protein pp71 also localized to ND10 rapidly after infection and in the absence of new protein synthesis (Hofmann *et al.*, 2002; Ishov *et al.*, 2002; Marshall *et al.*, 2002). All these findings are very reminiscent of those for HSV-1 and its equivalent proteins.

HCMV IE1 interacts directly with PML and the sequences required for this interaction have been mapped (Ahn *et al.*, 1998; Wilkinson *et al.*, 1998). As with HSV-1 ICP0, expression of IE1 by transfection caused the loss of SUMO-modified forms of PML, in this case an epitope-tagged PML-IV, and Sp100 (Müller and Dejean, 1999). However, unlike HSV-1, IE1 has no RING domain, and disruption of ND10 by HCMV infection was not dependent on proteasome activity (Xu *et al.*, 2001). Thus, the biochemical basis for ND10 disruption by IE1 is unclear. Both IE1 and IE2 are modified by SUMO-1 (Hofmann *et al.*, 2000; Spengler *et al.*, 2002; Xu *et al.*, 2001); however, for IE1 at least, this modification was not necessary for it to be targeted to ND10, nor for their disruption. Sumoylation seems to be a common feature of proteins that localize to ND10 but, with the possible exception of PML (see Section 8.2.2), this appears to be a consequence rather than a cause of such localization. In support of this model for IE1 sumoylation, overexpressing PML-VI prolonged the association of IE1 with ND10, delaying their dissociation (Ahn and Hayward, 2000), and enhanced IE1 sumoylation (Xu *et al.*, 2001). The association of pp71 with ND10 is mediated not directly by PML, but via Daxx. Pp71 bound to Daxx in yeast two-hybrid analysis (Hofmann *et al.*, 2002) and in cells from Daxx knockout mice, pp71 remained nucleoplasmic in the presence of otherwise normal ND10 (Ishov *et al.*, 2002).

Several findings suggest that these interactions of HCMV with ND10 are functionally significant. First, IE1 binding blocked the transcription repression activity of PML-VI (Xu *et al.*, 2001). Second, the overexpression of PML-VI that delayed ND10 disruption also led to impaired establishment of replication centres and reduced production of early and late proteins (Ahn and Hayward, 2000). Third, Daxx enhanced the activation of immediate-early gene expression by pp71 (Hofmann *et al.*, 2002) and, in a pp71-null virus infection, the viral transcription domains marked initially by IE2 localization were no longer associated with ND10 (Ishov *et al.*, 2002). Overall, these data support a model where, similar to HSV-1 infection, ND10 harbour inhibitors of HCMV infection but are nonetheless used as sites for the establishment of viral replication centres.

8.4.3   Epstein–Barr virus (EBV)

EBV, from the *gammaherpesvirus* subfamily, infects B lymphocytes and causes infectious mononucleosis (glandular fever). It is also a factor in several human cancers. In common with all herpesviruses, it has both lytic and latent phases to its life cycle but, in contrast to HSV-1, its latent phase is characterized by a distinct mechanism of viral DNA replication and by expression of a set of proteins that are necessary for establishment and maintenance of the latent state. The first report of an EBV interaction with ND10 concerned one of these proteins, EBNA-5 (also known as EBNA-LP), which co-localized with PML in EBV-positive lymphoblastoid cell lines by immunofluorescence without disrupting ND10 (Szekely *et al.*, 1996). More recently another EBV latency protein, EBNA-3C, was shown to localize to ND10 and to displace PML upon heterologous expression in HeLa cells (Rosendorff *et al.*, 2004). Other reports have focused more on the lytic phase of infection and its induction from the latent state. Unlike latent infection, an active lytic infection caused disruption of

ND10 (Bell *et al.*, 2000b); different ND10 components became dispersed at different times after activation of infection. Using *in situ* hybridization, these authors also showed a corresponding association of viral DNA and developing replication centres with ND10 in lytic infection, but no association of EBV replicons with ND10 in latent infection. The principal transcriptional inducer of the lytic pattern of EBV gene expression (BZLF-1) was sufficient to induce ND10 disruption (Adamson and Kenney, 2001). Like HCMV IE1, BZLF-1 was sumoylated and its expression resulted in reduced levels of SUMO-PML (Adamson and Kenney, 2001). By mutational analysis, there was a correlation between the ability of this protein to disperse Sp100 from ND10 and its ability to activate transcription via interaction with the ND10 component, CBP (Deng *et al.*, 2001). Thus, the targeting of ND10 during EBV lytic infection and the establishment there of replication centres seems to mimic findings for HSV-1 and HCMV; viruses from all three of the *Herpesvirus* subfamilies therefore interact with ND10 during infection and establish replication centres in their vicinity.

8.4.4   Other human herpesviruses

Human herpes virus 6 (HHV6, a betaherpesvirus) IE1 protein associated with ND10, both when transfected and expressed during infection, and became sumoylated (Gravel *et al.*, 2002, Stanton *et al.*, 2002). However, the effect on ND10 was disputed. While one study claimed some reduction in the number of ND10 during HHV6 infection (Gravel *et al.*, 2002), the other showed no rearrangement of PML in ND10, with co-localization of IE1 persisting into the late phase of infection (Stanton *et al.*, 2002). Human herpesvirus 8 (HHV8, a gammaherpesvirus also known as Kaposi's sarcoma herpesvirus) also encodes a protein, K8, that localizes to ND10, and establishes replication domains in association with these structures (Katano *et al.*, 2001; Wu *et al.*, 2001a). These results support the conclusion that ND10 targeting is a general feature of herpesvirus infections.

Parkinson and Everett (2000) surveyed ICP0 homologues from Varicella-zoster virus, Bovine herpesvirus 1, Equine herpesvirus 1 and Pseudorabies virus, all alphaherpesviruses, in comparison with HSV-1 and found that each caused degradation of endogenous Sp100 and PML, although only ICP0 itself caused degradation of overexpressed exogenous PML. This array of ICP0 homologues also all induced conjugated polyubiquitin upon transfection (Parkinson and Everett, 2001). These results suggest that broad parallels exist among the alphaherpesviruses as to the mechanism of ND10 targeting and disruption by their respective immediate-early proteins.

8.4.5   Adenoviruses

Adenoviruses are nuclear-replicating DNA viruses that are smaller than herpesviruses. The family comprises both mammalian and avian viruses, and within the former are more than 50 human serotypes, each with similar molecular organization, although they cause a variety of pathologies including respiratory, gut and eye infections. Human adenovirus type 5 (Ad5), from subgroup C, is the most widely studied model. In the context of ND10, the first relevant observation was of PML reorganization by Ad5 infection into what are now regarded as characteristic track-like structures (Puvion-Dutilleul *et al.*, 1995). These results were confirmed and extended, using a panel of mutant Ad5, to show that the Orf3 protein from the E4 transcription unit was responsible for this reorganization of ND10, during which PML and Sp100 become separated (Carvalho *et al.*, 1995; Doucas *et al.*, 1996). Moreover, Orf3 localized with PML in these reorganized tracks and exogenous expression of Orf3 alone was sufficient for both reorganization and co-localization (Carvalho *et al.*, 1995; Doucas *et al.*, 1996). Orf3 of Ad from subgroup A, B and D, also reorganized ND10 (Hosel *et al.*, 2001; Evans and Hearing, 2003), suggesting that this is a feature of all Orf3 proteins.

E4 Orf3 is not the only Ad protein that is found at ND10 or reorganized ND10 during infection. The E1A 13S protein, which serves as a transcriptional activator for other viral genes, was found in association with the ND10 following infection or transfection, although it did not appear to affect the integrity of the structures (Carvalho *et al.*, 1995). The larger E1B protein, E1B 55K, also localized transiently to reorganized ND10 during infection, and this localization was prolonged throughout infection when E4 Orf6, a known physical partner of E1B 55K, was absent (König *et al.*, 1999; Leppard and Everett, 1999). Furthermore, E4 Orf3 was required for E1B 55K to localize to reorganized ND10 (Lethbridge *et al.*, 2003), a result probably related to the ability of these two proteins to interact (Leppard and Everett, 1999). Thus, it appears that Orf3 mediates the localization of 55K to ND10 and that Orf6, an alternative partner protein for 55K, antagonizes this localization. Ultimately, E1B 55K/E4 Orf6 ends up in the periphery of virus replication centres (Ornelles and Shenk, 1991), a localization that is believed to be linked to its role in mediating viral RNA export from the nucleus (Dobner and Kzhyshkowska, 2001; Leppard, 1998). As with other proteins that show ND10 association, a fraction of E1B 55K was modified by SUMO-1 during infection (Endter *et al.*, 2001). The amount of SUMO-55K was increased in the absence of E4 Orf6, although this was true irrespective of whether the virus expressed E4 Orf3, discounting the possibility that modification correlated directly with the extent of ND10 association of the protein (Lethbridge *et al.*, 2003).

Ad5 establishes round or goblet-shaped replication domains within the nucleus of infected cells (Pombo *et al.*, 1994). Using an E4 Orf3 deletion mutant so that intact ND10 were retained during infection, it was possible to demonstrate that these replication domains were located adjacent to, or closely associated with, ND10 (Ishov and Maul, 1996). Similarly, monitoring the fate of input wild-type Ad5 genomes prior to ND10 disruption, it was shown that very early in infection (1.5 h post-infection) viral DNA locations had no obvious correlation with ND10, but that by 4 h post-infection, still well before the onset of DNA replication, ~ 50% of the input viral DNA was found close to ND10 (Ishov and Maul, 1996); this association was independent of new gene expression. As ND10 occupy less than 0.05% of the total nuclear volume, this result is highly significant and indicates a preferential association of viral DNA and establishing replication centres with ND10, similar to that seen for various other nuclear-replicating DNA viruses.

Of course, in an infection by wild-type virus, ND10 are being disrupted by the time replication begins, so what is the normal localization of ND10 components relative to the replication centres? While Sp100, PML and other ND10-associated proteins were initially redistributed to tracks through the action of Orf3, most of these proteins (notably Sp100) eventually segregated from the tracks and joined the replication domains as DNA replication proceeded (Ishov and Maul, 1996); of the antigens tested, only PML did not at any point join the replication domains. Thus, Ad5 replication domains are established adjacent to ND10 consequent upon the localization of input genomes and subsequently accumulate many former ND10 components.

There are two reports linking Ad structural proteins with PML. First, using a virus that lacked expression of the fibre protein, which normally forms the spikes at the vertices of the virion, accumulation of the other major capsid proteins was observed in amorphous nuclear inclusions that included PML and Sp100 (Puvion-Dutilleul *et al.*, 1999). The significance of these observations is uncertain, but it is interesting to note the more extensive data that links ND10 to papillomavirus assembly (Section 4.6). Second, the minor capsid protein IX associates with PML during the late phase of infection (Rosa-Calatrava *et al.*, 2001), with PML moving from the tracks formed earlier into spherical structures surrounded by protein IX (Rosa-Calatrava *et al.*, 2003), leading the authors to hypothesize that IX maintains the inactivation of PML, established earlier by Orf3, through the late phase of infection.

The mechanism by which Ad5 Orf3 causes ND10 disruption is still unclear. Western blotting for PML in samples from cells infected by various Ad5 mutants showed that E4 Orf3 expression correlated with a gradual loss of the high molecular weight SUMO-modified PML isoforms and the appearance of novel, infection-specific forms, one of which became a major species by late infection (Leppard and Everett, 1999; K. Lethbridge and K.N.Leppard, unpublished). Lack of either E1B 55K or E4 Orf6 proteins had no impact on these infection-induced changes to PML. It is tempting to hypothesize that this biochemical effect on PML and the effect of Orf3 on ND10 organization are manifestations of the same Orf3 activity, and that they might even be causally linked. However, expression of Ad5 Orf3 by transfection had no effect on the sumoylation state of exogenously overexpressed PML-IV (Müller and Dejean, 1999). Either the hypothesis that altered PML modification causes ND10 reorganization is incorrect, and Orf3 requires some other infection-specific factor in order to affect PML modification state, or else a PML isoform other than PML-IV is the significant target of Orf3-mediated modification changes.

Interestingly, the major novel PML species observed subsequent to Ad5 infection has marked similarity to a form of PML observed during the normal cell cycle at mitosis when the ND10 temporarily disperse (see Section 8.2.4). Indeed, the overall pattern of PML isoforms observed during late Ad5 infection is very reminiscent of the pattern seen in cells synchronized at mitosis (Everett *et al.*, 1999b). Possibly the Ad5 effect on ND10 is consequent upon changes to cell cycle regulation that the infection imposes, or perhaps Ad5 simply uses a similar molecular trigger to cause changes in PML organization, e.g. phosphorylation, as the cell uses during the cell cycle.

ND10 disruption is not essential for Ad replication as E4 Orf3-deficient viruses grow similarly to wild-type (Halbert *et al.*, 1985). However, in an otherwise completely E4-deficient virus, which grows very poorly, E4 Orf3 stimulates viral growth more than 2000-fold (Huang and Hearing, 1989). Enhancing ND10 size inhibits their disruption by E4 Orf3 and decreases viral DNA replication (Doucas *et al.*, 1996), suggesting that the ND10 reorganization function of E4 Orf3 is at least partly responsible for its ability to enhance viral growth. Until recently, it was thought that the key target of Orf3 during this reorganization was double strand DNA break repair (DSBR). DSBR is activated by Ad infection and components of the Mre11-Rad50-NBS (MRN) complex, which is required for DSBR, associate with ND10 when the pathway is stimulated (Lombard and Guarente, 2000). E4 Orf3, which disrupts MRN rearrangement, or an alternate function provided by E1B 55K / E4 Orf6, blocks DSBR (Stracker *et al.*, 2002); when both functions are missing, active DSBR ligates Ad genomes into dead-end products. However, it is now known that only Ad5 E4 Orf3 can target MRN (Stracker *et al.*, 2005) while, as already discussed, Orf3 from several serotypes can rearrange ND10. Moreover, the MRN and ND10 rearrangement functions have been separated by mutation and it is the former Orf3 activity that stimulates virus growth in culture (Evans and Hearing, 2005). The conserved effect of Orf3 on ND10 must play some other role that is only significant *in vivo*.

8.4.6 Papillomaviruses

Papillomaviruses (PV) are small DNA viruses that cause skin and genital warts, some of the latter infections being associated with the development of cervical cancer. They have biphasic replication cycles in epithelia. Infection initiates and is maintained in the basal cell layers of the epithelium, with expression of a subset of viral proteins and episomal DNA replication. However, vegetative DNA replication, late gene expression and progeny particle production only occur in differentiating cells as they progress upwards through the epithelium. Most of the data that relate PV infection to ND10 concern this latter phase of the life cycle. Both the major (L1) and minor (L2) capsid proteins of either bovine PV1 or human PV33 localized to ND10 when exogenously expressed (Day *et al.*, 1998; Florin *et al.*, 2002a). L1 localization was dependent on the presence of L2 (Day *et al.*, 1998) and L2 arrived at these structures earlier than L1 when co-expressed from vaccinia virus recombinants (Florin *et al.*, 2002a). These results suggest that ND10 serve as sites of PV assembly, although it has been reported recently that L2 forms nuclear foci, recruits L1 and forms virus-like particles in PML-null fibroblasts (Becker *et al.*, 2004). L2 expression also leads to changes in ND10 composition. Daxx was recruited to ND10, and Sp100 displaced and degraded, by expression of L2 (Florin *et al.*, 2002b). Finally, overexpression of the E4 protein (also a late infection phase product) formed nuclear inclusions to which PML relocalized and similar inclusions were seen in infected wart tissue (Roberts *et al.*, 2003). These structures may be related to those that develop from ND10 in the presence of L1 and L2.

The PV E1 and E2 proteins are involved in regulation of PV gene expression (E2) and viral DNA replication (E1, E2) and are required in both phases of the life cycle. L2 was reported to mediate the recruitment of bovine PV1 E2, but not E1, into ND10 (Day *et al.*, 1998). However, both E1 and E2 of human PV11 were found to localize adjacent to ND10 when expressed in the presence of PV replication origin-containing plasmid DNA, with localization of both proteins being to some extent dependent on the other (Swindle *et al.*, 1999); E1 and E2 are known to interact and to bind together to the PV origin of replication. Establishment of these active PV replication compartments was accompanied by some fragmentation of ND10 into a larger number of smaller bodies, judged by PML staining (Swindle *et al.*, 1999). In common with many other proteins that localize to ND10, human PV1 and PV18 E1 proteins were sumoylated *in vitro* but, in contrast to most other examples, this modification appeared to be necessary for E1 localization to ND10, since a non-sumoylatable mutant, E1, had altered localization and impaired replication capacity (Rangasamy *et al.*, 2000). Overall, it appears that PV replication centres, like those of herpesviruses and adenoviruses, are established in the vicinity of ND10, although in this case disruption to ND10 during infection is more subtle than for the other viruses.

Recent reports have linked the early PV proteins E6 and E7 to ND10. These proteins cause p53 degradation and inactivate the cell cycle regulator pRb and are important in maintaining infection in the basal cells of an epithelium. E6 and E7 from human PV11 partially co-localized with ND10, whereas the equivalent proteins from human PV18 showed diffuse nuclear staining (Guccione *et al.*, 2002). When specific PML isoforms were overexpressed, both these E6 proteins could associate with ND10 formed by some, but not all, of the isoforms and the HPV18 E6 could mediate degradation of PML-IV, an isoform associated with cellular senescence (Guccione *et al.*, 2004). It is tempting to speculate that differences between the interactions of the E6/E7 proteins of these two viruses and ND10/PML are related to the difference in oncogenicity of the two viruses in the genital tract.

8.4.7   Other DNA viruses

In common with HSV-1 and adenovirus, incoming genomes of the monkey polyoma-virus, SV40, were found to localize adjacent to ND10 and to begin to replicate in that location (Ishov and Maul, 1996). Subsequently it was shown that SV40 replication, but not transcription, required localization to ND10. SV40 plasmid replicons bearing a reporter gene were carried to ND10 only when they contained a minimal SV40 origin including T antigen binding site II and SV40 large T antigen was present (Tang *et al.*, 2000). Under such circumstances, reporter transcription was also localized to ND10 but equivalent levels of expression were also seen from replicons that could not localize to ND10, with transcription diffuse in the nucleoplasm. These results fit the general pattern for other mammalian DNA viruses, in which replication centres are established adjacent to ND10, consequent upon the initial targeting of input genomes to these sites.

Finally, two studies have been made of the interaction of insect baculoviruses with ND10. Using either heterologous expression of mammalian PML to create 'ND10' in insect cells, or else dead-end infection of mammalian cells, two baculovirus early proteins were found partially to co-localize with ND10 (Murges *et al.*, 2001). Similarly, baculovirus IE2 was found to localize to replication domains that were established adjacent to ND10 (Mainz *et al.*, 2002). It is important to appreciate that insects do not appear to possess a PML gene homologue and insect cells are not known to contain ND10. Thus, the significance of these findings is very unclear.

8.4.8 Retroviruses

Retroviruses are characterized by reverse transcription of their RNA genomes to form DNA that is inserted into the host chromosomes. Once infection is established in a cell, host cell transcription is used to produce progeny genomes. Viruses from three different genera of the *Retrovirus* family have been suggested to interact, in different ways, with ND10 or PML.

Human T cell lymphotropic virus (HTLV)-1, which causes tropical spastic para-paresis and adult T cell leukaemia, encodes a transcriptional activator (Tax) that is also the viral transforming protein. Tax interacts directly with the cellular protein Int-6, as judged by yeast and mammalian two-hybrid analysis and co-immunoprecipitation (Desbois *et al.*, 1996). Int-6, which is one of the targets of insertional mutagenesis in tumours caused by another retrovirus, MMTV (Marchetti *et al.*, 1995), normally co-localizes with PML in ND10 but is retained in the cytoplasm by its interaction with Tax (Desbois *et al.*, 1996). Interestingly, Tax inhibits the potentiation of the activity of steroid hormone receptors by PML-VI (Doucas and Evans, 1999), suggesting that the interaction of Tax with ND10 is functionally significant. One model based on these observations is that the transcription potentiation activity of overexpressed PML requires in some way the resulting enhanced Int-6 localization at ND10, and that Tax inhibits the activity by relocating Int-6 away from ND10.

Another retrovirus shown recently to target ND10 is human foamy virus (HFV). This virus is not known to be pathogenic in man and is now thought to be a zoonotic infection (reviewed by Linial, 2000). HFV encodes a transcriptional activator, Tas, which, unlike HTLV-1 Tax, operates through direct binding of elements in the two viral promoters. Overexpression of PML-III substantially inhibited HFV infection at an early stage through a direct interaction between Tas and the RING domain of PML, which caused localization of Tas to ND10 and inhibited its binding to viral promoter sequences (Regad and Chelbi-Alix, 2001). A similar effect on HFV infection was achieved by pre-treating the cells with interferon. This effect required PML, since interferon could not inhibit infection in PML-null cells (Regad and Chelbi-Alix, 2001), and might therefore be mediated by PML induction (see Section 8.2.5).

The third retrovirus on which studies relating to PML/ND10 have been reported is human immunodeficiency virus type 1 (HIV-1), the causative agent of AIDS. Turelli *et al*. (2001) reported that ND10 were transiently disturbed by HIV-1 infection, with PML being translocated to the cytoplasm at 2 h post-infection, where it associated with incoming viral pre-integration complexes. Movement of PML to the cytoplasm was sensitive to the inhibitor of nuclear export, leptomycin B, and treatment with arsenic trioxide, which would be expected to oppose movement of PML out of ND10, enhanced viral infection. This latter result suggested that PML was acting as part of an antiviral response, opposing the establishment of infection. However, findings from two other groups have contradicted these results. Bell *et al*. (2001) found no effect of HIV-1 on ND10 at any time post-infection and saw no co-localization of PML with preintegration complexes. Similarly, Berthoux *et al*. (2003) found no effect of HIV-1 on PML, Sp100 or Daxx staining. These authors did find an effect of arsenic trioxide on infection, but the effect was independent of PML, since a GFP-transducing virus infected normal and PML-null fibroblasts equally well and infectivity was enhanced by arsenic trioxide to the same extent in both cell types. Thus, the effect of HIV-1 on ND10 remains uncertain.

8.4.9   RNA viruses

As discussed above, a wide variety of viruses with DNA genomes that target the cell nucleus as part of their replication cycle have been found to have effects on ND10. More surprisingly, an increasing number of RNA viruses are being shown to make interactions with these structures. In particular, there is evidence that links members of several different families of negative-strand (segmented and non-segmented) and ambisense RNA viruses to ND10, as discussed below.

The best characterized of the RNA virus interactions with ND10 is that of the arenavirus, lymphocytic choriomeningitis virus (LCMV), which is studied as a model system in mice. PML appears to inhibit the replication of this virus, since PML knockout mice showed enhanced LCMV pathogenesis and fibroblasts from these mice supported enhanced virus growth (Bonilla *et al*., 2002; Djavani *et al*., 2001). As LCMV is sensitive to interferon, a plausible hypothesis in the light of these data was that this sensitivity might be mediated through induction of PML; however, PML and interferon acted additively in depressing virus growth, suggesting this was not the case (Djavani *et al*., 2001). LCMV infection disrupts ND10 via the action of a small non-structural protein, Z, which is a zinc-binding RING-domain protein. Z interacted with PML and relocated it to the cytoplasm (Borden *et al*., 1998b) although, from the biological data, it is clear that this is not sufficient to block the antiviral effect of PML. Indeed, Z action on ND10 may actively contribute to a negative regulation of virus growth, since Z has been shown to inhibit transcription and replication from minigenomes of either LCMV (Cornu and de la Torre, 2001) or the related Tacaribe virus (Lopez *et al.*, 2001).

Further studies of the action of Z suggest that it could affect translation. First, its cytoplasmic relocalization of PML seems selectively to spare the subset of PML that interacts with a nuclear fraction of ribosomal P proteins. While most Z and PML was cytoplasmic during an infection, a minor subset of both proteins remained co-localized with the P proteins in nuclear foci (Borden *et al.*, 1998a). Second, eIF4E, another protein involved in translation that partially co-localizes with PML in ND10 (Lai and Borden, 2000), also interacted with Z, with the result that translation of certain mRNA was impaired (Campbell Dwyer *et al.*, 2000). Both Z and PML were shown subsequently to inhibit directly the activity of eIF4E (Kentsis *et al.*, 2001). Possibly, these activities of Z serve to direct translation activity away from mRNA whose products would be detrimental to the outcome of infection.

Infection with Influenza A virus, an orthomyxovirus, caused an increase in the number and staining intensity of ND10 in HeLa cells, as defined by Sp100 antibodies (Guldner *et al.*, 1992), similar to that observed upon interferon treatment of cells. Moreover, a Chinese hamster ovary (CHO) cell line overexpressing PML-III was resistant to low multiplicity influenza A virus infection when compared with parental cells, although overexpressing Sp100 had no effect (Chelbi-Alix *et al.*, 1998). This effect did not involve either secreted interferon activity or the interferon-induced antiviral mediator, MxA, that is known to inhibit influenza A virus. More recently, three viral proteins, M1, NS1 and NS2, were found to co-localize with PML during infection (Sato *et al.*, 2003).

Productive infection by Vesicular stomatitis virus (VSV), a rhabdovirus, was also found to be inhibited by PML-III overexpression in CHO cells; viral RNA and protein synthesis were examined and found to be reduced (Chelbi-Alix *et al.*, 1998). Also, VSV infection had no effect on the localization of PML-III in infected, overexpressing CHO cells (Blondel *et al.*, 2002). In contrast, PML-III overexpression had no effect on growth of the related rabies virus in CHO cells. Infection induced a significant increase in PML staining intensity in ND10 and caused changes in their organization, revealed by electron microscopy. This was attributed to a direct and specific physical association between PML and the viral P and P3 proteins, the latter being particularly significant since, when expressed alone, it localized to ND10 (Blondel *et al.*, 2002). A tentative interpretation of these data is that ND10 exert an antiviral effect, which rabies virus counters successfully via its P proteins, while VSV does not.

There has been one report of an interaction between a hantavirus and ND10. Puumala virus N protein, the principal nucleocapsid protein, bound to the ND10 component Daxx in yeast and mammalian two-hybrid, GST pull-down and co-immunoprecipitation assays. The N protein also co-localized with overexpressed Daxx (but not endogenous Daxx) in nuclear foci (Li *et al.*, 2002). However, overexpression seems to direct many proteins to ND10; indeed, this has been used as evidence for ND10 functioning as nuclear depots for proteins not currently required (Negorev and Maul, 2001). Thus, concluding that there is a positive functional significance to such localizations is difficult.

Finally, hepatitis D virus has been reported to interact with ND10. This virus requires a second virus, hepatitis B, to grow, and exacerbates the liver disease caused by hepatitis B. It encodes only two sequence-related proteins, the S and L antigens, and is thought to employ host RNA polymerase II for its RNA synthesis in the nucleus. Using an artificial replication system, antigenome RNA (the template for genome synthesis), but not genome RNA, exactly co-localized with L antigen in nuclear foci and about half of these coincided with PML foci that were generally reduced in number (Bell *et al.*, 2000a). Altered Sp100, Daxx and SUMO-1 localization gave further evidence of an alteration to ND10 organization in these cells. This apparent establishment of replication centres adjacent to ND10, which are then altered as the molecular events of infection proceed, is very reminiscent of the interaction of many DNA viruses with ND10.

8.4.10   Viral non-interactions with ND10

In considering the general significance of ND10 targeting by viruses, cases where viruses do not apparently interact with these structures are just as important as those that do. Two reports concerning parvoviruses have suggested a lack of involvement with ND10. Virus H1 induced nuclear structures, presumptively replication centres, where viral DNA and the NS1 protein localized and to which the DNA polymerase processivity factor PCNA was recruited. However, unlike the case of DNA viruses such as HSV-1 (see Section 8.4.1), these replication centres were spatially distinct from ND10, which remained unaltered (Cziepluch *et al.*, 2000). The related virus MVM (minute virus of mice) produced similar results at a comparable time point, but with further progression of infection, co-localization of NS1 with PML in large aggregates was seen, along with gross disorganization of the nucleus; expression of NS1 alone had no effect on ND10 (Young *et al.*, 2002).

Finally, one study used the picornavirus EMCV, in comparison with influenza A and VSV (see Section 8.4.9) and found that, in contrast to these latter two viruses, EMCV growth was unaffected by the overexpression of PML in CHO cells (Chelbi-Alix *et al.*, 1998). This, however, is not proof of a 'non-interaction' and further work is needed to assess the relationship, if any, between picornaviruses and ND10. Certainly, picornaviruses do interact with the nucleus and other subnuclear structures (see Chapters 5 and 7).

8.5   Why do viruses interact with and modify ND10?

In the absence of evidence, two principal hypotheses could be advanced as to the purpose of virus interactions with ND10. First, ND10 might harbour components or provide function(s) that inhibit the growth of virus and which are therefore targeted and inactivated by a virus in order to maximize its growth potential. Alternatively,

ND10 might contain components or provide functions that are of positive value to the virus in furthering its growth cycle, and so a virus finds it advantageous to associate with them. Of course, there is no reason why both should not be true. It needs also to be allowed that, despite the temptation to generalize to a single basis for the diverse interactions between viruses and ND10 that are seen, individual viruses may target ND10 for very different reasons.

Among the viruses discussed in Section 8.4, there are many differences in the way ND10 are modified by infection. In some cases the bodies are completely dispersed, while in others only selected components are altered in localization. This diversity can be used to argue against there being any common component of ND10 that viruses require; indeed, the diversity of virus replication strategies would make this implausible anyway. There is also likely to be diversity in the mechanisms by which ND10 modification is achieved during infection. Although evidence so far is limited for most examples, it is probably safe to say that proteasome-mediated destruction of the core ND10 component, PML, which is clearly a feature of the interaction of HSV-1 with ND10, does not occur generally as a mechanism for ND10 alteration beyond the *alphaherpesvirus* subfamily. Thus, many other mechanisms remain to be fully elucidated.

So, if one were to assume that the plethora of viruses that target ND10 all do so for the same reason and were forced to speculate on that reason, attention should focus on functions of ND10 that these viruses might need to disable, and here thoughts would surely be drawn towards the role of ND10 in the interferon response. Activating the interferon pathway is understood to be a key early cell response to virus infection that has evolved precisely to limit virus growth. Not only are ND10 clearly implicated in that response, but there is evidence for several viruses that have distinct genome types and replication strategies, including HSV-1 (Section 8.4.1), HFV (Section 8.4.8) and influenza (Section 8.4.9), that ND10 or their components are important players in the interferon response as it applies to that virus. Moreover, arguing for the interferon response function of ND10 being the critical viral target explains most casily the diversity of viruses, of both nuclear and cytoplasmic replicating types, that cause changes to ND10 upon infection.

Apoptosis is also a known response to expression of certain viral proteins in cells, for example Ad5 E1A, via deregulation of the cell cycle and activation of a DNA damage response, and premature apoptosis of an infected cell would be sufficient to drive evolution of anti-apoptotic functions by the virus. Given the proposed role of ND10 in mediating apoptosis and the DNA damage response, could the primary target for ND10 disruption be these functions? In some cases, it might be so – p53 is a key ND10-localizing player in these processes and several viruses that target ND10 also target p53. However, accepting this hypothesis necessarily excludes the simple idea that all viruses target ND10 for the same reason – not all viruses have any need to block a DNA damage response. Moreover, in those viruses where there is a well-characterized p53 inactivating function, it is either distinct from the ND10 disrupting function (Ad5, PV) or there is no disruption of ND10 yet described (SV40).

One generalization that does emerge from Section 8.4 is that the nuclear-replicating DNA viruses all deposit their input genomes close to ND10 and, as a result, establish replication centres and even assembly sites in these locations. Moreover, there is evidence that progression into active infection is favoured when events begin close to ND10. On the face of it, these results suggest that the viruses concerned target ND10 in order to gain some general benefit. This seems to run counter to the idea that ND10 are fundamentally antiviral. However, the fact that all these incoming viral genomes are deposited at ND10 does not necessarily mean that this is by the design of the virus. It is equally possible that the prime driver for this localization comes from the host. If ND10 were actively involved not only in the interferon response but also in antigen processing, what would be more logical than that the host cell should draw these incoming viruses to ND10 as part of a damage-limitation exercise? Once there, the virus would not succeed unless it could rapidly disrupt the antiviral functions that were present. Faced with this challenge, many viruses would then have evolved early proteins to target ND10 and disrupt their function in the antiviral response. Following establishment of these aspects of the virus–host interaction, there would then have been scope for the interaction to evolve further, so that components and functions present at, or displaced from, ND10 might be found some more positive role in the virus life cycle and hence be actively recruited by the virus.

These ideas are based upon the proposition that there is some single explanation for the interaction between viruses and ND10, but perhaps this is not the case. As a greater range of viruses is examined in the same detail as HSV-1 has been, so it may be shown that blocking the interferon response is not a relevant consequence of virus interaction with ND10 in some cases. This may be the case for Ad5, for example, which has a function other than its ND10 disrupting protein, the absence of which leads to severe inhibition of the virus by interferon even in the presence of the ND10-disrupting Orf3 protein (Kitajewski *et al.*, 1986). Given the diversity of proteins that apparently reside at ND10, and the breadth of ND10 functions in the uninfected cell that have been suggested, it would certainly not be unreasonable to suggest that different viruses might be getting different things out of the changes they impose on ND10.

Even if there are different reasons why viruses around today target ND10, this would not necessarily negate the basic hypothesis that these interactions arose to counter interferon. Given that the original purpose of ND10 interaction or disruption by a virus was to interfere with an aspect of the interferon response, the subsequent evolution of other viral products might make that initial function relatively less significant. Analysis of the virus–ND10 interaction as it occurs today might then reveal it to be important not so much for its original purpose but for other reasons that were initially only secondary.

In summary, there is a rapidly growing body of information concerning the relationship between viruses of mammals and ND10 structures. This is already persuasive that the interactions are of importance to the outcome of infection in many cases. Based upon the available data, we have advanced various ideas as to how the generality of virus interactions with ND10 might be explained. Hopefully, the high intensity of study into these interactions that is currently ongoing will reveal whether or not these ideas are correct.

References

Adamson, A. L. and Kenney, S. (2001). Epstein–Barr virus immediate-early protein BZLF1 is SUMO-1 modified and disrupts promyelocytic leukemia bodies. *J Virol* **75**, 2388–2399.

Ahn, J. H., Brignole, E. J. and Hayward, G. S. (1998). Disruption of PML subnuclear domains by the acidic IE1 protein of human cytomegalovirus is mediated through interaction with PML and may modulate a RING finger-dependent cryptic transactivator function of PML. *Mol Cell Biol* **18**, 4899–4913.

Ahn, J. H. and Hayward, G. S. (1997). The major immediate-early proteins IE1 and IE2 of human cytomegalovirus colocalize with and disrupt PML-associated nuclear bodies at very early times in infected permissive cells. *J Virol* **71**, 4599–4613.

Ahn, J. H. and Hayward, G. S. (2000). Disruption of PML-associated nuclear bodies by IE1 correlates with efficient early stages of viral gene expression and DNA replication in human cytomegalovirus infection. *Virology* **274**, 39–55.

Ahn, J. H., Jang, W. J. and Hayward, G. S. (1999). The human cytomegalovirus IE2 and UL112–113 proteins accumulate in viral DNA replication compartments that initiate from the periphery of promyelocytic leukemia protein-associated nuclear bodies (PODs or ND10). *J Virol* **73**, 10458–10471.

Alcalay, M., Tomassoni, L., Colombo, E., Stoldt, S., Grignani, F., Fagioli, M., Szekely, L., Helin, K. and Pelicci, P. G. (1998). The promyelocytic leukemia gene product (PML) forms stable complexes with the retinoblastoma protein. *Mol Cell Biol* **18**, 1084–1093.

Anton, L. C., Schubert, U., Bacik, I., Princiotta, M. F., Wearsch, P. A., Gibbs, J., Day, P. M., Realini, C., Rechsteiner, M. C., Bennick, J. R. and Yewdell, J. W. (1999). Intracellular localization of proteasome degradation of a viral antigen. *J Cell Biol* **146**, 113–124.

Bailey, D. and O'Hare, P. (2002). Herpes simplex virus 1 ICP0 co-localizes with a SUMO-specific protease. *J Gen Virol* **83**, 2951–2964.

Becker, K. A., Florin, L., Sapp, C., Maul, G. G. and Sapp, M. (2004). Nuclear localization but not PML protein is required for incorporation of the papillomavirus minor capsid protein L2 into virus-like particles. *J Virol* **78**, 1121–1128.

Beech, S. J., Lethbridge, K. J., Killick, N., McGlincy, N. and Leppard, K. N. (2005). Isoforms of the promyelocytic leukemia protein differ in their effects on ND10 organization. *Exp Cell Res* **307**, 109–117.

Bell, P., Brazas, R., Ganem, D. and Maul, G. G. (2000a). Hepatitis delta virus replication generates complexes of large hepatitis delta antigen and antigenomic RNA that affiliate with and alter nuclear domain 10. *J Virol* **74**, 5329–5336.

Bell, P., Lieberman, P. M. and Maul, G. G. (2000b). Lytic but not latent replication of Epstein–Barr virus is associated with PML and induces sequential release of nuclear domain 10 proteins. *J Virol* **74**, 11800–11810.

Bell, P., Montaner, L. J. and Maul, G. G. (2001). Accumulation and intranuclear distribution of unintegrated human immunodeficiency virus type 1 DNA. *J Virol* **75**, 7683–7691.

Berthoux, L., Towers, G. J., Gurer, C., Salomoni, P., Pandolfi, P. P. and Luban, J. (2003). As2O3 enhances retroviral reverse transcription and counteracts Ref1 antiviral activity. *J Virol* **77**, 3167–3180.

Bischof, O., Kirsh, O., Pearson, M., Itahana, K., Pelicci, P. G. and Dejean, A. (2002). Deconstructing PML-induced premature senescence. *EMBO J* **21**, 3358–3369.

Blondel, D., Regad, T., Poisson, N., Pavie, B., Harper, F., Pandolfi, P. P., de Thé, H. and Chelbi-Alix, M. K. (2002). Rabies virus P and small P products interact directly with PML and reorganize PML nuclear bodies. *Oncogene* **21**, 7957–7970.

Boisvert, F. M., Hendzel, M. J. and Bazett-Jones, D. P. (2000). Promyelocytic leukemia (PML) nuclear bodies are protein structures that do not accumulate RNA. *J Cell Biol* **148**, 283–292.

Boisvert, F. M., Kruhlak, M. J., Box, A. K., Hendzel, M. J. and Bazett-Jones, D. P. (2001). The transcription coactivator CBP is a dynamic component of the promyelocytic leukemia nuclear body. *J Cell Biol* **152**, 1099–1106.

Bonilla, W. V., Pinschewer, D. D., Klenerman, P., Rousson, V., Gaboli, M., Pandolfi, P. P., Zinkernagel, R. M., Salvato, M. S. and Hengartner, H. (2002). Effects of promyelocytic leukemia protein on virus–host balance. *J Virol* **76**, 3810–3818.

Borden, K. L. B. (2000). RING domains: Master builders of molecular scaffolds? *J Mol Biol* **295**, 1103–1112.

Borden, K. L. B. (2002). Pondering the promyelocytic leukemia protein (PML) puzzle: possible functions for PML nuclear bodies. *Mol Cell Biol* **22**, 5259–5269.

Borden, K. L. B., Campbell Dwyer, E. J., Carlile, G. W., Djavani, M. and Salvato, M. S. (1998a). Two RING finger proteins, the oncoprotein PML and the arenavirus Z protein, colocalize with the nuclear fraction of the ribosomal P proteins. *J Virol* **72**, 3819–3826.

Borden, K. L. B., Dwyer, E. J. C. and Salvato, M. S. (1998b). An arenavirus RING (zinc-binding) protein binds to the oncoprotein promyelocytic leukemia protein (PML) and relocates PML nuclear bodies to the cytoplasm. *J Virol* **72**, 758–766.

Borden, K. L. B., Lally, J. M., Martin, S. R., O'Reilly, N. J., Solomon, E. and Freemont, P. S. (1996). *In vivo* and *in vitro* characterization of the B1 and B2 zinc-binding domains from the acute promyelocytic leukemia protooncoprotein PML. *Proc Natl Acad Sci USA* **93**, 1601–1606.

Boutell, C., Sadis, S. and Everett, R. D. (2002). Herpes simplex virus type 1 immediate-early protein ICP0 and its isolated RING finger domain act as ubiquitin E3 ligases *in vitro*. *J Virol* **76**, 841–850.

Burkham, J., Coen, D. M., Hwang, C. B. C. and Weller, S. K. (2001). Interactions of herpes simplex virus type 1 with ND10 and recruitment of PML to replication compartments. *J Virol* **75**, 2353–2367.

Campbell Dwyer, E. J., Lai, H. K., MacDonald, R. C., Salvato, M. S. and Borden, K. L. B. (2000). The lymphocytic choriomeningitis virus RING protein Z associates with eukaryotic initiation factor 4E and selectively represses translation in a RING-dependent manner. *J Virol* **74**, 3293–3300.

Carbone, R., Pearson, M., Minucci, S. and Pelicci, P. G. (2002). PML NBs associate with the hMre11 complex and p53 at sites of irradiation induced DNA damage. *Oncogene* **21**, 1633–1640.

Carvalho, T., Seeler, J. S., Öhman, K., Jordan, P., Pettersson, U., Akusjärvi, G., Carmofonseca, M. and Dejean, A. (1995). Targeting of adenovirus E1A and E4-ORF3 proteins to nuclear matrix-associated PML bodies. *J Cell Biol* **131**, 45–56.

Chee, A. V., Lopez, P., Pandolfi, P. P. and Roizman, B. (2003). Promyelocytic leukemia protein mediates interferon-based anti-herpes simplex virus 1 effects. *J Virol* **77**, 7101–7105.

Chelbi-Alix, M. K. and de Thé, H. (1999). Herpes virus induced proteasome-dependent degradation of the nuclear bodies-associated PML and Sp100 proteins. *Oncogene* **18**, 935–941.

Chelbi-Alix, M. K., Quignon, F., Pelicano, L., Koken, M. H. M. and de Thé, H. (1998). Resistance to virus infection conferred by the interferon-induced promyelocytic leukemia protein. *J Virol* **72**, 1043–1051.

Cornu, T. I. and de la Torre, J. C. (2001). RING finger Z protein of lymphocytic choriomeningitis virus (LCMV) inhibits transcription and RNA replication of an LCMV S-segment minigenome. *J Virol* **75**, 9415–9426.

Cziepluch, C., Lampel, S., Grewenig, A., Grund, C., Lichter, P. and Rommelaere, J. (2000). H-1 parvovirus-associated replication bodies: a distinct virus-induced nuclear structure. *J Virol* **74**, 4807–4815.

Day, P. M., Roden, R. B. S., Lowy, D. R. and Schiller, J. T. (1998). The papillomavirus minor capsid protein, L2, induces localization of the major capsid protein, L1, and the viral transcription/replication protein, E2, to PML oncogenic domains. *J Virol* **72**, 142–150.

Deng, Z., Chen, C. J., Zerby, D., Delecluse, H. J. and Lieberman, P. M. (2001). Identification of acidic and aromatic residues in the Zta activation domain essential for Epstein–Barr virus reactivation. *J Virol* **75**, 10334–10347.

Desbois, C., Rousset, R., Bantignies, F. and Jalinot, P. (1996). Exclusion of Int-6 from PML nuclear bodies by binding to the HTLV-I tax oncoprotein. *Science* **273**, 951–953.

Djavani, M., Rodas, J., Lukashevich, I. S., Horejsh, D., Pandolfi, P. P., Borden, K. L. B. and Salvato, M. S. (2001). Role of the promyelocytic leukemia protein PML in the interferon sensitivity of lymphocytic choriomeningitis virus. *J Virol* **75**, 6204–6208.

Dobner, T. and Kzhyshkowska, J. (2001). Nuclear export of adenovirus RNA. *Curr Topics Microb Immunol* **259**, 25–54.

Doucas, V. and Evans, R. M. (1999). Human T-cell leukemia retrovirus-Tax protein is a repressor of nuclear receptor signaling. *Proc Natl Acad Sci USA* **96**, 2633–2638.

Doucas, V., Ishov, A. M., Romo, A., Juguilon, H., Weitzman, M. D., Evans, R. M. and Maul, G. G. (1996). Adenovirus replication is coupled with the dynamic properties of the PML nuclear structure. *Genes Dev* **10**, 196–207.

Doucas, V., Tini, M., Egan, D. A. and Evans, R. M. (1999). Modulation of CREB binding protein function by the promyelocytic (PML) oncoprotein suggests a role for nuclear bodies in hormone signaling. *Proc Natl Acad Sci USA* **96**, 2627–2632.

Duprez, E., Saurin, A. J., Desterro, J. M., Lallemand-Breitenbach, V., Howe, K., Boddy, M. N., Solomon, E., de Thé, H., Hay, R. T. and Freemont, P. S. (1999). SUMO-1 modification of the acute promyelocytic leukaemia protein PML: implications for nuclear localisation. *J Cell Sci* **112**, 381–393.

Dyck, J. A., Maul, G. G., Miller, W. J., Chen, J. D., Kakizuka, A. and Evans, R. M. (1994). A novel macromolecular structure is a target of the promyelocyte-retinoic acid receptor oncoprotein. *Cell* **76**, 333–343.

Eidson, K. M., Hobbs, W. E., Manning, B. J., Carlson, P. and DeLuca, N. A. (2002). Expression of herpes simplex virus ICP0 inhibits the induction of interferon-stimulated genes by viral infection. *J Virol* **76**, 2180–2191.

Endter, C., Kzhyshkowska, J., Stauber, R. and Dobner, T. (2001). SUMO-1 modification required for transformation by adenovirus type 5 early region 1B 55 kDa oncoprotein. *Proc Natl Acad Sci USA* **98**, 11312–11317.

Engelhardt, O. G., Ullrich, E., Kochs, G. and Haller, O. (2001). Interferon-induced antiviral Mx1 GTPase is associated with components of the SUMO-1 system and promyelocytic leukemia protein nuclear bodies. *Exp Cell Res* **271**, 286–295.

Evans, J. D. and Hearing, P. (2003). Distinct roles of the adenovirus E4 ORF3 protein in viral DNA replication and inhibition of genome concatenation. *J Virol* **77**, 5295–5304.

Evans, J. D. and Hearing, P. (2005). Relocalization of the Mre11-Rad50-Nbs1 complex by the adenovirus E4 ORF3 protein is required for viral replication. *J Virol* **79**, 6207–6215.

Everett, R., Sourvinos, G. and Orr, A. (2003). Recruitment of herpes simplex virus type 1 tran-scriptional regulatory protein ICP4 into foci juxtaposed to ND10 in live, infected cells. *J Virol* **77**, 3680–3689.

Everett, R. D. (2000a). ICP0 induces the accumulation of colocalizing conjugated ubiquitin. *J Virol* **74**, 9994–10005.

Everett, R. D. (2000b). ICP0, a regulator of herpes simplex virus during lytic and latent infec-tion. *Bioessays* **22**, 761–770.

Everett, R. D., Earnshaw, W. C., Findlay, J. and Lomonte, P. (1999a). Specific destruction of kinetochore protein CENP-C and disruption of cell division by herpes simplex virus imme-diate-early protein Vmw110. *EMBO J* **18**, 1526–1538.

Everett, R. D., Freemont, P., Saitoh, H., Dasso, M., Orr, A., Kathoria, M. and Parkinson, J. (1998). The disruption of ND10 during herpes simplex virus infection correlates with the Vmw110- and proteasome-dependent loss of several PML isoforms. *J Virol* **72**, 6581–6591.

Everett, R. D., Lamonte, P., Sternsdorf, T., van Driel, R. and Orr, A. (1999b). Cell cycle regu-lation of PML modification and ND10 composition. *J Cell Sci* **112**, 4581–4588.

Everett, R. D. and Maul, G. G. (1994). HSV-1 IE protein Vmw110 causes redistribution of PML. *EMBO J* **13**, 5062–5069.

Everett, R. D., Meredith, M. and Orr, A. (1999c). The ability of herpes simplex virus type 1 immediate-early protein Vmw110 to bind to a ubiquitin-specific protease contributes to its roles in the activation of gene expression and stimulation of virus replication. *J Virol* **73**, 417–426.

Everett, R. D., Meredith, M., Orr, A., Cross, A., Kathoria, M. and Parkinson, J. (1997). A novel ubiquitin-specific protease is dynamically associated with the PML nuclear domain and binds to a herpesvirus regulatory protein. *EMBO J* **16**, 1519–1530.

Fabunmi, R. P., Wigley, W. C., Thomas, P. J. and DeMartino, G. N. (2001). Interferon gamma regulates accumulation of the proteasome activator PA28 and immunoproteasomes at nuclear PML bodies. *J Cell Sci* **114**, 29–36.

Fagioli, M., Alcalay, M., Pandolfi, P. P., Venturini, L., Mencarelli, A., Simeone, A., Acampora, D., Grignani, F. and Pelicci, P. G. (1992). Alternative splicing of PML transcripts predicts expression of several carboxyterminally different protein isoforms. *Oncogene* **7**, 1083–1091.

Fagioli, M., Alcalay, M., Tomassoni, L., Ferrucci, P. F., Mencarelli, A., Riganelli, D., Grignani, F., Pozzan, T., Nicoletti, I., Grignani, F. and Pelicci, P. G. (1998). Cooperation between the RING + B1-B2 and coiled-coil domains of PML is necessary for its effects on cell survival. *Oncogene* **16**, 2905–2913.

Flenghi, L., Fagioli, M., Tomassoni, L., Pileri, S., Gambacorta, M., Pacini, R., Grignani, F., Casini, T., Ferrucci, P. F., Martelli, M. F., Pelicci, P. G. and Falini, B. (1995). Characteriza-tion of a new monoclonal antibody (PG-M3) directed against the aminoterminal portion of the PML gene-product—immunocytochemical evidence for high expression of PML proteins on activated macrophages, endothelial cells, and epithelia. *Blood* **85**, 1871–1880.

Florin, L., Sapp, C., Streeck, R. E. and Sapp, M. (2002a). Assembly and translocation of papil-lomavirus capsid proteins. *J Virol* **76**, 10009–10014.

Florin, L., Schafer, F., Sotlar, K., Streeck, R. E. and Sapp, M. (2002b). Reorganization of nuclear domain 10 induced by papillomavirus capsid protein L2. *Virology* **295**, 97–107.

Fogal, V., Gostissa, M., Sandy, P., Zacchi, P., Sternsdorf, T., Jensen, K., Pandolfi, P. P., Will, H., Schneider, C. and Del Sal, G. (2000). Regulation of p53 activity in nuclear bodies by a specific PML isoform. *EMBO J* **19**, 6185–6195.

Freemont, P. S. (1993). The RING finger. A novel protein sequence motif related to the zinc finger. *Ann NY Acad Sci* **684**, 174–192.

Gottifredi, V. and Prives, C. (2001). P53 and PML: new partners in tumor suppression. *Trends Cell Biol* **11**, 184–187.

Gravel, A., Gosselin, J. and Flamand, L. (2002). Human herpesvirus 6 immediate-early 1 protein is a sumoylated nuclear phosphoprotein colocalizing with promyelocytic leukemia protein-associated nuclear bodies. *J Biol Chem* **277**, 19679–19687.

Grobelny, J. V., Godwin, A. K. and Broccoli, D. (2000). ALT-associated PML bodies are present in viable cells and are enriched in cells in the G(2)/M phase of the cell cycle. *J Cell Sci* **113**, 4577–4585.

Grötzinger, T., Jensen, K. and Will, H. (1996). The interferon (IFN)-stimulated gene Sp100 promoter contains an IFN-gamma activation site and an imperfect IFN-stimulated response element which mediate type I IFN inducibility. *J Biol Chem* **271**, 25253–25260.

Guccione, E., Lethbridge, K. J., Killick, N., Leppard, K. N. and Banks, L. (2004). HPV E6 proteins interact with specific PML isoforms and allow distinctions to be made between different POD structures. *Oncogene* **23** (in press).

Guccione, E., Massimi, P., Bernat, A. and Banks, L. (2002). Comparative analysis of the intra-cellular location of the high- and low-risk human papillomavirus oncoproteins. *Virology* **293**, 20–25.

Guldner, H. H., Szostecki, C., Grötzinger, T. and Will, H. (1992). IFN enhances expression of Sp100, an autoantigen in primary biliary cirrhosis. *J Immunol* **149**, 4067–4073.

Guo, A., Salomoni, P., Luo, J. Y., Shih, A., Zhong, S., Gu, W. and Pandolfi, P. P. (2000). The function of PML in p53-dependent apoptosis. *Nat Cell Biol* **2**, 730–736.

Halbert, D. N., Cutt, J. R. and Shenk, T. (1985). Adenovirus early region 4 encodes functions required for efficient DNA replication, late gene expression, and host-cell shutoff. *J Virol* **56**, 250–257.

Hofmann, H., Floss, S. and Stamminger, T. (2000). Covalent modification of the transactivator protein IE2-p86 of human cytomegalovirus by conjugation to the ubiquitin-homologous proteins SUMO-1 and hSMT3b. *J Virol* **74**, 2510–2524.

Hofmann, H., Sindre, H. and Stamminger, T. (2002). Functional interaction between the pp71 protein of human cytomegalovirus and the PML-interacting protein human Daxx. *J Virol* **76**, 5769–5783.

Hosel, M., Schroer, J., Webb, D., Jaroshevskaja, E. and Doerfler, W. (2001). Cellular and early viral factors in the interaction of adenovirus type 12 with hamster cells: the abortive response. *Virus Res* **81**, 1–16.

Huang, M.-M. and Hearing, P. (1989). Adenovirus early region 4 encodes 2 gene products with redundant effects in lytic infection. *J Virol* **63**, 2605–2615.

Hutchinson, I., Whiteley, A., Browne, H. and Elliott, G. (2002). Sequential localization of two herpes simplex virus tegument proteins to punctate nuclear dots adjacent to ICP0 domains. *J Virol* **76**, 10365–10373.

Ishov, A. M. and Maul, G. G. (1996). The periphery of nuclear domain 10 (ND10) as site of DNA virus deposition. *J Cell Biol* **134**, 815–826.

Ishov, A. M., Sotnikov, A. G., Negorev, D., Vladimirova, O. V., Neff, N., Kamitani, T., Yeh, E. T. H., Strauss, J. F. and Maul, G. G. (1999). PML is critical for ND10 formation and recruits the PML-interacting protein Daxx to this nuclear structure when modified by SUMO-1. *J Cell Biol* **147**, 221–233.

Ishov, A. M., Stenberg, R. M. and Maul, G. G. (1997). Human cytomegalovirus immediate early interaction with host nuclear structures: definition of an immediate transcript environment. *J Cell Biol* **138**, 5–16.

Ishov, A. M., Vladimirova, O. V. and Maul, G. G. (2002). Daxx-mediated accumulation of human cytomegalovirus tegument protein pp71 at ND10 facilitates initiation of viral infection at these nuclear domains. *J Virol* **76**, 7705–7712.

Jensen, K., Shiels, C. and Freemont, P. S. (2001). PML protein isoforms and the RBCC/TRIM motif. *Oncogene* **20**, 7223–7233.

Jiang, W. Q. and Ringertz, N. (1997). Altered distribution of the promyelocytic leukemia-associated protein is associated with cellular senescence. *Cell Growth Different* **8**, 513–522.

Kamei, H. (1996). Reversible large-body formation from nuclear bodies upon amino acid(s) starvation in T24 cells. *Exp Cell Res* **224**, 302–311.

Kamitani, T., Kito, K., Nguyen, H. P., Wada, H., Fukuda-Kamitani, T. and Yeh, E. T. H. (1998). Identification of three major sentrinization sites in PML. *J Biol Chem* **273**, 26675–26682.

Katano, H., Ogawa-Goto, K., Hasegawa, H., Kurata, T. and Sata, T. (2001). Human-herpesvirus-8-encoded K8 protein colocalizes with the promyelocytic leukemia protein (PML) bodies and recruits p53 to the PML bodies. *Virology* **286**, 446–455.

Kelly, C., van Driel, R. and Wilkinson, G. W. G. (1995). Disruption of PML-associated nuclear bodies during human cytomegalovirus infection. *J Gen Virol* **76**, 2887–2893.

Kentsis, A., Dwyer, E. C., Perez, J. M., Sharma, M., Chen, A., Pan, Z. Q. and Borden, K. L. B. (2001). The RING domains of the promyelocytic leukemia protein PML and the arenaviral protein Z repress translation by directly inhibiting translation initiation factor eIF4E. *J Mol Biol* **312**, 609–623.

Kiesslich, A., von Mikecz, A. and Hemmerich, P. (2002). Cell cycle-dependent association of PML bodies with sites of active transcription in nuclei of mammalian cells. *J Struct Biol* **140**, 167–179.

Kitajewski, J., Schneider, R. J., Safer, B., Munemitsu, S. M., Samuel, C. E., Thimmappaya, B. and Shenk, T. (1986). Adenovirus VAI RNA antagonizes the antiviral action of interferon by preventing activation of the interferon-induced eIF-2-alpha kinase. *Cell* **45**, 195–200.

Koken, M. H. M., Linarescruz, G., Quignon, F., Viron, A., Chelbi-Alix, M. K., Sobczakthepot, J., Juhlin, L., Degos, L., Calvo, F. and de Thé, H. (1995). The PML growth-suppressor has an altered expression in human oncogenesis. *Oncogene* **10**, 1315–1324.

Koken, M. H. M., Puvion-Dutilleul, F., Guillemin, M. C., Viron, A., Linarescruz, G., Stuurman, N., Dejong, L., Szostecki, C., Calvo, F., Chomienne, C., Degos, L., Puvion, E. and de Thé, H. (1994). The t(1517) translocation alters a nuclear body in a retinoic acid-reversible fashion. *EMBO J* **13**, 1073–1083.

König, C., Roth, J. and Dobbelstein, M. (1999). Adenovirus type 5 E4orf3 protein relieves p53 inhibition by E1B 55 kDa protein. *J Virol* **73**, 2253–2262.

Korioth, F., Maul, G. G., Plachter, B., Stamminger, T. and Frey, J. (1996). The nuclear domain 10 (ND10) is disrupted by the human cytomegalovirus gene product IE1. *Exp Cell Res* **229**, 155–158.

LaFemina, R. L., Pizzorno, M. C., Mosca, J. D. and Hayward, G. S. (1989). Expression of the acidic immediate-early protein (IE1) of human cytomegalovirus in stable cell lines and its preferential association with metaphase chromosomes. *Virology* **172**, 584–600.

Lai, H. K. and Borden, K. L. B. (2000). The promyelocytic leukemia (PML) protein suppresses cyclin D1 protein production by altering the nuclear cytoplasmic distribution of cyclin D1 mRNA. *Oncogene* **19**, 1623–1634.

Lallemand-Breitenbach, V., Zhu, J., Puvion, F., Koken, M., Honore, N., Doubeikovsky, A., Duprez, E., Pandolfi, P. P., Puvion, E., Freemont, P. and de Thé, H. (2001). Role of promyelocytic leukemia (PML) sumolation in nuclear body formation, 11S proteasome recruitment, and As2O3-induced PML or PML/retinoic acid receptor alpha degradation. *J Exp Med* **193**, 1361–1371.

LaMorte, V. J., Dyck, J. A., Ochs, R. L. and Evans, R. M. (1998). Localization of nascent RNA and CREB binding protein with the PML-containing nuclear body. *Proc Natl Acad Sci USA* **95**, 4991–4996.

Langley, E., Pearson, M., Faretta, M., Bauer, U.-M., Frye, R. A., Minucci, S., Pelicci, P. G. and Kouzarides, T. (2002). Human SIR2 deacetylates p53 and antagonizes PML/p53-induced cellular senescence. *EMBO J* **21**, 2383–2396.

Lavau, C., Marchio, A., Fagioli, M., Jansen, J., Falini, B., Lebon, P., Grosveld, F., Pandolfi, P. P., Pelicci, P. G. and Dejean, A. (1995). The acute promyelocytic leukemia-associated PML gene is induced by interferon. *Oncogene* **11**, 871–878.

Lehembre, F., Muller, S., Pandolfi, P. P. and Dejean, A. (2001). Regulation of Pax3 transcriptional activity by SUMO-1-modified PML. *Oncogene* **20**, 1–9.

Leib, D. A., Harrison, T. E., Laslo, K. M., Machalek, M. A., Moorman, N. J. and Virgin, H. W. (1999). Interferons regulate the phenotype of wild-type and mutant herpes simplex viruses *in vivo. J Exp Med* **189**, 663–672.

Leppard, K. N. (1998). Regulated RNA processing and RNA transport during adenovirus infection. *Semin Virol* **8**, 301–307.

Leppard, K. N. and Everett, R. D. (1999). The adenovirus type 5 E1b 55K and E4 Orf3 proteins associate in infected cells and affect ND10 components. *J Gen Virol* **80**, 997–1008.

Lethbridge, K. J., Scott, G. E. and Leppard, K. N. (2003). Nuclear matrix localization and SUMO-1 modification of adenovirus type 5 E1b 55K protein are controlled by E4 Orf6 protein. *J Gen Virol* **84**, 259–268.

Li, H., Leo, C., Zhu, J., Wu, X. Y., O'Neil, J., Park, E. J. and Chen, J. D. (2000). Sequestration and inhibition of Daxx-mediated transcriptional repression by PML. *Mol Cell Biol* **20**, 1784–1796.

Li, X. D., Makela, T. P., Guo, D. Y., Soliymani, R., Koistinen, V., Vapalahti, O., Vaheri, A. and Lankinen, H. (2002). Hantavirus nucleocapsid protein interacts with the Fas-mediated apoptosis enhancer Daxx. *J Gen Virol* **83**, 759–766.

Lin, R., Noyce, R. S., Collins, S. E., Everett, R. D. and Mossman, K. L. (2004). The herpes simplex virus ICP0 RING finger domain inhibits IRF3- and IRF7-mediated activation of interferon-responsive genes. *J Virol* **78**, 1675–1684.

Linial, M. (2000). Why aren't foamy viruses pathogenic? *Trends Microbiol* **8**, 284–289.

Lombard, D. B. and Guarente, L. (2000). Nijmegen breakage syndrome disease protein and MRE11 at PML nuclear bodies and meiotic telomeres. *Cancer Res* **60**, 2331–2334.

Lopez, N., Jacamo, R. and Franze-Fernandez, M. T. (2001). Transcription and RNA replication of tacaribe virus genome and antigenome analogs require N and L proteins: Z protein is an inhibitor of these processes. *J Virol* **75**, 12241–12251.

Lopez, P., Jacob, R. J. and Roizman, B. (2002). Overexpression of promyelocytic leukemia protein precludes the dispersal of ND10 structures and has no effect on accumulation of infectious herpes simplex virus 1 or its proteins. *J Virol* **76**, 9355–9367.

Mainz, D., Quadt, I. and Knebel-Morsdorf, D. (2002). Nuclear IE2 structures are related to viral DNA replication sites during baculovirus infection. *J Virol* **76**, 5198–5207.

Marchetti, A., Buttitta, F., Miyazaki, S., Gallahan, D., Smith, G. H. and Callahan, R. (1995). Int-6, a highly conserved, widely expressed gene, is mutated by mouse mammary tumor virus in mammary preneoplasia. *J Virol* **69**, 1932–1938.

Marshall, K. R., Rowley, K. V., Rinaldi, A., Nicholson, I. P., Ishov, A. M., Maul, G. G. and Preston, C. M. (2002). Activity and intracellular localization of the human cytomegalovirus protein pp71. *J Gen Virol* **83**, 1601–1612.

Maul, G. G. and Everett, R. D. (1994). The nuclear location of PML, a cellular member of the C3HC4 zinc binding domain protein family, is rearranged during herpes simplex virus infection by the C3HC4 viral protein ICP0. *J Gen Virol* **75**, 1223–1233.

Maul, G. G., Guldner, H. H. and Spivack, J. G. (1993). Modification of discrete nuclear domains induced by herpes simplex virus type 1 immediate early gene product ICP0. *J Gen Virol* **74**, 2679–2690.

Maul, G. G., Ishov, A. M. and Everett, R. D. (1996). Nuclear domain 10 as preexisting potential replication start sites of herpes simplex virus type-1. *Virology* **217**, 67–75.

Maul, G. G., Jensen, D. E., Ishov, A. M., Herlyn, M. and Rauscher III, F. J. (1998). Nuclear redistribution of BRCA1 during viral infection. *Cell Growth Diff* **9**, 743–755.

Maul, G. G., Yu, E., Ishov, A. M. and Epstein, A. L. (1995). Nuclear domain 10 (ND10) associated proteins are also present in nuclear bodies and redistribute to hundreds of nuclear sites after stress. *J Cell Biochem* **59**, 498–513.

Mossman, K. L., Saffran, H. A. and Smiley, J. R. (2000). Herpes simplex virus ICP0 mutants are hypersensitive to interferon. *J Virol* **74**, 2052–2056.

Mu, Z. M., Le, X. F., Vallian, S., Glassman, A. B. and Chang, K. S. (1997). Stable overexpression of PML alters regulation of cell cycle progression in HeLa cells. *Carcinogenesis* **18**, 2063–2069.

Müller, S. and Dejean, A. (1999). Viral immediate-early proteins abrogate the modification by SUMO-1 of PML and Sp100 proteins, correlating with nuclear body disruption. *J Virol* **73**, 5137–5143.

Müller, S., Matunis, M. J. and DeJean, A. (1998). Conjugation of the ubiquitin-related modifier SUMO-1 regulates the partitioning of PML within the nucleus. *EMBO J* **17**, 61–70.

Muratani, M., Gerlich, D., Janicki, S. M., Gebhard, M., Eils, R. and Spector, D. L. (2002). Metabolic-energy-dependent movement of PML bodies within the mammalian cell nucleus. *Nature Cell Biology* **4**, 106–110.

Murges, D., Quadt, I., Schroer, J. and Knebel-Morsdorf, D. (2001). Dynamic nuclear localization of the baculovirus proteins IE2 and PE38 during the infection cycle: the promyelocytic leukemia protein colocalizes with IE2. *Exp Cell Res* **264**, 219–232.

Negorev, D., Ishov, A. M. and Maul, G. G. (2001). Evidence for separate ND10-binding and homo-oligomerization domains of Sp100. *J Cell Sci* **114**, 59–68.

Negorev, D. and Maul, G. G. (2001). Cellular proteins localized at and interacting within ND10/PML nuclear bodies/PODs suggest functions of a nuclear depot. *Oncogene* **20**, 7234–7242.

Ornelles, D. A. and Shenk, T. (1991). Localization of the adenovirus early region 1b 55 kDa protein during lytic infection: association with nuclear viral inclusions requires the early region 4 34 kDa protein. *J Virol* **65**, 424–439.

Parkinson, J. and Everett, R. D. (2000). Alphaherpesvirus proteins related to herpes simplex virus type 1 ICP0 affect cellular structures and proteins. *J Virol* **74**, 10006–10017.

Parkinson, J. and Everett, R. D. (2001). Alphaherpesvirus proteins related to herpes simplex virus type 1 ICP0 induce the formation of colocalizing, conjugated ubiquitin. *J Virol* **75**, 5357–5362.

Pearson, M. and Pelicci, P. G. (2001). PML interaction with p53 and its role in apoptosis and replicative senescence. *Oncogene* **20**, 7250–7256.

Pombo, A., Ferreira, J., Bridge, E. and Carmo-Fonseca, M. (1994). Adenovirus replication and transcription sites are spatially separated in the nucleus of infected cells. *EMBO J* **13**, 5075–5085.

Puvion-Dutilleul, F., Chelbi-Alix, M. K., Koken, M., Quignon, F., Puvion, E. and de Thé, H. (1995). Adenovirus infection induces rearrangements in the intranuclear distribution of the nuclear body-associated PML protein. *Exp Cell Res* **218**, 9–16.

Puvion-Dutilleul, F., Legrand, V., Mehtali, M., Chelbi-Alix, M. K., de Thé, H. and Puvion, E. (1999). Deletion of the fiber gene induces the storage of hexon and penton base proteins in PML/Sp100-containing inclusions during adenovirus infection. *Biol Cell* **91**, 617–628.

Rangasamy, D., Woytek, K., Khan, S. A. and Wilson, V. G. (2000). SUMO-1 modification of bovine papillomavirus E1 protein is required for intranuclear accumulation. *J Biol Chem* **275**, 37999–38004.

Regad, T. and Chelbi-Alix, M. K. (2001). Role and fate of PML nuclear bodies in response to interferon and viral infections. *Oncogene* **20**, 7274–7286.

Reymond, A., Meroni, G., Fantozzi, A., Merla, G., Cairo, S., Luzi, L., Riganelli, D., Zanaria, E., Messali, S., Cainarca, S., Guffanti, A., Minucci, S., Pelicci, P. G. and Ballabio, A. (2001). The tripartite motif family identifies cell compartments. *EMBO J* **20**, 2140–2151.

Roberts, S., Hillman, M. L., Knight, G. L. and Gallimore, P. H. (2003). The ND10 component promyelocytic leukemia protein relocates to human papillomavirus type 1 E4 intranuclear inclusion bodies in cultured keratinocytes and in warts. *J Virol* **77**, 673–684.

Rockel, T. D. and von Mikecz, A. (2002). Proteasome-dependent processing of nuclear proteins is correlated with their subnuclear localization. *J Struct Biol* **140**, 189–199.

Rosa-Calatrava, M., Grave, L., Puvion-Dutilleul, F., Chatton, B. and Kedinger, C. (2001). Functional analysis of adenovirus protein IX identifies domains involved in capsid stability, transcriptional activity, and nuclear reorganization. *J Virol* **75**, 7131–7141.

Rosa-Calatrava, M., Puvion-Dutilleul, F., Lutz, P., Dreyer, D., de Thé, H., Chatton, B. and Kedinger, C. (2003). Adenovirus protein IX sequesters host-cell promyelocytic leukaemia protein and contributes to efficient viral proliferation. *EMBO Rep* **4**, 969–975.

Rosendorff, A., Illanes, D., David, G., Lin, J., Kieff, E. and Johannsen, E. (2004). EBNA3C coactivation with EBNA2 requires a SUMO homology domain. *J Virol* **78**, 367–377.

Sato, Y., Yoshioka, K., Suzuki, C., Awashima, S., Hosaka, Y., Yewdell, J. and Kuroda, K. (2003). Localization of influenza virus proteins to nuclear dot 10 structures in influenza virus-infected cells. *Virology* **310**, 29–40.

Shiels, C., Islam, S. A., Vatcheva, R., Sasieni, P., Sternberg, M. J. E., Freemont, P. S. and Sheer, D. (2001). PML bodies associate specifically with the MHC gene cluster in interphase nuclei. *J Cell Sci* **114**, 3705–3716.

Sourvinos, G. and Everett, R. D. (2002). Visualization of parental HSV-1 genomes and replication compartments in association with ND10 in live infected cells. *EMBO J* **21**, 4989–4997.

Spengler, M. L., Kurapatwinski, K., Black, A. R. and Azizkhan-Clifford, J. (2002). SUMO-1 modification of human cytomegalovirus IE1/IE72. *J Virol* **76**, 2990–2996.

Stanton, R., Fox, J. D., Caswell, R., Sherratt, E. and Wilkinson, G. W. G. (2002). Analysis of the human herpesvirus-6 immediate-early 1 protein. *J Gen Virol* **83**, 2811–2820.

Sternsdorf, T., Jensen, K. and Will, H. (1997). Evidence for covalent modification of the nuclear dot-associated proteins PML and Sp100 by PIC1/SUMO-1. *J Cell Biol* **139**, 1621–1634.

Stracker, T. H., Carson, C. T. and Weitzman, M. D. (2002). Adenovirus oncoproteins inactivate the Mre11-Rad50-NBS1 DNA repair complex. *Nature* **418**, 348–352.

Stracker, T. H., Lee, D. V., Carson, C. T., Araujo, F. D., Ornelles, D. A. and Weitzman, M. D.
(2005). Serotype-specific reorganization of the Mre11 complex by adenoviral E4orf3
proteins. *J Virol* **79**, 6664–6673.

Swindle, C. S., Zou, N. X., Van Tine, B. A., Shaw, G. M., Engler, J. A. and Chow, L. T.
(1999). Human papillomavirus DNA replication compartments in a transient DNA replication system. *J Virol* **73**, 1001–1009.

Szekely, L., Pokrovskaja, K., Jiang, W. Q., de Thé, H., Ringertz, N. and Klein, G. (1996). The
Epstein–Barr virus-encoded nuclear antigen EBNA-5 accumulates in PML-containing
bodies. *J Virol* **70**, 2562–2568.

Szostecki, C., Guldner, H. H., Netter, H. J. and Will, H. (1990). Isolation and characterization of cDNA encoding a human nuclear antigen predominantly recognized by
autoantibodies from patients with primary biliary cirrhosis. *J Immunol* **145**, 4338–4347.

Tang, Q. Y., Bell, P., Tegtmeyer, P. and Maul, G. G. (2000). Replication but not transcription
of simian virus 40 DNA is dependent on nuclear domain 10. *J Virol* **74**, 9694–9700.

Terris, B., Baldin, V., Dubois, S., Degott, C., Flejou, J. F., Henin, D. and Dejean, A. (1995). PML
nuclear bodies are general targets for inflammation and cell proliferation. *Cancer Res* **55**, 1590–1597.

Turelli, P., Doucas, V., Craig, E., Mangeat, B., Klages, N., Evans, R., Kalpana, G. and Trono,
D. (2001). Cytoplasmic recruitment of INI1 and PML on incoming HIV preintegration
complexes: interference with early steps of viral replication. *Mol Cell* **7**, 1245–1254.

Vallian, S., Chin, K. V. and Chang, K. S. (1998). The promyelocytic leukemia protein interacts
with Sp1 and inhibits its transactivation of the epidermal growth factor receptor promoter.
Mol Cell Biol **18**, 7147–7156.

Wang, Z. G., Delva, L., Gaboli, M., Rivi, R., Giorgio, M., Cordon-Cardo, C., Grosveld, F. and
Pandolfi, P. P. (1998a). Role of PML in cell growth and the retinoic acid pathway. *Science*
279, 1547–1551.

Wang, Z. G., Ruggero, D., Ronchetti, S., Zhong, S., Gaboli, M., Rivi, R. and Pandolfi, P. P.
(1998b). PML is essential for multiple apoptotic pathways. *Nat Genet* **20**, 266–272.

Weis, K., Rambaud, S., Lavau, C., Jansen, J., Carvalho, T., Carmo-Fonseca, M., Lamond, A. and
Dejean, A. (1994). Retinoic acid regulates aberrant nuclear-localization of PML-RAR-α in
acute promyelocytic leukemia cells. *Cell* **76**, 345–356.

Wiesmeijer, K., Molenaar, C., Bekeer, I., Tanke, H. J. and Dirks, R. W. (2002). Mobile foci of
Sp100 do not contain PML: PML bodies are immobile but PML and Sp100 proteins are not.
J Struct Biol **140**, 180–188.

Wilkinson, G. W. G., Kelly, C., Sinclair, J. H. and Rickards, C. (1998). Disruption of PML-
associated nuclear bodies mediated by the human cytomegalovirus major immediate early
gene product. *J Gen Virol* **79**, 1233–1245.

Wilson, V. G. and Rangasamy, D. (2001). Viral interaction with the host cell sumoylation
system. *Virus Res* **81**, 17–27.

Wu, F. Y., Ahn, J. H., Alcendor, D. J., Jang, W. J., Xiao, J. S., Hayward, S. D. and Hayward, G. S.
(2001a). Origin-independent assembly of Kaposi's sarcoma-associated herpesvirus DNA
replication compartments in transient cotransfection assays and association with the ORF-
K8 protein and cellular PML. *J Virol* **75**, 1487–1506.

Wu, W. S., Vallian, S., Seto, E., Yang, W. M., Edmondson, D., Roth, S. and Chang, K. S.
(2001b). The growth suppressor PML represses transcription by functionally and physically
interacting with histone deacetylases. *Mol Cell Biol* **21**, 2259–2268.

Xu, Y. X., Ahn, J. H., Cheng, M. F., Ap Rhys, C. M., Chiou, C. J., Zong, J. H., Matunis, M. J. and Hayward, G. S. (2001). Proteasome-independent disruption of PML oncogenic domains (PODs), but not covalent modification by SUMO-1, is required for human cytomegalovirus immediate-early protein IE1 to inhibit PML-mediated transcriptional repression. *J Virol* **75**, 10683–10695.

Yang, S. T., Kuo, C., Bisi, J. E. and Kim, M. K. (2002). PML-dependent apoptosis after DNA damage is regulated by the checkpoint kinase hCds1/Chk2. *Nat Cell Biol* **4**, 865–870.

Young, P. J., Jensen, K. T., Burger, L. R., Pintel, D. J. and Lorson, C. L. (2002). Minute virus of mice NS1 interacts with the SMN protein, and they colocalize in novel nuclear bodies induced by parvovirus infection. *J Virol* **76**, 3892–3904.

Zhong, S., Muller, S., Ronchetti, S., Freemont, P. S., Dejean, A. and Pandolfi, P. P. (2000a). Role of SUMO-1-modified PML in nuclear body formation. *Blood* **95**, 2748–2753.

Zhong, S., Salomoni, P. and Pandolfi, P. P. (2000b). The transcriptional role of PML and the nuclear body. *Nat Cell Biol* **2**, E85–E90.

9 Viruses and the Cell Cycle

Crisanto Gutierrez[1], Brian Dove[2] and Julian A. Hiscox[2]

[1] Centro de Biologia Molecular 'Severo Ochoa', Consejo Superior de Investigaciones Cientificas, Universidad Autonoma de Madrid, Cantoblanco, Madrid, Spain
[2] School of Biochemistry and Microbiology, University of Leeds, Leeds, UK

9.1 Introduction: an overview of the eukaryotic cell cycle

The evolutionary trends of DNA replication, cell cycle and growth control are mechanistically well conserved among eukaryotes, from yeast to humans, from flies to plants. The function of the cell cycle itself can be seen as a number of checkpoints and phases in order to provide conditions suitable for the proper function of that cell and for DNA replication. Needless to say, viruses can interact with the cell cycle. As we shall see in this chapter, interaction of viruses with the cell cycle is not restricted to DNA and retroviruses and an emerging paradigm is that RNA viruses can also usurp host cell cycle functions to facilitate virus can replication. The cell cycle itself was reviewed in detail in Chapter 2 and the reader is referred that chapter for a greater understanding of the eukaryotic system; here we briefly recap the most salient points.

The cell cycle can be divided into a number of separate events [(Pines, 1999); DNA replication (S phase), nuclear division (mitosis, M)] and cell division (cytokinesis), separated by the two gap periods (G_1 and G_2). Quiescent cells are described as being in G_0 (Pardee, 1989). Progression through each phase and from one phase to the next is tightly regulated and highly orchestrated (Pines, 1999) (Figure 9.1). So-called checkpoints are found in each phase in order to ensure that intracellular conditions are favourable for the onset and progression of the subsequent phase (Rupes, 2002). This is especially important during the S and M phases, where errors in DNA replication

Viruses and the Nucleus   Edited by Julian A. Hiscox
© 2006 John Wiley & Sons, Ltd

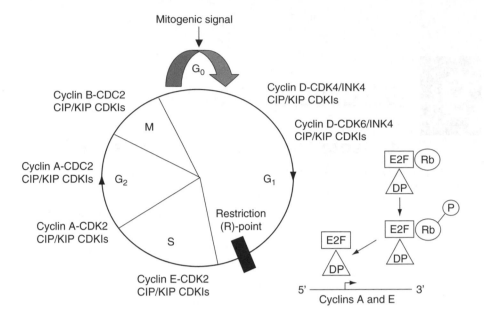

Figure 9.1 Schematic diagram of the cell cycle showing the interactions of various molecules involved in this process at each different stage. Highlighted is the restriction point in G_1. Many viruses can be the cause of a mitogenic stimulus. Adapted (with permission) from Brooks *et al.* (1998)

and nuclear division can cause uncontrolled cell growth (Hartwell and Kastan, 1994) and genetic instability (Hollander and Fornace, 2002). Such delays enable repair mechanisms to be activated. In the G_1 phase cells monitor both their internal and external environments to determine whether to divide or to remain in a state of quiescence. One such control point in mammalian cells is called the restriction or R-point. In mammalian cells the R-point is widely believed to be analogous to START, which has been defined in yeast cell cycle experiments (Simanis *et al.*, 1987). Once yeast cells have passed START they are committed to cell division (Cross and Roberts, 2001).

Cell cycle progression is regulated by different cyclin-dependent kinases (cdks), whose activity depends on cellular localization, phosphorylation state and association with activators (cyclins) or inhibitors (cdk inhibitors) (Pines, 1993) (Figure 9.1). This interconnecting web of control has been aptly described as 'four-dimensional' (Pines, 1999). Cytokinesis can be viewed as the final stage to the cell cycle and is the programmed division of a cell into two daughter cells, each containing one nucleus. Once nuclear division has occurred, a division site is chosen (in animal cells this is at the equator) and a cleavage furrow, composed of a contractile ring formed from actin, myosin and other proteins, is assembled at the division site. Cytokinesis in eukaryotes has recently been reviewed by Guertin *et al.* (2002).

9.2   DNA viruses and the cell cycle

9.2.1   Animal DNA viruses and the cell cycle

A variety of host cell functions are required for the completion of a productive viral infectious cycle, a characteristic typical of viruses derived from the strict cellular dependence of viruses. The interactions are required to sustain expression, and duplication of the viral genome and macromolecular trafficking are among the best studied. In addition, the proliferative state of the infected cell is, in most cases, crucial for a productive infection. Consequently, DNA viruses have evolved a rich variety of interactions with the host cell to modify and interfere with the cell cycle regulatory machinery. In some cases viruses are adapted to multiply in resting cells, while in others they induce proliferation of arrested cells or just wait until the infected cell replicates. As discussed briefly below, two main strategies for interfering with host cell cycle control can be distinguished.

One has generally evolved in viruses with large genomes with the potential to encode many proteins, including those required for viral DNA replication. A typical example of this strategy is found in herpesviruses, which have large and complex genomes. Recent studies, which have been comprehensively discussed (Castillo and Kowalik, 2002; Flemington, 2001), have provided clear insights into the mechanisms used by alpha- (HSV, herpes simplex virus), beta- (CMV, cytomegalovirus; Kalejta and Shenk, 2002) and gamma-herpesviruses (EBV, Epstein–Barr virus) to block cell cycle progression, preventing entry into the S-phase. In general, some of the viral immediate-early proteins are responsible for this effect: HSV ICP0 protein (Hobbs and DeLuca, 1999; Lomonte and Everett, 1999; Castillo and Kowalik, 2002), CMV UL69 (a ICP27 homologue; Lu and Shenk, 1999) and IE2 (Wiebush *et al.*, 1999) proteins, and EBV protein Zta (Cayrol and Flemington, 1996) and BZLF1 induces a G_2 and M phase block (Mauser *et al.*, 2002). For example, HCMV causes a large increase in the amount of cyclin E and redistributes Cdk2 from the cytoplasm to the nucleus. In addition to this, the abundance of both p21 and p27 is reduced (Chen *et al.*, 2001). In addition, Cdk1/cyclin B1 activity is maintained in HCMV-infected cells, by accumulation of factors that promote Cdk1 activity, accumulation of cyclin B1 and inactivation of negative regulatory pathways for Cdk1 (Sanchez *et al.*, 2003). These viruses also activate some cell cycle functions (reviewed in Flemington, 2001), suggesting that although an arrested cell environment may be helpful to avoid competition with cellular DNA replication, some functions and/or activities present in cycling cells seem to be required, or at least beneficial, for the viral replication cycle. In this context, EBV DNA replication requires the function of the origin recognition complex (ORC) and is inhibited by geminin (Dhar *et al.*, 2001), a six-subunit multimeric complex required for activation of cellular DNA replication origins (Bell and Dutta, 2002). Since the activity of ORC, as well as other pre-replication complex components, is strictly regulated in connection to cell cycle progression, this association with EBV DNA replication origins point to a complex and fine interference with host cell cycle regulatory pathways. Another example comes after the finding that homologues of D-type cyclins are encoded by some oncogenic gamma-herpesviruses (Hardwick, 2000): the herpesvirus Saimiri cyclin V, the Kaposi sarcoma-associated herpesvirus cyclin K, and the murine gamma-herpesvirus-68 cyclin M. These cyclins bind to CDK4/6, are resistant to CDK inhibitors and broaden the range of substrates of CDK6 with respect to those phosphorylated when it is normally bound to cyclin D (Laman *et al.*, 2000; Schulze-Gahmen and Kim, 2002; Swanton *et al.*, 1999).

The other strategy to impinge on cell cycle control is more typical of viruses with small genomes. Here, viral-encoded proteins, which are not homologues of any known cellular protein, interfere directly with the normal function of cell cycle regulatory components to subvert their activity. Typical examples are the viral oncoproteins, which sequester the retinoblastoma (RB) tumour suppressor protein as a first step in inducing S-phase entry by activating the expression of E2F-regulated genes.

The molecular mechanisms evolved by animal DNA tumour viruses to interfere with cell cycle regulatory pathways have been largely delineated and discussed comprehensively in different reviews for polyomaviruses (Ludlow, 1993; Simmons, 2000; Sullivan and Pipas, 2002), adenoviruses (Ben-Israel and Kleinberger, 2002; Moran, 1993; Op De Beeck and Caillet-Fauquet, 1997) and papillomaviruses (DiMaio and Mattoon, 2001; Vousden, 1993). S phase is fundamental for efficient viral replication for many of these viruses, because the infected cells themselves are epithelial cells (and therefore in G_0) or, if not already, would otherwise exit the cell cycle into G_0. For example, the human papillomavirus type 16 E7 oncoprotein inactivates both the Rb protein (Giarre *et al.*, 2001) and p21 (Helt *et al.*, 2002a) in order to inhibit cell cycle arrest and therefore induce S phase. Interestingly, adenovirus-associated virus (AAV) NS1 protein exhibits an anti-proliferative effect on the adenovirus-induced inactivation and degradation of RB, most likely through interaction of NS1 with adenovirus E1A protein (Batchu *et al.*, 2002).

The strategy used by parvoviruses, although they have a small genome, is not to induce cell proliferation but instead to wait until cells began to replicate (Op De Beeck and Caillet-Fauquet, 1997). Cellular factors associated with S phase are required for the initial transformation of the single-stranded DNA genome into a double-stranded form, which is then transcriptionally active and used for viral DNA replication. In addition, different parvoviruses are able to arrest the cell cycle, most frequently, although not exclusively, in G_2, by inducing accumulation of the CDK inhibitor $p21^{Kip}$ (Op De Beeck *et al.*, 2001) and suppressing the nuclear accumulation of cyclin B (Morita *et al.*, 2001). In some cases of cells infected with Aleutian mink disease parvovirus (ADV), two populations of cells arise (Oleksiewicz and Alexandersen, 1997). The first population are arrested in late S phase or G_2 and fail to undergo cytokinesis. In these cells levels of DNA replication are high and, from the host organism's perspective, levels of production of the major viral non-structural protein 1 (NS1) are high. This protein is the main viral protein responsible for the cytotoxic and cytostatic state observed during ADV infection. The second population of cells forms after mitosis and becomes arrested in G_0/G_1 phase. In this population levels of DNA replication are low, but high levels of non-structural proteins are expressed. The two populations of cells led Oleksiewicz and Alexandersen (1997) to suggest that the second population arose from the first due to leakage of cells through the S phase block caused by sub-threshold levels of ADV gene products.

9.2.2   Plant DNA viruses and the cell cycle

Functional interactions between viral proteins and the cell cycle machinery are not restricted to animal DNA viruses. DNA viruses that infect plant cells have also established a rich variety of interactions with the host cell to interfere with cell cycle regulation, among other cellular processes. Although these interactions have been the subject of several review articles (Boulton, 2002; Gutierrez, 2000, 2002; Gutierrez *et al.*, 2002; Hanley-Bowdoin *et al.*, 1999; Palmer and Rybicki, 1998), recent developments justify an update in the context of a more general discussion on both animal and plant DNA viruses. Moreover, this topic has been less well covered in general virology journals. The new data are contributing to both better understanding of the original observations in molecular terms and to widening our perspective of previously unidentified interactions between viral proteins and cell cycle regulatory components.

The family *Geminiviridae* include a large number of plant viruses that frequently produce significant reductions in economically important crops of both monocotyledonous and dicotyledonous plants (Moffat, 1999). They are non-enveloped viruses with a single-stranded DNA (ssDNA) genome (either one or two circular molecules, 2.5–3.0 kb in length) that encodes a few proteins. The transcriptional regulation of viral genes has been reviewed (Boulton, 2002; Hanley-Bowdoin *et al.*, 1999). The geminivirus genome has a specific organization in each of the four genera currently recognized.

Members of the genus *Mastrevirus*, which includes about a dozen species, are characterized by having a monopartite genomic ssDNA and generally infect monocotyledonous plants (*Maize streak virus*, MSV, and *Wheat dwarf virus*, WDV, being the most commonly studied viruses). Some of them are adapted to infect dicotyledonous plants, e.g. *Bean yellow dwarf virus*, BeYDV). Of the four proteins encoded, RepA and Rep are relevant for the discussion below: RepA, a regulatory protein, and Rep, the replication initiator protein. While RepA is unique to this genus, Rep is encoded by all geminiviruses, although it has specific functions in some cases, as described below. *Beet curly top virus* (BCTV; Genus *Curtovirus*) has a monopartite genome that encodes seven proteins, of which Rep and C4 are relevant for the present discussion.

Tomato pseudo-curly top virus (TPCTV; Genus *Topocuvirus*) also has a monopartite genome with a genetic organization similar to that of curtoviruses.

Members of the genus *Begomovirus* (*Bean golden mosaic virus*, BGMV) infect exclusively dicotyledonous plants, e.g. *Tomato golden mosaic virus* (TGMV), *African cassava mosaic virus* (ACMV), *Tomato yellow leaf curl virus* (TYLCV) and *Squash leaf curl virus* (SLCV). Most begomoviruses have a bipartite genome (two ssDNA circles named A and B) and component A encodes Rep and REn, a replication enhancer protein (also named AC3 or AL3), which interfere with cell cycle regulators.

Different from caulimoviruses, the other large plant DNA virus family, the geminivirus replicative cycle occurs within the nucleus of the infected cell and relies entirely on DNA intermediates. The DNA replication cycle (Gutierrez, 1999; Hanley-Bowdoin *et al.*, 1999) can be divided into several functionally distinct stages. The first stage is the conversion of the genomic ssDNA into a super-coiled, covalently-closed circular dsDNA intermediate; the second stage leads to the amplification of viral DNA through a rolling-circle mechanism; and the final stage includes the transport of genomic ssDNA to neighbouring cells and encapsidation to form mature viral particles. Only the dsDNA intermediates are transcriptionally active, therefore it is important to keep in mind that the first replication stage relies entirely on host cell proteins.

Geminivirus-encoded proteins do not possess DNA polymerase activity or accessory functions. Consistent with a requirement for the cellular DNA replication machinery, replication intermediates are significantly more abundant in cells undergoing S-phase (Accotto *et al.*, 1993). However, paradoxically, geminiviruses do not infect actively proliferating cells, located in the meristems of the plant body or in developing organs. Instead they are normally present in arrested, fully differentiated cells that lack most of the DNA replication. These, together with other indirect observations, supported the hypothesis that geminivirus infection might be linked to a process whereby the infected cell physiology, in particular its proliferation competence, was modified by one or more viral proteins in order to induce a cellular state permissive for viral DNA replication. This would be similar to what animal oncoviruses do.

Early studies showed that proliferating cell nuclear antigen (PCNA), a protein with multiples roles in cell cycle progression, DNA replication and repair, accumulated in TGMV-infected cells while it was undetectable in the absence of virus infection (Nagar *et al.*, 1995). The finding that the WDV RepA protein was able to interact with human tumour suppressor retinoblastoma (RB) protein (Xie *et al.*, 1995) and plant RB-related (RBR) protein (Xie *et al.*, 1996) provided a first clue strongly suggesting a link between geminiviral proteins and the cell cycle regulatory network (Gutierrez, 1998). As discussed earlier in this chapter, in animal cells, progression through G_1 and transition to S phase is largely controlled by the activity of RB and other family members on the E2F/DP family of transcription factors. Inhibition of E2F/DP activity is relieved after sequential phosphorylation of pocket proteins by CDK/cyclin complexes (Mittnacht, 1998). A similar situation seems to occur in plant cells, where RBR protein is phosphorylated by CDK/cyclin complexes late in G_1 (Boniotti and Gutierrez, 2001; Nakagami *et al.*, 2002).

Mastrevirus RepA protein interacts with plant RBR through a typical LXCXE amino acid motif (Xie *et al.*, 1996), while mastrevirus Rep, although containing the same motif, does not interact efficiently with RBR (Gutierrez *et al.*, 2002; Horvath *et al.*, 1998; Liu *et al.*, 1999). A possible reason is that the LXCXE motif is hindered in Rep by its C-terminal domain, as suggested by secondary structure predictions and deletion analysis (Gutierrez *et al.*, 2002). Other geminiviruses do not encode any protein homologous to mastrevirus RepA. However, begomovirus Rep does interact with plant RBR (Ach *et al.*, 1997) in a LXCXE-independent manner (Kong *et al.*, 2000). Recently, it has been shown that accumulation of PCNA in TGMV Rep-expressing cells depends on transcriptional activation of the PCNA gene by overcoming E2F-mediated repression (Egelkrout *et al.*, 2001).

Point mutations in the RBR-binding motif of TGMV Rep that reduce RBR binding lead to a decrease in the amount of viral DNA that accumulates after infection (Kong *et al.*, 2000). Interestingly, viruses expressing mutant Rep proteins also have an altered tissue specificity (Kong *et al.*, 2000). Point mutations that affect WDV

RepA–RBR interaction have a deleterious effect on viral DNA accumulation in cultured cells (Xie *et al.*, 1995). However, in the case of dicotyledon-infecting mastreviruses, e.g. *Bean yellow dwarf virus* (BeYDV), mutants with different RBR-binding efficiencies are able to replicate in tobacco protoplasts and to infect and produce systemic symptoms in susceptible *N. benthamiana* plants (Liu *et al.*, 1999). Therefore, the requirements for cell cycle interference and the functional relevance of the interaction between viral proteins and the G_1–S control may differ depending on the virus. Mastreviruses adapted to replicate in dicotyledonous plants may have evolved a LXCXE-independent mechanism of RBR interaction as it occurs with begomoviruses, e.g. TGMV. These possibilities need to be further addressed. Similar complex interactions between viral proteins and the RB pathway also occur with animal oncoviruses, as discussed elsewhere (Gutierrez, 2000, 2002).

Whatever the activation mechanism, at least some S-phase functions occur, and viral proteins require the participation of cellular DNA replication factors to amplify viral DNA through a rolling-circle process. Recent studies have aimed at identifying a possible role of viral proteins in recruiting these cellular factors for viral replication. Recently Luque *et al.* (2002) found that WDV Rep interacts both in the yeast two-hybrid system and *in vitro* with purified proteins with the large subunit of replication factor C (RFC-1). The WDV Rep–RFC-1 interaction results are consistent with a model in which the viral protein stimulates the recruitment of a RFC-1–PCNA–DNA polymerase δ to the 3′-OH primer generated during the initiation reaction by the Rep protein (Luque *et al.*, 2002). Based on the similarity among Rep proteins of different geminiviruses, it is tempting to speculate that a similar mechanism may apply to other geminiviruses, perhaps with slight variations.

Induction of S phase, or at least some S-phase functions, seems to be a characteristic feature of geminivirus infection. TGMV Rep interacts with several cellular proteins, one of which is geminivirus Rep-interacting motor protein (GRIMP), a protein that localizes to the spindle apparatus and binds to CDC2a (Kong and Hanley-Bowdoin, 2002). Interestingly, Rep and CDC2a bind to the same region of GRIMP, leading to the suggestion that Rep may prevent the normal function of GRIMP at the mitotic apparatus (Kong and Hanley-Bowdoin, 2002). Furthermore, MSV promoter activity has been shown to be cell cycle-specific: while the coat protein gene promoter has a high activity in G_2 cells, the c-sense promoter exhibits two peaks of activity in S and G_2 cells (Nikovics *et al.*, 2001). This provides further support for the idea of a Rep-mediated effect helping to lock infected cells in S-phase (or G_2). These data, together with the fact that infected cells do not proceed through mitosis (Nagar *et al.*, 1995) and the lack of geminivirus-induced cell proliferation, has led to speculation that TGMV Rep may induce a switch to endoreplication cycles (Kong and Hanley-Bowdoin, 2002). An alternative, or perhaps complementary, possibility that has not yet been explored is that the geminivirus-induced S phase-like state is detected as abnormal, triggering a defence response. In this context, it is interesting that inhibition of post-transcriptional gene silencing (PTGS), also known in animal cells as RNA interference (reviewed by Voinnet, 2002), is dependent on the begomovirus transcriptional activator protein (TrAP; Voinnet *et al.*, 1999). The functional relevance of such a silencing suppression is not yet known but the use of geminivirus-derived vectors could be of significant help (Peele *et al.*, 2001)

9.3 Retroviruses and the cell cycle

Unlike most animal retroviruses, primate lentiviruses [HIV-1, HIV-2 and the simian immunodeficiency virus (SIV)] are able to replicate efficiently in non-dividing cells. Their infections are characterized by a progressive depletion of CD4$^+$ T cells in peripheral blood and lymphoid organs. Several mechanisms have been proposed to account for this loss, including the activation and proliferation of naïve CD4 T cells in primary lymphoid organs during viral infection, and abnormalities in differentiation and disruption of precursor CD4$^+$ T cells during T cell development.

Central to the ability of HIV (and SIV) to replicate in non-dividing cells is Vpr (Vpx in SIV), a ~ 14 kDa, 96-amino acid protein, conserved amongst the primate lentiviruses. Two main functions have been ascribed to Vpr; the first is a role in enhancing nuclear migration of the pre-initiation complex (Sherman and Greene, 2002) and the second is to prevent mitosis (Bukrinsky and Adzhubei, 1999). HIV-infected cells, or cells transfected with an expression plasmid encoding Vpr, do not proliferate but accumulate in the G$_2$ phase of the cell cycle (He *et al.*, 1995; Re *et al.*, 1995). The delay is caused by preventing cyclin B/cdc2 activation and its upstream regulator cdc25 (Re and Luban, 1997). Hrimech *et al.* (2000) suggested that Vpr enhanced the nuclear import of protein phosphatase 2A and also modulating its catalytic activity towards active phosphorylated nuclear Cdc25A phosphatase. This delay in the cell cycle results in an increase in virus production (Re *et al.*, 1995), probably because long terminal repeat-directed transcription is more active in G$_2$ (Goh *et al.*, 1998; Gummuluru and Emerman, 1999). SIV also encodes a protein similar to HIV Vpr that shares 31% amino acid identity, and it too induces cell cycle arrest in G$_2$ (Planelles *et al.*, 1996). However, while arrest in the G$_2$ phase of the cell cycle is necessary for the induction of apoptosis by HIV Vpr via the mitochondria, induction of apoptosis by SIV Vpr is, in contrast, cell cycle-independent (Zhu *et al.*, 2001).

HIV-1 Vpr also shuttles between the nucleus and cytoplasm (Sherman *et al.*, 2001) and induces changes in the nuclear envelope caused by the successive projection and retraction of nuclear envelope herniations. Mutational analysis of Vpr revealed a relationship between cell cycle arrest and changes in the nuclear lamina (de Noronha *et al.*, 2001). Vpr present in the virus particle alone is sufficient for cell cycle arrest (Poon *et al.*, 1998) and, interestingly, Henklein *et al.* (2000) demonstrated that extracellular added Vpr also localized to the nucleus and induced G$_2$ phase arrest, and suggested that free Vpr in the serum of HIV-infected patients may pre-program cells in order to facilitate replication of HIV in infected individuals.

Consistent with the perturbation of the cell cycle observed in tissue culture infected cells, lymphocytes from HIV-infected individuals show high levels of cyclin B and also nucleolar proteins (whose expression is linked to the cell cycle) (Cannavo *et al.*, 2001). Nucleolin has a fundamental role in T lymphocyte-mediated apoptotic cell death and in the target cell is cleaved by granzyme A (Pasternack *et al.*, 1991). Apoptosis is induced by autolytic endonucleases that fragment DNA. The endonucleases are themselves activated by cleaved nucleolin. Although protein profiles in HIV-infected cells are characteristic of cells in the G_2 phase of the cell cycle, the actual metabolic profile and DNA content are consistent with cells that are in the G_0 phase (Piedimonte *et al.*, 1999). Interestingly, treatment of HIV-infected individuals with interleukin-2 normalized intracellular protein turnover and corrected cell cycle abnormalities (Paiardini *et al.*, 2001).

Several other lentivirus proteins interact with the cell cycle. Nef has also been shown to delay $CD4^+$ T cells through the G_1/S phase of the cell cycle, probably through the downregulation of cyclin D1 and cyclin A and an upregulation of p21 and p27, thus possibly contributing to the observed depletion of $CD4^+$ T cells measured during SIV infection (Ndolo *et al.*, 2002). HIV Tat is a potent stimulator of virus transcription and is required for viral replication. Tat *trans*-activation is TAR (tat-response element)-dependent in the G_1 phase of the cell cycle whereas it is TAR-independent in G_2 (Kashanchi *et al.*, 2000). Indeed, Kashanchi *et al.* (2000) hypothesized that transcription in G_1 phase was important for HIV genomic and viral mRNA synthesis, whereas transcription in G_2 phase might activate cellular genes, such as cytokines, important for activating neighbouring cells prior to virus infection. These predictions were subsequently shown to have some basis, in that Tat protein-mediated IL-8 production is regulated in a cell cycle-dependent manner, with maximum synthesis in S phase (Mahieux *et al.*, 2001). Also, Kootstra *et al.* (2000) reported that intracellular pools of nucleotides and cellular factors that are found in late G_1-phase of the cell cycle may play a role in efficient reverse transcription and nuclear localization of the HIV genome.

Human T cell leukaemia virus type 1 (HTLV-1) affects the cell cycle. HTLV-1 causes clonal malignancy of CD4-bearing T lymphocytes. One of the major gene products of this virus, Tax1, causes an activation of transcription of the p21 promoter. This is in comparison with the Tax2 protein from the closely related HTLV-2 (Sieburg *et al.*, 2004). The consequence of expressing Tax1 is that HTLV-1 causes the cell cycle to be pushed through the G_1/S checkpoint by deregulating cyclin and cdk complexes and pushing p21 away from the cyclin E/cdk2 complex, thus allowing this protein complex to phosphorylate Rb (Kehn *et al.*, 2004).

9.4 RNA viruses and the cell cycle

The rationale for the interaction of RNA viruses with the cell cycle might, on first inspection, appear uncertain. For DNA viruses, the rationale is fairly obvious, i.e. the inhibition of cellular DNA synthesis and the promotion of conditions to favour intracellular conditions for viral DNA synthesis. In the main, RNA virus replication (especially positive-strand replication) is cytoplasmic and does not conflict with cellular DNA synthesis, yet several examples of RNA viral interactions with the cell cycle have been described. In the case of positive-strand RNA viruses, interactions with the cell cycle have been known for several years but are probably not widely appreciated.

9.4.1   Positive-strand RNA viruses and the cell cycle

Several examples have been described of different picornaviruses interacting with the cell cycle (Eremenko *et al.*, 1972; Mallucci *et al.*, 1985; Suarez *et al.*, 1975). More recently, Feuer *et al.* (2002) have shown that cells arrested in the G_1 or G_1/S phase produced higher levels of infectious virus and viral polyproteins when compared to cells in the G_0 phase or cells blocked at the G_2–M boundary. Feuer *et al.* (2002) suggested that persistence of *Coxsackie B3 virus* (CVB3) *in vivo* might be dependent on infection of G_0 cells, which do not support replication. If such cells were to re-enter the cell cycle, then the virus may reactive and, in the case of CVB3, trigger chronic viral or immune-mediated pathology in the host. Such findings suggest that locally delivered cell cycle inhibitors could form part of an antiviral therapy, similar to the example of IL-2 used to correct cell cycle aberrations in HIV-infected individuals.

Positive-strand RNA viruses whose genome expression is under the control of internal ribosome entry sites (IRESs), such as the picornaviruses, may take advantage of the cell cycle stage of the infected cell as part of a strategy to promote virus translation. Cells arrested in G_2–M phase translate capped mRNAs 75% less than other interphase cells and this has been attributed to an inhibition in translation initiation. Certainly, translation of the poliovirus genome was shown to be unaffected by the general inhibition of translation in G_2–M phase (Bonneau and Sonenberg, 1987). Translation of a number of cellular genes is under the control of IRES elements, presumably for efficient operation during G_2 and M phase (Hellen and Sarnow, 2001; Sachs, 2000).

Infection of both primary and continuous cells with the avian coronavirus infectious bronchitis virus (IBV) results in an apparent arrest in the G_2/M phase of the cell cycle and inhibition of cytokinesis (Chen *et al.*, 2002; Wurm *et al.*, 2001). These observations may be attributable to the coronavirus nucleoprotein (N protein; the viral RNA binding protein), where overexpression of N proteins from several different coronaviruses causes a delay in cell growth (Chen *et al.*, 2002) and inhibition of cytokinesis and promotion of endoreduplication (Wurm *et al.*, 2001; Figure 9.2). Incidentally, endoreduplication can also be caused by DNA viruses, e.g. papillomavirus (Garner-Hamrick and Fisher, 2002) and Gemini viruses, and Rb would seem to be a target of many viral proteins to cause this effect (Garner-Hamrick and Fisher, 2002; Grafi *et al.*, 1996; Helt *et al.*, 2002b). Again regard to coronaviruses, Chen and Makino (2004) have shown that in cells infected with the murine coronavirus, mouse hepatitis virus (MHV), the cells become arrested in the G_0/G_1 phase of the cell cycle and this results in reduced cellular proliferation.

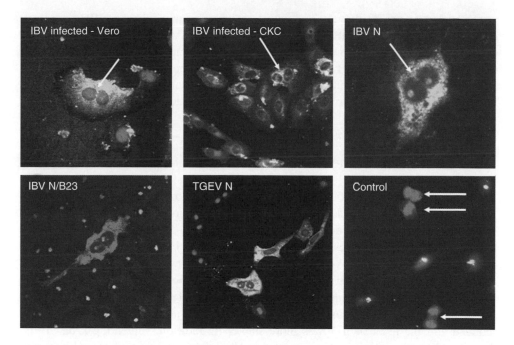

Figure 9.2 An example of endoreduplication in coronavirus infected continuous (IBV infected, Vero) and primary (IBV infected, CKC) cell lines. This phenomena of nuclear division in the absence of cellular division is common in many virus infections. Certainly in the case of coronaviruses this has been shown to be induced by the virus nucleoprotein (IBV N, IBVN/B23 (avian coronaviruses) and TGEV (porcine coronavirus) N). The panel marked control, shows normal cellular cytokinesis, i.e. nucleoli are absent

Hepatitis C virus (HCV), on the other hand, has been reported to promote cell growth through the action of the NS5a protein repressing transcription of the cyclin-dependent kinase inhibitor gene, p21^{WAF1}, while in human hepatoma cells and murine fibroblasts it activates the proliferating cell nuclear antigen gene (Ghosh *et al.*, 1999). Apart from NS5a protein, HCV core protein also interacts with p21^{WAF1}. However, whether this causes up- or downregulation of p21 is uncertain (Honda *et al.*, 2000; Jung *et al.*, 2001; Kwun *et al.*, 2001; Nguyen *et al.*, 2003; Otsuka *et al.*, 2000; Ruggieri *et al.*, 2004). Indeed, Scholle *et al.* (2004) have suggested that reports of different cell cycle effects attributed to the same HCV protein may be due to different experimental systems – including different viral sequences and cell lines and overexpression of only partial segments of the HCV polyprotein. To overcome this, they expressed the full-length HCV polyprotein in cells and analysed cell cycle profiles using flow cytometry, and found that there were no apparent changes in cell cycle profiles compared to control cells. Interestingly, Scholle *et al.* (2004) found that viral RNA synthesis was increased in proliferating cells or those cells in S phase compared to cells which were serum-starved and, presumably in G$_0$, confluent or poorly proliferating.

In contrast to the results reported with translation of the poliovirus IRES and cell cycle status, data suggests that translations from the HCV and encephalomyocarditis (EMCV) virus IRESs are lowest in the G_2/M phases compared to the other phases of the cell cycle (Venkatesan *et al.*, 2003) and that translation from these IRESs is stimulated in S phase. This observation may help explain the results of Scholle *et al.* (2004), in that presumably more viral proteins are available for replication of the viral genome.

9.4.2   Cell cycle and the Mononegavirales

There are several examples of virus–cell cycle interactions in the *Mononegavirales*. Measles virus can cause disease and death by suppression of the immune response. This is thought to occur because measles virus induces G_1 arrest and thus prevents the proliferation of lymphocytes (McChesney *et al.*, 1987, 1988). Naniche *et al.* (1999) showed that in measles virus-infected cells, expression of cyclins D3 and E decreased and Rb protein is not inactivated, resulting in the block in the G_0–G_1 transition. Engelking *et al.* (1999) found that, in peripheral blood lymphocytes infected with UV-inactivated measles virus, the activity of Cdk4-cyclin D and Cdk2-cyclin E were reduced and the level of p27 was more stable. In addition, Engelking *et al.* (1999) described how the activity and accumulation levels of Cdk2 were restricted in mouse spleen cells stimulated with a mitogen and infected with UV-inactivated measles virus. Engelking *et al.* (1999) suggested that the G_1–S phase progression was prevented because measles virus induced deregulation of the relevant Cdk–cyclin complexes. Studies using the cotton rat as a model for measles virus infection indicated that cell cycle delay may prevent the initiation of an effective immune response against infection (Niewiesk *et al.*, 1999). Simian virus 5 also interacts with the cell cycle, in this case protein V slows progression of the cell cycle by delaying the transition from G_1 to S phase, prolonging S phase, and some infected cells were arrested in G_2 or M phase (Lin and Lamb, 2000). *Borna disease virus* (BDV), which replicates in the nucleus/nucleolus and infects the central nervous system, causes a reduction in proliferation in host cells which is due to a delayed G_2–M phase transition (Planz *et al.*, 2003). The cell cycle perturbations observed in BDV-infected cells are due to the viral nucleoprotein forming a complex with Cdc2–cyclin B1, perhaps leading to its inactivation (Planz *et al.*, 2003).

9.4.3   Double-stranded RNA viruses and the cell cycle

Reoviruses are non-enveloped, and have double-stranded segmented RNA genomes. Replication occurs in the cytoplasm of an infected cell, although nuclear functions are altered during virus infection (Tyler *et al.*, 2001). Infection of cultured cells by reovirus results in inhibition of cellular proliferation (Cox and Shaw, 1974) and apoptosis (Debiasi *et al.*, 1999). Serotype 3 σ1s protein causes hyperphosphorylation of Cdc2 (Poggioli *et al.*, 2001) and thus accounts for an observed arrest in G_2/M phase (Poggioli *et al.*, 2000).

9.5   Viruses, the nucleolus and cell cycle control

Apart from direct interactions with cell cycle factors, viruses may also interact with the nucleolus and other subnuclear structures to induce cell cycle perturbations. The nucleolus has been reviewed in detail in the previous chapter, together with its interaction with the cell cycle. Briefly, the nucleolus and associated proteins are also implicated in (and regulated by) the cell cycle (Carmo-Fonseca $et al.$, 2000; Lyon and Lamond, 2000; Olson $et al.$, 2000). For example, the nucleolus sequesters cell cycle regulatory complexes. Hyperphosphorylated Rb protein is sequestered by B23 (a nucleolar protein) to the nucleolus in late S-phase (Takemura $et al.$, 2002). During interphase in higher eukaryotic cells, the number of nucleoli varies depending on the stage of the cell cycle, and the nucleolus disappears at the start of mitosis. During G_1 cells can contain more than one nucleolus. This is probably reflected in the fact that these cells are translationally active, and therefore require more ribosomes. As the cells progress through S phase and into G_2, only single nucleoli may be present. The nucleolus then disperses during mitosis (Olson $et al.$, 2002). Cajal bodies, which contain nucleolar proteins and associate with nucleoli (Ogg and Lamond, 2002), also sequester cell cycle regulatory complexes, including CDK2–cyclin E (Liu $et al.$, 2000)

Therefore, at a gross level virally induced changes in nucleolar architecture may induce cell cycle changes by displacing cell cycle factors and/or, at a more complex level, defined viral proteins may target the nucleolus to target specific cell cycle factors. Certainly there are some examples of viral proteins that localize to the nucleolus and cause cell cycle perturbations, but no formal relationship has been established between the two processes. For example, the coronavirus nucleoprotein has been shown to localize to the nucleolus and cause a reduction in cellular proliferation, as well as a G_2/M phase arrest (Chen $et al.$, 2002). Likewise HIV-1 Vpr protein can localize to the nucleolus (Sherman $et al.$, 2001) and also cause a G_2/M phase arrest (He $et al.$, 1995; Henklein $et al.$, 2000; Re $et al.$, 1995). Indeed, HIV-1 Vpr can induce changes in nuclear architecture which are related to its cell cycle function (de Noronha $et al.$, 2001). The interaction of HIV-1 Rev protein with the nucleolar protein B23 may also impinge on the cell cycle; certainly, expression of Rev protein causes cells to accumulate in either the G_2 phase of the cell cycle, prophase, or mitotic cells which have failed cytokinesis (Miyazaki $et al.$, 1996). In a somewhat similar manner, Rous sarcoma virus-transforming protein pp60v-src localizes to the nucleus/nucleolus and causes a redistribution of nucleolar proteins in infected cells and also interacts with cell cycle components (David-Pfeuty and Nouvian-Dooghe, 1995).

9.6   Viral interaction with activator protein-1 (AP-1)

The activator protein-1 (AP-1) is a collective term referring to a series of homo- and heterodimeric transcription factors composed of Jun, Fos and activating transcription factor (ATF) subunits. These dimers bind to a common DNA binding site, the AP-1 binding site. The DNA binding proteins forming AP-1 dimers are induced by a wide variety of both physiological and pathological stimuli. These then activate transcription of many genes that play essential roles in several physiological functions, including cell proliferation, neoplastic transformation and cell cycle control (Milde-Langosch *et al.*, 2000; Shaulian and Karin, 2001, 2002). AP-1 proteins, especially those that belong to the Jun group, control cell proliferation through their ability to regulate the expression and function of cell cycle regulators such as cyclin D1, p53 and p21. A variety of both RNA and DNA and retroviruses have been shown to either induce or suppress AP-1 expression within infected cells and therefore may have concomitant cell cycle effects.

HCV non-structural protein NS5A protein plays a key role in virus replication and has also been demonstrated as being capable of manipulation of host cell physiology, including inhibition of AP-1 function (Macdonald and Harris, 2004). AP-1 activity was inhibited by perturbation of the ras–ERK signalling pathway and it was postulated that downregulation of the ERK MAPK pathway by NS5A could be a mechanism to induce the upregulation of P13-kinase, a regulator of cell proliferation and apoptosis, to protect HCV-infected cells against apoptotic stimulation (Macdonald *et al.*, 2004; Street *et al.*, 2004). Recently, the nucleocapsid (N) protein of the severe acute respiratory virus (SARS) coronavirus has been shown to induce the upregulation of various AP-1 factors when expressed within cells (He *et al.*, 2003), although the status of AP-1 in infected cells is unknown. It is conceivable that the coronavirus N protein could be manipulating the cell cycle of infected host cells through interactions with AP-1 transcription factors, such as the downregulation of p21.

Epstein–Barr virus latent membrane protein 1 (LMP1) has been shown to activate the AP-1 pathway and cause an arrest at the G_2/M phase of the cell cycle (Deng *et al.*, 2003). In contrast, the simian virus 40 small t antigen also activates AP-1 to cause cell cycle progression (Howe *et al.*, 1998); in this case the protein must also inhibit protein phosphatase 2A (PP2A) and also causes induction of cyclin D1 through an AP-1-mediated effect (Watanabe *et al.*, 1996). Transcription of the E6 and E7 oncoproteins of human papillomavirus is regulated by AP-1 (Kyo *et al.*, 1997); these viral proteins are intimately involved in cellular proliferation and also maintaining an efficient infection.

Similar to how free HIV-1 Vpr has been shown to enter cells and promote progression of the cell cycle into the G_2 phase, the envelope glycoprotein 120 (gp120) might also have the same effect. Studies on heat-inactivated virus suggested that when added to cells these particles could stimulate the AP-1 pathway and cause cells to enter G_2 phase (Briant *et al.*, 1996).

9.7   Summary

Viral interactions with the cell cycle appear to be pan-virus phenomena (Figure 9.3). By comparing the variety of mechanisms used by different viruses to subvert the host

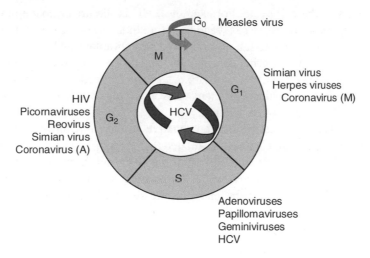

Figure 9.3 Diagram showing the interaction of a number of viruses discussed in this chapter with the cell cycle. Most of the literature discussing DNA viruses and the cell cycle reports arrests in either the G_1 or S phases depending on the replication mechanism. In comparison, there are very few examples of DNA viruses causing an arrest in G_2/M. A number of RNA viruses can arrest at multiple cell cycle stages, e.g. simian virus. Hepatitis C virus (HCV) has been reported to stimulate cellular proliferation with replication of the viral genome being more efficient in the S phase of the cell cycle. Viruses of the same family can cause different cell cycle perturbations. For example the avian coronavirus (coronavirus A) can arrest cells in G_2, where as the murine coronavirus (coronavirus M) arrests cell in G_1

cell cycle, a number of different reasons can be elucidated as to why viruses would do this. In the case of DNA viruses, the integration of virus replication with different phases of the host cell cycle, especially host cell DNA replication (depending on the type of virus), is logical. In the case of RNA viruses, the validity of such an explanation is less apparent. Positive-strand RNA viruses may alter the host cell cycle in order to promote favourable conditions for translation of genomes. In common with retroviruses, both negative- and positive-strand RNA viruses may alter the host cell cycle in order to modulate host immune responses. Indeed, induction of cell cycle changes by free-floating HIV Vpr provides an excellent example of how viruses could pre-programme cells prior to virus infection in order to generate optimum conditions, perhaps by inducing genes to generate appropriate metabolic profiles. In yeast, estimates suggest that one-eighth of all genes are cell cycle-regulated (Lord *et al.*, 2000); a similar proportion may be true for other eukaryotic cells. Like retroviruses, both DNA and RNA viruses can stimulate cells from a quiescent state, e.g. the human cytomegalovirus (HCMV) pp71 protein contains a retinoblastoma protein (Rb)-binding motif (LxCxD), and stimulates DNA synthesis in quiescent cells (Kalejta, 2004); likewise, hepatitis C virus would appear to replicate better in cells which are in S phase (Scholle *et al.*, 2004).

Indeed, it may be a common paradigm for viruses that infect quiescent cells to stimulate these cells into G_1 (via activation of AP-1) and then arrest these cells at different stages of the cell cycle. For example, cells in the uppermost epithelial layer in, say, the respiratory tract are in G_0 and therefore quiescent. Viruses will infect these cells, and may through the action of viral protein(s) stimulated AP-1 signal transduction. This results in cells entering the cell cycle and proceeding to G_1. Viral translation occurs in G_1, although host cell translation may be reduced. Cells become arrested before mitosis in order to allow efficient virus assembly. Some viruses will arrest in G_2 to allow IRES-dependent translation.

Intriguingly, as part of their studies on negative-strand RNA virus interactions with the cell cycle, Lin and Lamb (2000) postulated that enveloped viruses might delay the cell cycle in order to promote suitable conditions for virus assembly, especially if the virus life cycle is longer than the cell cycle. Intracellular sites of assembly for enveloped viruses include the Golgi and the endoplasmic reticulum, which are disorganized during cell division.

References

Accotto, G. P., Mullineaux, P. M., Brown, S. C. and Marie, D. (1993). Digitaria streak geminivirus replicative forms are abundant in S-phase nuclei of infected cells. *Virology* **195**, 257–259.

Ach, R. A., Durfee, T., Miller, A. B., Taranto, P., Hanley-Bowdoin, L., Zambryski, P. C. and Gruissem, W. (1997). *RRB1* and *RRB2* encode maize retinoblastoma-related proteins that interact with a plant D-type cyclin and geminivirus replication protein. *Mol Cell Biol* **17**, 5077–5086.

Batchu, R. B., Shammas, M. A., Wang, J. Y., Freeman, J., Rosen, N. and Munshi, N. C. (2002). Adeno-associated virus protects the retinoblastoma family of proteins from adenoviral-induced functional inactivation. *Cancer Res* **62**, 2982–2985.

Bell, S. P. and Dutta, A. (2002). DNA replication in eukaryotic cells. *Annu Rev Biochem* **71**, 333–374.

Ben-Israel, H. and Kleinberger, T. (2002). Adenovirus and cell cycle control. *Front Biosci* **7**, d1369–1395.

Boniotti, M. B. and Gutierrez, C. (2001). A cell-cycle-regulated kinase activity phosphorylates plant retinoblastoma protein and contains, in *Arabidopsis*, a CDKA/cyclin D complex. *Plant J* **28**, 341–350.

Bonneau, A. M. and Sonenberg, N. (1987). Involvement of the 24 kDa cap-binding protein in regulation of protein synthesis in mitosis. *J Biol Chem* **262**, 11134–11139.

Boulton, M. I. (2002). Functions and interactions of mastrevirus gene products. *Physiol Mol Plant Pathol* **60**, 243–255.

Briant, L., Coudronniere, N., Robert-Hebmann, V., Benkirane, M. and Devaux, C. (1996). Binding of HIV-1 virions or gp120-anti-gp120 immune complexes to HIV-1-infected quiescent peripheral blood mononuclear cells reveals latent infection. *J Immunol* **156**, 3994–4004.

Brooks, G., Poolman, R. A. and Li, J. M. (1998). Arresting developments in the cardiac myocyte cell cycle: role of cyclin-dependent kinase inhibitors. *Cardiovasc Res* **39**, 301–311.

Bukrinsky, M. and Adzhubei, A. (1999). Viral protein R of HIV-1. *Rev Med Virol* **9**, 39–49.

Cannavo, G., Paiardini, M., Galati, D., Cervasi, B., Montroni, M., De Vico, G., Guetard, D., Bocchino, M. L., Picerno, I., Magnani, M., Silvestri, G. and Piedimonte, G. (2001). Abnormal intracellular kinetics of cell-cycle-dependent proteins in lymphocytes from patients infected with human immunodeficiency virus: a novel biologic link between immune activation, accelerated T-cell turnover, and high levels of apoptosis. *Blood* **97**, 1756–1764.

Carmo-Fonseca, M., Mendes-Soares, L. and Campos, I. (2000). To be or not to be in the nucleolus. *Nat Cell Biol* **2**, E107–E112.

Castillo, J. P. and Kowalik, T. F. (2002). Human cytomegalovirus immediate early proteins and cell growth control. *Gene* **290**, 19–34.

Cayrol, C. and Flemington, E. K. (1996). The Epstein-Barr virus bZIP transcription factor Zta causes G_0/G_1 cell cycle arrest through induction of cyclin-dependent kinase inhibitors. *EMBO J* **15**, 2748–2759.

Chen, C. J. and Makino, S. (2004). Murine coronavirus replication induces cell cycle arrest in G_0/G_1 phase. *J Virol* **78**, 5658–5669.

Chen, H., Wurm, T., Britton, P., Brooks, G. and Hiscox, J. A. (2002). Interaction of the coronavirus nucleoprotein with nucleolar antigens and the host cell. *J Virol* **76**, 5233–5250.

Chen, Z., Knutson, E., Kurosky, A. and Albrecht, T. (2001). Degradation of p21cip1 in cells productively infected with human cytomegalovirus. *J Virol* **75**, 3613–3625.

Cox, D. C. and Shaw, J. E. (1974). Inhibition of the initiation of cellular DNA synthesis after reovirus infection. *J Virol* **13**, 760–761.

Cross, F. R. and Roberts, J. M. (2001). Retinoblastoma protein: combating algal bloom. *Curr Biol* **11**, 824–827.

David-Pfeuty, T. and Nouvian-Dooghe, Y. (1995). Highly specific antibody to Rous sarcoma virus *src* gene product recognizes nuclear and nucleolar antigens in human cells. *J Virol* **69**, 1699–1713.

de Noronha, C. M. C., Sherman, M. P., Lin, H., Cavrois, M. V. and Greene, W. C. (2001). HIV-1 Vpr induces dynamic disruptions in nuclear envelope architecture and integrity. *Science* **294**, 1105–1108.

Debiasi, R. L., Squier, M. K., Pike, B., Wynes, M., Dermody, T. S., Cohen, J. J. and Tyler, K. L. (1999). Reovirus-induced apoptosis is preceded by increased cellular calpain activity and is blocked by calpain inhibitors. *J Virol* **73**, 695–701.

Deng, L., Yang, J., Zhao, X. R., Deng, X. Y., Zeng, L., Gu, H. H., Tang, M. and Cao, Y. (2003). Cells in G_2/M phase increased in human nasopharyngeal carcinoma cell line by EBV-LMP1 through activation of NF-κB and AP-1. *Cell Res* **13**, 187–194

Dhar, S. K., Yoshida, K., Machida, Y., Khaira, P., Chaudhuri, B., Wohlschlegel, J. A., Leffak, M., Yates, J. and A, D. (2001). Replication from oriP of Epstein–Barr virus requires human ORC and is inhibited by geminin. *Cell* **106**, 287–296.

DiMaio, D. and Mattoon, D. (2001). Mechanisms of cell transformation by papillomavirus E5 proteins. *Oncogene* **20**, 7866–7873.

Egelkrout, E. M., Robertson, D. and Hanley-Bowdoin, L. (2001). Proliferating cell nuclear antigen transcription is repressed through an E2F consensus element and activated by geminivirus infection in mature leaves. *Plant Cell* **13**, 1437–1452.

Engelking, O., Fedorov, L., Lilischkis, R., ter Meulen, V. and Schneider-Schaulies, S. (1999). Measles virus-induced immunosuppression *in vitro* is associated with deregulation of G(1) cell cycle control proteins. *J Gen Virol* **80**, 1599–1608.

Eremenko, T., Benedetto, A. and Volpe, P. (1972). Virus infection as a function of the host cell life cycle: replication of poliovirus RNA. *J Gen Virol* **16**, 61–68.

Feuer, R., Mena, I., Pagarigan, R., Slifka, M. K. and Whitton, J. L. (2002). Cell cycle status affects Coxsackievirus replication, persistence, and reactivation *in vitro*. *J Virol* **76**, 4430–4440.

Flemington, E. K. (2001). Herpesvirus lytic replication and the cell cycle: arresting new developments. *J Virol* **75**, 4475–4481.

Garner-Hamrick, P. A. and Fisher, C. (2002). HPV episomal copy number closely correlates with cell size in keratinocyte monolayer cultures. *Virology* **301**, 334–341.

Ghosh, A. K., Steele, R., Meyer, K., Ray, R. and Ray, R. B. (1999). Hepatitis C virus NS5a protein modulates cell cycle regulatory genes and promotes cell growth. *J Gen Virol* **80**, 1179–1183.

Giarre, M., Caldeira, S., Malanchi, I., Ciccolini, F., Leao, M. J. and Tommasino, M. (2001). Induction of pRb degradation by the human papillomavirus type 16 E7 protein is essential to efficiently overcome p16INK4a-imposed G_1 cell cycle arrest. *J Virol* **75**, 4705–4712.

Goh, W. C., Rogel, M. E., Kinsey, C. M., Michael, S. F., Fultz, P. N., Nowak, M. A., Hahn, B. H. and Emerman, M. (1998). HIV-1 Vpr increases viral expression by manipulation of the cell cycle: a mechanism for selection of Vpr *in vivo*. *Nat Med* **4**, 65–71.

Grafi, G., Burnett, R. J., Helentjaris, T., Larkins, B. A., DeCaprio, J. A., Sellers, W. R. and Kaelin, W. G. Jr (1996). A maize cDNA encoding a member of the retinoblastoma protein family: involvement in endoreduplication. *Proc Natl Acad Sci USA* **93**, 8962–8967.

Guertin, D. A., Trautmann, S. and McCollum, D. (2002). Cytokinesis in eukaryotes. *Microbiol Mol Biol Rev* **66**, 155–178.

Gummuluru, S. and Emerman, M. (1999). Cell cycle- and Vpr-mediated regulation of human immunodeficiency virus type 1 expression in primary and transformed T-cell lines. *J Virol* **73**, 5422–5430.

Gutierrez, C. (1998). The retinoblastoma pathway in plant cell cycle and development. *Curr Opin Plant Biol* **1**, 492–497.

Gutierrez, C. (1999). Geminivirus DNA replication. *Cell Mol Life Sci* **56**, 313–329.

Gutierrez, C. (2000). DNA replication and cell cycle in plants: learning from geminiviruses. *EMBO J* **19**, 792–799.

Gutierrez, C. (2002). Strategies for geminivirus DNA replication and cell cycle interference. *Physiol Mol Plant Pathol* **60**, 219–230.

Hanley-Bowdoin, L., Settlage, S. B., Orozco, B. M., Nagar, S. and Robertson, D. (1999). Geminiviruses: models for plant DNA replication, transcription and cell cycle regulation. *Crit Rev Plant Sci* **18**, 71–106.

Hardwick, J. M. (2000). Cyclin on the viral path to destruction. *Nat Cell Biol* **2**, 203–204.

Hartwell, L. H. and Kastan, M. B. (1994). Cell cycle control and cancer. *Science* **266**, 1821–1828.

He, J., Choe, S., Walker, R., Di Marzio, P., Morgan, D. O. and Landau, N. R. (1995). Human immunodeficiency virus type 1 protein R (Vpr) blocks cells in the G_2 phase of the cell cycle by inhibiting $p34^{cdc2}$ activity. *J Virol* **69**, 6705–6711.

He, R., Leeson, A., Andonov, A., Li, Y., Bastien, N., Cao, J., Osiowy, C., Dobie, F., Cutts, T., Ballantine, M. and Li, X. (2003). Activation of AP-1 signal transduction pathway by SARS coronavirus nucleocapsid protein. *Biochem Biophys Res Commun* **311**, 870–876.

Hellen, C. U. T. and Sarnow, P. (2001). Internal ribosome entry sites in eukaryotic mRNA molecules. *Genes Dev* **15**, 1593–1612.

Helt, A. M., Funk, J. O. and Galloway, D. A. (2002a). Inactivation of both the retinoblastoma tumor suppressor and p21 by the human papillomavirus type 16 E7 oncoprotein is necessary to inhibit cell cycle arrest in human epithelial cells. *J Virol* **76**, 10559–10568.

Helt, A. M., Funk, J. O. and Galloway, D. A. (2002b). Inactivation of both the retinoblastoma tumor suppressor and p21 by the human papillomavirus type 16 E7 oncoprotein is necessary to inhibit cell cycle arrest in human epithelial cells. *J Virol* **76**, 10559–10568.

Henklein, P., Bruns, K., Sherman, M. P., Tessmer, U., Licha, K., Kopp, J., de Noronha, C. M. C., Greene, W. C., Wray, V. and Schubert, U. (2000). Functional and structural characterization of synthetic HIV-1 Vpr that transduces cells, localizes to the nucleus and induces G_2 cell cycle arrest. *J Biol Chem* **275**, 32016–32026.

Hollander, M. C. and Fornace, A. J. Jr (2002). Genomic instability, centrosome amplification, cell cycle checkpoints and Gadd45a. *Oncogene* **21**, 6228–6233.

Honda, M., Kaneko, S., Shimazaki, T., Matsushita, E., Kobayashi, K., Ping, L. H., Zhang, H. C. and Lemon, S. M. (2000). Hepatitis C virus core protein induces apoptosis and impairs cell-cycle regulation in stably transformed Chinese hamster ovary cells. *Hepatology* **31**, 1351–1359.

Horvath, G. V., Pettko-Szandtner, A., Nikovics, K., Bilgin, M., Boulton, M., Davies, J. W., Gutierrez, C. and Dudits, D. (1998). Prediction of functional regions of the maize streak virus replication-associated proteins by protein–protein interaction analysis. *Plant Mol Biol* **38**, 699–712.

Howe, A. K., Gaillard, S., Bennett, J. S. and Rundell, K. (1998). Cell cycle progression in monkey cells expressing simian virus 40 small t antigen from adenovirus vectors. *J Virol* **72**, 9637–9644.

Hrimech, M., Yao, X.-J., Branton, P. E. and Cohen, E. A. (2000). Human immunodeficinecy virus type 1 Vpr-mediated G_2 cell cycle arrest: Vpr interferes with cell cycle signaling cascades by interacting with the B subunit of serine/threonine protein phosphatase 2A. *EMBO J* **19**, 3956–3967.

Jung, E. Y., Lee, M. N., Yang, H. Y., Yu, D.-Y. and Jang, K. L. (2001). The repressive activity of hepatitis C virus core protein on the transcription of p21waf1 is regulated by protein kinase A-mediated phosphorylation. *Virus Res* **79**, 109–115.

Kalejta, R. F. (2004). Human cytomegalovirus pp71: a new viral tool to probe the mechanisms of cell cycle progression and oncogenesis controlled by the retinoblastoma family of tumor suppressors. *J Cell Biochem* **93**, 37–45.

Kalejta, R. F. and Shenk, T. (2002). Manipulation of the cell cycle by human cytomegalovirus. *Front Biosci* **7**, d295–306.

Kashanchi, F., Agbottah, E. T., Pise-Masison, C. A., Mahieux, R., Duvall, J., Kumar, A. and Brady, J. N. (2000). Cell cycle-regulated transcription by the human immunodeficiency virus type 1 Tat transactivator. *J Virol* **74**, 652–660.

Kehn, K., Deng, L., de la Fuente, C., Strouss, K., Wu, K., Maddukuri, A., Baylor, S., Rufner, R., Pumfery, A., Bottazzi, M. E. and Kashanchi, F. (2004). The role of cyclin D2 and p21/waf1 in human T-cell leukemia virus type 1 infected cells. *Retrovirology* **1**, 6.

Kong, L. J. and Hanley-Bowdoin, L. (2002). A geminivirus replication protein interacts with a protein kinase and a motor protein that display different expression patterns during plant development and infection. *Plant Cell* **14**, 1817–1832.

Kong, L. J., Orozco, B. M., Roe, J. L., Nagar, S., Ou, S., Feiler, H. S., Durfee, T., Miller, A. B., Gruissem, W., Robertson, D. and Hanley-Bowdoin, L. (2000). A geminivirus replication protein interacts with retinoblastoma through a novel domain to determine symptoms and tissue specificity of infection in plants. *EMBO J* **19**, 3485–3495.

Kootstra, N. A., Zwart, B. M. and Schuitemaker, H. (2000). Diminished human immunodeficiency virus type 1 reverse transcription and nuclear transport in primary macrophages arrested in early G_1 phase of the cell cycle. *J Virol* **74**, 1712–1717.

Kwun, H. J., Jung, E. Y., Ahn, J. Y., Lee, M. N. and Jang, K. L. (2001). p53-dependent transcriptional repression of p21waf1 by hepatitis C virus NS3. *J Gen Virol* **82**, 2235–2241.

Kyo, S., Klumpp, D. J., Inoue, M., Kanaya, T. and Laimins, L. A. (1997). Expression of AP1 during cellular differentiation determines human papillomavirus E6/E7 expression in stratified epithelial cells. *J Gen Virol* **78** (2), 401–411.

Laman, H., Mann, D. J. and Jones, N. C. (2000). Viral-encoded cyclins. *Curr Opin Genet Dev* **10**, 70–74.

Lin, G. Y. and Lamb, R. A. (2000). The paramyxovirus simian virus 5 V protein slows progression of the cell cycle. *J Virol* **74**, 9152–9166.

Liu, J.-L., Hebert, M. B., Ye, Y., Templeton, D. J., King, H.-J. and Matera, A. G. (2000). Cell cycle-dependent localization of the CDK2–cyclin E complex in Cajal (coiled) bodies. *J Cell Sci* **113**, 1543–1552.

Liu, L., Saunders, K., Thomas, C. L., Davies, J. W. and Stanley, J. (1999). Bean yellow dwarf virus RepA, but not Rep, binds to maize retinoblastoma protein, and the virus tolerates mutations in the consensus binding motif. *Virology* **256**, 270–279.

Lord, M., Yang, M. C., Mischke, M. and Chant, J. (2000). Cell cycle programs of gene expression control morphogenetic protein localization. *J Cell Biol* **151**, 1501–1512.

Lu, M. and Shenk, T. (1999). Human cytomegalovirus UL69 protein induces cells to accumulate in G_1 phase of the cell cycle. *J Virol* **73**, 676–683.

Ludlow, J. W. (1993). Interactions between SV40 large-tumor antigen and the growth suppressor proteins pRB and p53. *FASEB J* **7**, 866–871.

Luque, A., Sanz-Burgos, A. P., Ramirez-Parra, E., Castellano, M. M. and Gutierrez, C. (2002). Interaction of geminivirus Rep protein with replication factor C and its potential role during geminivirus DNA replication. *Virology* **302**, 83–94.

Lyon, C. E. and Lamond, A. I. (2000). The nucleolus. *Curr Biol* **10**, R323.

Macdonald, A., Crowder, K., Street, A., McCormick, C. and Harris, M. (2004). The hepatitis C virus NS5A protein binds to members of the Src family of tyrosine kinases and regulates kinase activity. *J Gen Virol* **85**, 721–729.

Macdonald, A. and Harris, M. (2004). Hepatitis C virus NS5A: tales of a promiscuous protein. *J Gen Virol* **85**, 2485–2502.

Mahieux, R., Lambert, P. F., Agbottah, E., Halanski, M. A., Deng, L., Kashanchi, F. and Brady, J. N. (2001). Cell cycle regulation of human interleukin-8 gene expression by the human immunodeficiency virus type 1 Tat protein. *J Virol* **75**, 1736–1743.

Mallucci, L., Wells, V. and Beare, D. (1985). Cell cycle position and expression of encephalomyocarditis virus in mouse embryo fibroblasts. *J Gen Virol* **66**, 1501–1506.

Mauser, A., Holley-Guthrie, E., Simpson, D., Kaufmann, W. and Kenney, S. (2002). The Epstein–Barr virus immediate-early protein BZLF1 induces both a G_2 and a mitotic block. *J Virol* **76**, 10030–10037.

McChesney, M. B., Altman, A. and Oldstone, M. B. A. (1988). Suppression of T lymphocyte function by measles virus is due to cell cycle arrest in G_1. *J Immunol* **140**, 1269–1273.

McChesney, M. B., Kehrl, J. H., Valsamakis, A., Fauci, A. S. and Oldstone, M. B. A. (1987). Measles virus infection of B lymphocytes permits cellular activation but blocks progression through the cell cycle. *J Virol* **61**, 3441–3447.

Milde-Langosch, K., Bamberger, A. M., Methner, C., Rieck, G. and Loning, T. (2000). Expression of cell cycle-regulatory proteins rb, p16/MTS1, p27/KIP1, p21/WAF1, cyclin D1 and cyclin E in breast cancer: correlations with expression of activating protein-1 family members. *Int J Cancer* **87**, 468–472.

Mittnacht, S. (1998). Control of pRB phosphorylation. *Curr Opin Genet Dev* **8**, 21–27.

Miyazaki, Y., Nosaka, T. and Hatanaka, M. (1996). The post-transcriptional regulator Rev of HIV: implications for its interaction with the nucleolar protein B23. *Biochimie* **78**, 1081–1086.

Moffat, A. S. (1999). Geminiviruses emerge as serious crop threat. *Science* **286**, 1835.

Moran, E. (1993). Interactions of adenoviral proteins with pRB and p53. *FASEB J* **7**, 880–885.

Morita, E., Tada, K., Chisaka, H., Asao, H., Sato, H., Yaegashi, N. and Sugamura, K. (2001). Human parvovirus B19 induces cell cycle arrest at G(2) phase with accumulation of mitotic cyclins. *J Virol* **75**, 7555–7563.

Nagar, S., Pedersen, T. J., Carrick, K. M. and Hanley-Bowdoin, L. (1995). A geminivirus induces expression of a host DNA synthesis protein in terminally differentiated plant cells. *Plant Cell* **7**, 705–719.

Naniche, D., Reed, S. I. and Oldstone, M. B. A. (1999). Cell cycle arrest during measles virus infection: a G_0-like block leads to suppression of retinoblastoma protein expression. *J Virol* **73**, 1894–1901.

Ndolo, T., Dhillon, N. K., Nguyfen, H., Guadalupe, M., Mudryj, M. and Dandekar, S. (2002). Simian immunodeficiency virus nef protein delays the progression of CD4$^+$ T cells through G_1/S phase of the cell cycle. *J Virol* **76**, 3587–3595.

Nguyen, H., Mudryj, M., Guadalupe, M. and Dandekar, S. (2003). Hepatitis C virus core protein expression leads to biphasic regulation of the p21 cdk inhibitor and modulation of hepatocyte cell cycle. *Virology* **312**, 245–253.

Niewiesk, S., Ohnimus, H., Schnorr, J.-J., Gotzelmann, M., Schneider-Schaulies, S., Jassoy, C. and ter Meulen, V. (1999). Measles virus-induced immunosuppression in cotton rats is associated with cell cycle retardation in uninfected lymphocytes. *J Gen Virol* **80**, 2023–2029.

Nikovics, K., Simidjieva, J., Peres, A., Ayaydin, F., Pasternak, T., Davies, J. W., Boulton, M. I., Dudits, D. and Horvath, G. V. (2001). Cell-cycle, phase-specific activation of Maize streak virus promoters. *Mol Plant Microbe Interact* **14**, 609–617.

Ogg, S. C. and Lamond, A. I. (2002). Cajal bodies and coilin-moving towards function. *J Cell Biol* **159**, 17–21.

Oleksiewicz, M. B. and Alexandersen, S. (1997). S-phase-dependent cell cycle disturbances caused by Aleutian mink disease parvovirus. *J Virol* **71**, 1386–1396.

Olson, M. O., Dundr, M. and Szebeni, A. (2000). The nucleolus: an old factory with unexpected capabilities. *Trends Cell Biol* **10**, 189–196.

Olson, M. O., Hingorani, K. and Szebeni, A. (2002). Conventional and nonconventional roles of the nucleolus. *Int Rev Cytol* **219**, 199–266.

Op De Beeck, A. and Caillet-Fauquet, P. (1997). Viruses and the cell cycle. In *Progress in Cell Cycle Research*, pp. 1–19, L. Meijer, S. Guidet and M. Philippe (eds). Plenum: New York.

Op De Beeck, A., Sobczak-Thepot, J., Sirma, H., Bourgain, F., Brechot, C. and Caillet-Fauquet, P. (2001). NS1 and minute virus of mice-induced cell cycle arrest: involvement of p53 and p21^{cip1}. *J Virol* **75**, 11071–11078.

Otsuka, M., Kato, N., Lan, K., Yoshida, H., Kato, J., Goto, T., Shiratori, Y. and Omata, M. (2000). Hepatitis C virus core protein enhances p53 function through augmentation of DNA binding affinity and transcriptional ability. *J Biol Chem* **275**, 34122–34130.

Paiardini, M., Galati, D., Cervasi, B., Cannavo, G., Galluzzi, L., Montroni, M., Guetard, D., Magnani, M., Piedimonte, G. and Silvestri, G. (2001). Exogenous interleukin-2 administration corrects the cell cycle perturbation of lymphocytes from human immunodeficiency virus-infected individuals. *J Virol* **75**, 10843–10855.

Palmer, K. E. and Rybicki, E. P. (1998). The molecular biology of mastreviruses. *Adv Virus Res* **50**, 183–234.

Pardee, A. B. (1989). G_1 events and regulation of cell proliferation. *Science* **246**, 603–608.

Pasternack, M. S., Bleier, K. J. and McInerney, T. N. (1991). Granzyme A binding to target cell proteins. Granzyme A binds to and cleaves nucleolin *in vitro*. *J Biol Chem* **266**, 14703–14708.

Peele, C., Jordan, C. V., Muangsan, N., Turnage, M., Egelkrout, E., Eagle, P., Hanley-Bowdoin, L. and Robertson, D. (2001). Silencing of a meristematic gene using geminivirus-derived vectors. *Plant J* **27**, 357–366.

Piedimonte, G., Corsi, D., Paiardini, M., Cannavo, G., Ientile, R., Picerno, I., Montroni, M., Silvestri, G. and Magnani, M. (1999). Unscheduled cyclin B expression and p34 cdc2 activation in T lymphocytes from HIV-infected patients. *AIDS* **13**, 1159–1164.

Pines, J. (1993). Cyclins and cyclin-dependent kinases: take your partners. *Trends Biochem Sci* **18**, 195–197.

Pines, J. (1999). Four-dimensional control of the cell cycle. *Nature Cell Biol* **1**, 73–79.

Planelles, V., Jowett, J. B. M., Li, Q. X., Xie, Y., Hahn, B. and Chen, I. S. Y. (1996). Vpr-induced cell cycle arrest is conserved among primate lentiviruses. *J Virol* **70**, 2516–2524.

Planz, O., Pleschka, S., Oesterle, K., Berberich-Siebelt, F., Ehrhardt, C., Stitz, L. and Ludwig, S. (2003). Borna disease virus nucleoprotein interacts with the Cdc2–cyclin B1 complex. *J Virol* **77**, 11186–11192.

Poggioli, G. J., Dermody, T. S. and Tyler, K. L. (2001). Reovirus-induced sigma1s-dependent G(2)/M phase cell cycle arrest is associated with inhibition of p34(cdc2). *J Virol* **75**, 7429–7434.

Poggioli, G. J., Keefer, C., Connolly, J. L., Dermody, T. S. and Tyler, K. L. (2000). Reovirus-induced G_2/M cell cycle arrest requires σ1s and occurs in the absence of apoptosis. *J Virol* **74**, 9562–9570.

Poon, B., Grovit-Ferbans, K., Stewart, S. A. and Chen, I. S. Y. (1998). Cell cycle arrest by Vpr in HIV-1 virions and insensitivity to antiretroviral agents. *Science* **281**, 266–269.

Re, F., Braaten, D., Franke, E. K. and Luban, J. (1995). Human immunodeficiency virus type 1 Vpr arrests the cell cycle in G_2 by inhibiting the activation of $p34^{cdc2}$-cyclin B. *J Virol* **69**, 6859–6864.

Re, F. and Luban, J. (1997). HIV-1 Vpr: G_2 cell cycle arrest, macrophages and nuclear transport. *Progr Cell Cycle Res* **3**, 21–27.

Ruggieri, A., Murdolo, M., Harada, T., Miyamura, T. and Rapicetta, M. (2004). Cell cycle perturbation in a human hepatoblastoma cell line constitutively expressing hepatitis C virus core protein. *Arch Virol* **149**, 61–74.

Rupes, I. (2002). Checking cell size in yeast. *Trends Genet* **18**, 479–485.

Sachs, A. B. (2000). Cell cycle-dependent translation initiation: IRES elements prevail. *Cell* **101**, 243–245.

Sanchez, V., McElroy, A. K. and Spector, D. H. (2003). Mechanisms governing maintenance of Cdk1/cyclin B1 kinase activity in cells infected with human cytomegalovirus. *J Virol* **77**, 13214–13224.

Scholle, F., Li, K., Bodola, F., Ikeda, M., Luxon, B. A. and Lemon, S. M. (2004). Virus–host cell interactions during hepatitis C virus RNA replication: impact of polyprotein expression on the cellular transcriptome and cell cycle association with viral RNA synthesis. *J Virol* **78**, 1513–1524.

Schulze-Gahmen, U. and Kim, S. H. (2002). Structural basis for CDK6 activation by a virus-encoded cyclin. *Nature Struct Biol* **9**, 177–181.

Shaulian, E. and Karin, M. (2001). AP-1 in cell proliferation and survival. *Oncogene* **20**, 2390–2400.

Shaulian, E. and Karin, M. (2002). AP-1 as a regulator of cell life and death. *Nat Cell Biol* **4**, E131–136.

Sherman, M. P., de Noronha, C. M. C., Heusch, M. I., Greene, S. and Greene, W. C. (2001). Nucleocytoplasmic shuttling by human immunodeficiency virus type 1 Vpr. *J Virol* **75**, 1522–1532.

Sherman, M. P. and Greene, W. C. (2002). Slipping through the door: HIV entry into the nucleus. *Microb Infect* **4**, 67–73.

Sieburg, M., Tripp, A., Ma, J.-W. and Feuer, G. (2004). Human T-cell leukemia virus type 1 (HTLV-1) and HTLV-2 Tax oncoproteins modulate cell cycle progression and apoptosis. *J Virol* **78**, 10399–10409.

Simanis, V., Hayles, J. and Nurse, P. (1987). Control over the onset of DNA synthesis in fission yeast. *Phil Trans R Soc Lond B Biol Sci* **317** (1187), 507–516.

Simmons, D. T. (2000). SV40 large T antigen functions in DNA replication and transformation. *Adv Virus Res* **55**, 75–134.

Street, A., Macdonald, A., Crowder, K. and Harris, M. (2004). The hepatitis C virus NS5A protein activates a phosphoinositide 3-kinase-dependent survival signaling cascade. *J Biol Chem* **279**, 12232–12241.

Suarez, M., Contreras, G. and Friedlender, B. (1975). Multiplication of Coxsackie B1 virus in synchronized HeLa cells. *J Virol* **16**, 1337–1339.

Sullivan, C. S. and Pipas, J. M. (2002). T antigens of simian virus 40: molecular chaperones for viral replication and tumor genesis. *Microbiol Mol Biol Rev* **66**, 179–202.

Swanton, C., Card, G. L., Mann, D., McDonald, N. and Jones, N. (1999). Overcoming inhibitions: subversion of CKI function by viral cyclins. *Trends Biochem Sci* **24**, 116–120.

Takemura, M., Ohoka, F., Perpelescu, M., Ogawa, M., Matsushita, H., Takaba, T., Akiyama, T., Umekawa, H., Furuichi, Y., Cook, P. R. and Yoshida, S. (2002). Phosphorylation-dependent migration of retinoblastoma protein into the nucleolus triggered by binding to nucleophosmin/B23. *Exp Cell Res* **276**, 233–241.

Tyler, K. L., Clarke, P., DeBiasi, R. L., Kominsky, D. and Poggioli, G. J. (2001). Reoviruses and the host cell. *Trends Microbiol* **9**, 560–564.

Venkatesan, A., Sharma, R. and Dasgupta, A. (2003). Cell cycle regulation of hepatitis C and encephalomyocarditis virus internal ribosome entry site-mediated translation in human embryonic kidney 293 cells. *Virus Res* **94**, 85–95.

Voinnet, O. (2002). RNA silencing: small RNAs as ubiquitous regulators of gene expression. *Curr Opin Plant Biol* **5**, 444–451.

Voinnet, O., Pinto, Y. M. and Baulcombe, D. C. (1999). Suppression of gene silencing: a general strategy used by diverse DNA and RNA viruses of plants. *Proc Natl Acad Sci USA* **96**, 14147–14152.

Vousden, K. (1993). Interactions of human papillomavirus transforming proteins with the products of tumor suppressor genes. *FASEB J* **7**, 872–879.

Watanabe, G., Howe, A., Lee, R. J., Albanese, C., Shu, I. W., Karnezis, A. N., Zon, L., Kyriakis, J., Rundell, K. and Pestell, R. G. (1996). Induction of cyclin D1 by simian virus 40 small tumor antigen. *Proc Natl Acad Sci USA* **93**, 12861–12866.

Wurm, T., Chen, H., Britton, P., Brooks, G. and Hiscox, J. A. (2001). Localisation to the nucleolus is a common feature of coronavirus nucleoproteins and the protein may disrupt host cell division. *J Virol* **75**, 9345–9356.

Xie, Q., Sanz-Burgos, A. P., Hannon, G. J. and Gutierrez, C. (1996). Plant cells contain a novel member of the retinoblastoma family of growth regulatory proteins. *EMBO J* **15**, 4900–4908.

Xie, Q., Suárez-López, P. and Gutierrez, C. (1995). Identification and analysis of a retinoblastoma binding motif in the replication protein of a plant DNA virus: requirement for efficient viral DNA replication. *EMBO J* **14**, 4073–4082.

Zhu, Y., Gelbard, H. A., Roshal, M., Pursell, S., Jamieson, B. D. and Planelles, V. (2001). Comparison of cell cycle arrest, transactivation, and apoptosis induced by the simian immunodeficiency virus SIVagm and human immunodeficiency virus type 1 *vpr* genes. *J Virol* **75**, 3791–3801.

Index

Note: Page numbers in *italics* refer to figures and tables.

Index compiled by Jill Halliday